KU-325-523

DISPOSED OF
BY LIBRARY
HOUSE OF LORDS

French Constitutional Law

French Constitutional Law

JOHN BELL

CLARENDON PRESS · OXFORD
1992

Oxford University Press, Walton Street, Oxford OX2 6DP
Oxford New York Toronto
Delhi Bombay Calcutta Madras Karachi
Kuala Lumpur Singapore Hong Kong Tokyo
Nairobi Dar es Salaam Cape Town
Melbourne Auckland Madrid
and associated companies in
Berlin Ibadan

Oxford is a trade mark of Oxford University Press

Published in the United States
by Oxford University Press Inc., New York

British Library Cataloguing in Publication Data
Data available

Library of Congress Cataloging-in-Publication Data
Bell, John, 1953–
French constitutional law / John Bell.
Includes bibliographical references (p.) and index.
1. France. Conseil constitutionnel. 2. Constitutional courts—France. 3. France—Constitutional law. I. Title.
KJV4390.B45 1992 342.44—dc20 [344.402] 92–16016
ISBN 0-19-825271-4

Typeset by Best-set Typesetter Ltd., Hong Kong
Printed in Great Britain by
Bookcraft Ltd.
Midsomer Norton, Avon

PREFACE

THE Preacher warns us that the writing of books is an endless process (Eccles. 12: 12). This book is an illustration of that maxim. Arising out of lectures given in the University of Oxford, it was developed by research conducted in 1985–6, when I was *professeur associé* at the Universities of Paris 1 and Paris 2. But it has taken a number of years to bring the project to final completion, not least because the subject of French constitutional, law, even just focused on the Conseil constitutionnel, seems to mushroom at every turn.

The purpose of the book is to provide an introduction to the legal side of France's Fifth Republic Constitution, notably the contribution of the Conseil constitutionnel. By supplying translations of basic texts, it is hoped that the subject might become accessible, in particular to students of law and politics who do not have the necessary linguistic skills to tackle the difficult structure of French legal texts. The translations are my own, and apologies are given to the reader for any inaccuracies and infelicities of language that are not due to the notorious problems associated with the almost incomprehensible structure of French court judgments.

My thanks are due to many people for their help in the production of this book. Most of all, they are due to my wife and son for their endurance and support during the long process of writing and preparing the manuscript. My thanks are also due to Christine Taylor, Christine Wigley, Rachael Haist, and Barbara Goodison in the Faculty of Law at Leeds of their efforts in typing and producing the manuscript, as well as to Hilary Walford and Nicola Pike at the Press. On the academic side, thanks are due in particular to Jean Gicquel, Jacques Debû, Nicole Gain, and Louis Favoreu for their patient explanations of the workings of the Conseil constitutionnel, and to Roger Errera and David Goldie for comments on early drafts of sections of this work. The errors this book contains are none of their doing, but it would have contained more but for their help.

The book tries to state the law as at 31 October 1991, though some later points have been added.

J.B.

Leeds
February 1992

CONTENTS

CONVENTIONS AND GLOSSARY OF TERMS

DECISIONS of the Conseil constitutionnel are conventionally referred to by both decision number and date. For the convenience of the common lawyer, the author has given names to the decisions, and these form the principal basis for indexing in the Table of Decisions. There are three series of numbers given to decisions, corresponding to the three categories of decision that the Conseil gives, and which are identified in the *ordonnance* of 7 November 1958, relevant extracts of which are reproduced in Part Two, *Constitutional Texts*, Section IV. The three series are 'DC' (*déclarations de conformité*), which covers the review of *lois* passed by Parliament, parliamentary standing orders, and treaties; 'L' (*textes en forme législative*), which covers applications by the Government to amend or repeal provisions of *lois* on the ground that they fall within its legislative province; and 'FNR' (*fins de non-recevoir*), which covers motions of inadmissibility raised against private members' bills or parliamentary amendments. Thus, DECISION 1 is cited as 'CC decision no. 71-44 DC of 16 July 1971, *Associations Law*'; DECISION 6 is cited as 'CC decision no. 69-55 L of 26 June 1969, *Protection of Beauty Spots and Monuments*'; and DECISION 5 as 'CC decision no. 59-1 FNR of 27 November 1959, *Prices of Agricultural Leases*'.

The case-law of French courts is referred to in the usual French way, by court and date. The reference frequently includes the names of the parties. Thus, a decision of the Assemblée du contentieux of the Conseil d'État would be cited as 'CE Ass., 20 December 1985, *Établissement Outters*'; and a decision of the Assemblée plénière of the Cour de cassation as 'Cass. Ass. plén., 19 May 1978, *Dame Roy*'.

French courts divide into administrative and ordinary courts, reflecting the French division between public law and private law. The administrative courts are the Tribunaux administratifs (at first instance), the Cours administratives d'appel (on appeal in many matters since 1989), and the Conseil d'État. The Conseil d'État may sit in a number of formations. Normally, decisions are taken by one or more subsections, and are referred to simply as decisions of the Conseil (cited 'CE'). Occasionally, the decision is taken by the Section du contentieux ('Sect.') or by the Assemblée du contentieux ('Ass.'). The ordinary courts (*tribunaux judiciaires*) have civil and criminal jurisdictions. The first-instance civil courts include the Tribunaux d'instance, Tribunaux de grande instance, and the Conseils de prud'hommes (labour courts). The first-instance criminal courts are the Tribunaux de police, and Tribunaux correctionnels, and the Cours d'assises, depending on the seriousness of the offence. Preliminary criminal proceedings are conducted by the *procureur de la République* (the prosecutor) and the investigating magistrate (the *juge d'instruction*), and, on appeal, by the Chambre d'accusation. Appeals from decisions of lower courts are heard by the Cour d'appel, and that decision may be quashed by the Cour de cassation. (Exceptionally, there is no appeal against decisions of the Cour d'assises, which may only be quashed by the Cour de cassation.) The Cour de cassation sits in a number of formations. Normally, it will sit in three civil chambers ('civ.'), one commercial ('com.'), one social ('soc.'), and one criminal ('crim.'). It may sit with members of more than one chamber ('Chambres mixtes'), or even as a full court, the Assemblée plénière ('Ass. plén.'). Arbitrating on the boundaries of jurisdiction between the administrative and ordinary courts, sits the Tribunal des conflits ('TC').

Court personnel, apart from judges, include the public representative, styled the *commissaire du gouvernement* in administrative courts, or the *procureur* or *avocat général* in ordinary courts. These persons, like the Advocate General before the European Court of Justice, give reasoned arguments, called *conclusions*, for a solution to the case, taken from the perspective of the public interest in the proper administration of justice, rather than from that of the parties.

Law reports are found in a number of series. The most important for the Conseil constitutionnel are the *Recueil des décisions du Conseil constitutionnel* (cited '*Rec.*'), which contains all the decisions given in that year. A parallel series, the *Recueil Lebon* (cited '*Leb.*'), is published for decisions of the Conseil d'État. Both are official series. Private publishing houses do produce periodicals that publish decisions of the courts. In public law the most important are *Actualité juridique, droit administratif* (cited as '*AJDA*'), and *la Revue française de droit administratif* (cited as '*RFDA*'). More generally, reports are found in the *Recueil Dalloz-Sirey* (cited as 'D.'), which combines two previous periodicals, *Dalloz* (cited as 'D.') and *Sirey* (cited as 'S.'), and in part II of the *Juris-Classeur Périodique* (*JCP*). In these unofficial series, as in the case of other periodicals, the year of publication follows the reference to the periodical, and then the page numbers follow, e.g. '*RFDA* 1986, 513', referring to page 513 of the 1986 issue of *la Revue française de droit administratif*.

ABBREVIATIONS

AJDA	*Actualité juridique, droit administratif*
Avis et débats	*Avis et débats du Comité consultatif constitutionnel* (Paris, 1961)
Avril & Gicquel	P. Avril and J. Gicquel, *Droit parlementaire* (Paris, 1988)
Cass.	Cour de cassation
CC	Conseil constitutionnel
CE	Conseil d'État
CSA	Conseil supérieur de l'audiovisuel
D.	*Recueil Dalloz (-Sirey)*
Droits et libertés	F. Luchaire, *La Protection constitutionnelle des droits et libertés* (Paris, 1987)
Franck	C. Franck, *Les Fonctions juridictionnelles du Conseil constitutionnel et du Conseil d'État dans l'ordre constitutionnel* (Paris, 1974)
GA	M. Long, P. Weil, and G. Braibant, *Les Grands Arrêts de la jurisprudence administrative* (9th edn., Paris, 1990)
GD	L. Favoreu and L. Philip, *Les Grandes Décisions du Conseil constitutionnel* (5th edn., Paris, 1989)
Genevois	B. Genevois, *La Jurisprudence du Conseil constitutionnel* (Paris, 1988)
GT	D. Maus, *Les Grands Textes de la pratique institutionnelle de la Ve République* (2nd revised edn., Paris, 1985)
JCP	*Juris-Classeur Périodique*
JO	*Journal officiel de la République*
Leb.	*Recueil Lebon* (decisions of the Conseil d'État)
Le Domaine de la loi	L. Favoreu (ed.), *Le Domaine de la loi et du règlement* (2nd edn., Paris, 1981)
Luchaire & Conac	F. Luchaire and G. Conac, *La Constitution de la République française* (2nd edn., Paris, 1987)
Nicholas	B. Nicholas, 'Fundamental Rights and Judicial Review in France' [1978] *Public Law* 82, 155
RDP	*Revue de droit public*
Rec.	*Recueil des décisions du Conseil constitutionnel*
RFDA	*Revue française de droit administratif*
RFDC	*Revue française de droit constitutionnel*
RFSP	*Revue française de sciences politiques*

RPR	Rassemblement pour la République
S.	*Recueil Sirey*
SB	O. Kahn-Freund, C. Lévy, and B. Rudden, *A Source-Book of French Law* (3rd edn. by B. Rudden, Oxford, 1991)
TC	Tribunal des conflits
UDC	Union démocratique du centre
UDF	Union pour la démocratie française

TABLE OF DECISIONS

DECISIONS OF THE CONSEIL CONSTITUTIONNEL LISTED BY NAME

DECISIONS OF THE CONSEIL CONSTITUTIONNEL LISTED BY NUMBER

Déclarations de conformité (DC)

Election cases

Fin de non-recevoir (FNR)

Declassification (L)

DECISIONS OF THE CONSEIL D'ÉTAT

DECISIONS OF THE COUR DE CASSATION

INTRODUCTION

FRANCE has had a total of fifteen constitutional instruments in the two hundred years since the Revolution of 1789. In Britain, in the three hundred years since the Revolution of 1688, there has been no single document purporting to lay down a comprehensive list of constitutional norms. Yet at present in the United Kingdom there is considerable interest in constitutional reform, especially in the protection of fundamental rights, relations between central and local government, and the electoral system. Connected with such interest is the suggestion that many constitutional norms should be reduced to writing. Of course, most common law jurisdictions have lived with written constitutions for many years, notably the United States and Australia, and countries such as Eire and Canada have seen constitutional litigation grow significantly in recent years. Because of ease of communication and familiarity with the legal system of those countries, United Kingdom lawyers have tended to look to other common law jurisdictions for examples of the ways in which our constitutional system might be changed and what the potential consequences of this might be.

This book offers an exposition of an alternative model that is, in many ways, far closer to the United Kingdom than many common law jurisdictions, particularly those of federal countries. France is a unitary State that, until recently, was a fervent believer in parliamentary sovereignty and the unacceptability of judicial review of legislation. Protection of fundamental rights and principles of good government were matters of political obligation and could not be legally enforced against Parliament. The rule of law involved simply the subjection of the executive and citizens to the laws made by Parliament, and did not involve legally enforceable limits on Parliament's powers. Yet since 1958, and especially since 1971, parliamentary sovereignty has come to be understood in a new way. Parliament now has a designated sphere of lawmaking power, and residuary legislative power lies with the Government. In addition, constitutional values, especially in the area of fundamental rights, impose restrictions on the legislation that Parliament may pass. The new boundaries of Parliament's powers are policed by the Conseil constitutionnel, a body that is best seen as a form of constitutional court. Such developments in very recent years demonstrate the way in which a country wedded to many of the fundamental constitutional principles that currently prevail in the United Kingdom can move in new directions. To some extent, this was the result of a break with the past, marked by the adoption of the Constitution of the new Fifth Republic in 1958. But many of the changes have really come about as a result of new attitudes and new interpretations of the French republican constitutional tradition. In both the establishment of a new formal constitutional settlement and in changes in attitudes, there are lessons for other countries, notably the United Kingdom. This book seeks to draw out some of these lessons by examining developments since 1958 in important areas of constitutional law.

In many ways, the distinction commonly made between a written and an unwritten constitution is unhelpful. As will be seen in this book, even where there is a written text or several texts, there will be additional unwritten rules and principles that form part of both constitutional law and constitutional conventions. In any case, a text has to be interpreted, and the very need for interpretation gives rise to the possibility of developing the constitution in different directions. The history of the Constitution of the United States this century is a good illustration of this. In France it is clearly acknowledged that there is a constitutional tradition that has endured through the various Constitutions that have existed since 1789. The idea of a constitution as a tradition is also taken up by jurisprudential writers in the United States.[1] Such a tradition is a set of practices and ideas that provide a justification for constitutional texts and institutions, a set of values that can be used as an interpretative scheme for reading the laws and conventions of the system.[2] This book will explore the importance of such a tradition in interpreting new constitutional texts, and also the way in which a new constitutional settlement affected the tradition.

If there is a constitutional tradition in any system, it may almost be accidental whether provisions are found in the written texts or are unwritten. What matters is how constitutional norms are going to be interpreted. The central features of a constitution are *what we do*—how government is conducted—*the values we proclaim*—the principles by which government is conducted—and *the institutions that we establish*—who conducts government. Whether a constitution is a *written document*, and whether it establishes *enforcement mechanisms* to ensure compliance with the principles of government are subsidiary features of a constitutional system. The United Kingdom has the central features without a written text or a constitutional court.

So what is the magic in a written constitution? After all, first and

[1] See e.g. M. Perry, 'The Authority of the Text, Tradition, and Reason' (1985) 58 *Southern California Law Review* 561.

[2] On the importance of texts and interpretation here, see R. M. Dworkin, *Law's Empire* (London, 1986), ch. 9; S. Fish, *Doing What Comes Naturally* (Oxford, 1989), esp. chs. 4, 16.

foremost a constitution is, as Hanna Pitkin writes, what we *do*, how we run the government of a country.[3] We can say that a minister is acting unconstitutionally in refusing to explain his or her actions to a parliamentary committee without needing to refer to a written document. What we mean by saying something is 'unconstitutional' is not that a rule has been broken, merely that this behaviour does not conform to the way in which we expect ministers to behave towards Parliament. Such an appeal to a vague set of constitutional ideas and values is typical of most citizens and politicians, who never have the time or inclination to study the specific wording of the constitutional texts, be they contained in written instruments called 'constitutions', or treatises on constitutional law, conventions, and practice.

So why are provisions written down in texts? Such instruments can have two functions, declaratory and constitutive. A *declaratory* text records in written form our common understandings about the way in which the government of the country should work. Specific violations may encourage us to write these down. This might be the best way to understand the English Bill of Rights of 1689, which really only tries to set out what was thought to be the result of the 1649 Revolution, but which James II had violated. As will be seen, the French Constitution of 1958 is really declaratory in terms of the fundamental values to which it expresses attachment. This is why there was no effort made to enumerate and consolidate the diverse provisions from the past, and why the concept of 'fundamental principles recognized by the laws of the Republic' was retained without specifying the content. A declaratory text can be filled out by reference back to the elements of the tradition; a continuity between past and present is taken for granted.

By contrast, a *constitutive* text seeks to make a new beginning, and does attempt to be exhaustive. Writers of the early French Constitutions, especially those of 1791 and 1793, certainly thought that they were setting out a charter for a new régime with new values. Such a role for the constitution was obvious in the German Basic Law of 1949. The institutional features of the Fifth Republic Constitution, particularly the roles of the President, Parliament, and the executive, were intended to mark a significant departure from the past. Therefore, to an important extent, these institutional arrangements are set out exhaustively in the provisions of the 1958 Constitution and the texts enacted under it. These need to be interpreted, but this is done primarily by treating them as a self-contained and novel body of provisions and values. All the same, as will be seen in relation to *loi* and *règlement*, the practice of interpretation has seen less of a break with

[3] H. Pitkin, 'The Idea of a Constitution' (1987) 37 *Journal of Legal Education* 167.

the past than might have been anticipated. This merely points to the rarity of a fully constitutive constitution, and a natural desire in most countries to retain what was good in the past. As Ian Harden has pointed out,[4] a constitution may be viewed as 'an official map of public power', setting out political institutions, and as 'a set of principles of legitimate government'. This book will argue that these two features of the Fifth Republic's Constitution demonstrate different degrees of discontinuity with the past, and thus that it is both constitutive and declaratory at the same time.

Ian Harden also draws a valuable distinction between 'fundamental constitutional principles binding on Parliament and Government', and 'preservation mechanisms' that enforce and protect such principles.[5] Clearly, a constitution, like that of the United Kingdom, can have fundamental principles of this kind with no formal enforcement mechanism such as a constitutional court. The interest of the French Constitution of 1958 is not simply in the principles it sets out for lawmaking, parliamentary proceedings, and fundamental rights, but also in the preservation mechanisms that it has created, notably the Conseil constitutionnel. To a great extent, this book will consider the principles as they are protected by this new body. Of course, there are other valuable ways in which the French Constitution can be studied, and excellent works have been written by political scientists.[6] The reason for the focus of this book is the concern to examine the role of *law* in the Constitution, and the way in which this affects the structure of government. It is thus predominantly a lawyer's book, but I hope it is not without interest to others.

The book is divided into two parts: Part One sets out the basic institutions and principles of the 1958 Constitution; Part Two consists of a selection of constitutional texts and materials, together with a number of important decisions of the Conseil constitutionnel. The text of Part One provides a systematic introduction to French constitutional law. It starts in Chapter 1 with an outline of the history and basic features of the 1958 Constitution, in particular the place of constitutional review. Chapter 2 introduces the various sources of the Constitution and how they relate to each other. The remaining chapters look at a number of basic principles of government. Chapter 3 examines the division of lawmaking powers between Parliament and Government, one of the major changes introduced in 1958. Chapter 4 examines another

[4] 'The Constitution and its Discontents' (1991) *British Journal of Political Science* 489.

[5] Ibid. 508–9.

[6] See esp. V. Wright, *The Government and Politics of France* (3rd edn., London, 1989), and J. E. S. Hayward, *Governing France: The One and Indivisible Republic* (2nd edn., London, 1983).

major change, the principles governing parliamentary procedure and thus the political power of Parliament. Chapters 5 and 6 examine the protection of fundamental rights: Chapter 5 considers the extent to which fundamental freedoms are now protected through the Conseil constitutionnel; and Chapter 6 then discusses the doctrine of equality, and how far it is a general principle of the Constitution. The texts, materials, and decisions in Part Two of the book illustrate the issues described in Part One. They are especially useful for showing the reasoning processes of the Conseil constitutionnel, and provide a basis for class discussion. Further useful extracts on French constitutional law can be found in O. Kahn-Freund, C. Lévy, and B. Rudden, *A Source-Book on French Law* (3rd edn. by B. Rudden, Oxford 1991), where materials are up to date but are in French, and in A. von Mehren and J. Gordley, *The Civil Law System* (2nd edn., Boston, 1977), where materials are more dated, but in English. The best book for following decisions of the Conseil constitutionnel is L. Favoreu and L. Philip, *Les Grands Décisions du Conseil constitutionnel* (5th edn., Paris, 1989). The bibliography contains further suggestions for reading, though it is not exhaustive. Current developments are best followed in the recently started journal, *La Revue française de droit constitutionnel*, and in the regular 'Chronique constitutionelle française' by P. Avril and J. Gicquel in the review *Pouvoirs*. The best textbook on the subject is written by the current Secretary-General of the Conseil constitutionnel, Bruno Genevois, *La Jurisprudence du Conseil constitutionnel: Principes directeurs* (Paris, 1988).

PART ONE
Constitutional Law

1

CONSTITUTIONAL REVIEW WITHIN THE FIFTH REPUBLIC

THE idea of constitutional law is not novel in France. More novel is the role of law in the Fifth Republic's Constitution. From 1789 the very act of making constitutions defining the competence of governmental institutions introduced law as a regulator of public power. But the concern was predominantly with the division of institutional competence—who does what, and how each institution should operate. The Fifth Republic introduces two new dimensions: limits on the omnicompetence of the legislature, and a supervisory body akin to a court that can make binding rulings on the validity of Parliament's legislation. As a result, almost all decisions of organs of government are now subject to some form of judicial review. The growth of judicial review of the executive power is documented elsewhere.[1] Since, as will be seen, the executive has always had some autonomous legislative power, judicial review of legislation is not totally novel. The form and scope of judicial review of legislation in the Fifth Republic is far more extensive than before. Such a development was both planned and unplanned: planned to the extent that a new supervisory organ, the Conseil constitutionnel, was created to review the exercise of legislative power by Parliament and the executive; unplanned to the extent that the kind of judicial review exercised by this organ has exceeded the desires of those who drafted the Constitution. This book seeks to map out the extent of these developments primarily through decisions of the Conseil constitutionnel.

The nature of judicial review of legislation in France can only be understood by relating the role of the Conseil constitutionnel to that of other organs of government. It had a specific place in the new institutions created in 1958, but has evolved in symbiosis with changes in other institutions of government. In this chapter I shall try to highlight the place of the Conseil constitutionnel in the arrangements of the Fifth Republic's Constitution and in the State tradition of France.

[1] See L. N. Brown and J. Bell, *French Administrative Law* (4th edn., Oxford, 1993).

I.I. What Is Special about the Fifth Republic?

Origins

The Fourth Republic was established in 1946, and attempted to restore the liberal democratic tradition of constitutional government after the events following the fall of France in 1940. Congenital weaknesses compounded by problems caused by the ending of France's colonial role explain its early demise. Three major problems can be identified: the existence of major political divisions that weakened support for the system, governmental instability, and military unrest in the wake of decolonization.[2]

Political Division Political division was nothing new. The Third Republic (1870–1940) had to contend with strong anti-republican and anti-deomocratic forces for much of its existence. Many of these rallied around the Pétain Government of 1940–44, and were never reconciled to the Fourth Republic. The rebirth of the extreme right in the 1950s (the Poujadist movement) continued to provide a strong focus for this tendency. In addition, the Communist party grew in importance after the First World War, especially as the result of the Resistance movement, in which it was a leading member. It was at the peak of its power in the late 1940s, and captured over 25 per cent of votes in the 1946 and 1956 elections. Although it took part in the Governments from 1944 to 1947, it was by no means reconciled to the liberal democratic regime of the Fourth Republic. The Cold War was the final straw, and, under pressure from the United States, the Communists were expelled from the Government in May 1947. Once in opposition, the Communists provided a strong and active force of obstruction to the liberal democratic regime both inside and outside Parliament, and were treated with suspicion by it.[3]

Although not part of the anti-democratic movements, the followers of General de Gaulle also remained unreconciled to the Fourth Republic's regime. The General had made no secret of his disdain for politicians and their power plays when he resigned in January 1946. The parliamentarian system of the Fourth Republic provided scope for these, without the strong authority of the executive that he believed necessary. The man of destiny presented himself as swept aside by the squabbles of politicians. His Bayeux declaration of June 1946 set out an alternative, more presidential form of government, but it had little

[2] See generally V. Wright, *The Government and Politics of France* (3rd edn., London, 1989), 1-7.

[3] P. M. Williams, *Crisis and Compromise: Politics in the Fourth Republic* (3rd edn., London, 1964), 218-19.

influence on the Second Constituent Assembly. He founded a popular party (the Rassemblement du peuple français) in 1947 with significant success, but the electoral reforms introduced in time for the 1951 elections prevented his party (and the Communists) from benefiting from their popular support. For most of the 1950s the General withdrew from active politics, and his parliamentary followers were loath to give support to the various Governments that came to power. These anti-regime forces could count on the votes of over one-third of the electorate, and in 1946 and 1956 they took some 40 per cent of the seats in the National Assembly.

The Fourth Republic was ill-fated from the start. The First Constituent Assembly presented a draft Constitution to a referendum in May 1946, but it was defeated. The left-wing dominated Assembly had tried to draw up a new charter of rights and to establish a parliamentary regime similar to that under the latter part of the Third Republic, though with only one chamber. In October 1946 the second draft only secured approval by a margin of a mere one million votes among an electorate of twenty million, with 31.1 per cent of electors voting against, and 31.2 per cent abstaining. The lukewarm reception of the Constitution, and the strong contrary forces that existed within it, meant that it did not have the broad-based political support necessary to carry it through any crisis.

Governmental Instability Like the Third Republic, the Fourth was a parliamentary regime, with the real power lying in the Assemblies, which chose the President, and to whom the Government was responsible, and upon whose support it relied. Again like the Third Republic, the institutions chosen produced a significant degree of governmental instability, which brought the politicians into disrepute and failed to create the consensus necessary to maintain the system in a crisis.

Although the Fourth Republic formally increased the power of the National Assembly, the proportional representation electoral system undermined its effectiveness. Coalitions were inevitable; but with one-third of seats going to the parties that were antipathetic to the regime, Governments could only function if there was support from the large majority of the remaining parliamentary groups. Since these disagreed significantly, long-term agreement was unlikely. In consequence, the country lurched from one Government crisis to the next. The combination of anti-regime groups and the centre parties that were disaffected from time to time was able to bring down a Government without being able to offer a constructive alternative. As a result, France got through twenty-three Governments in twelve years, the most enduring being that of Guy Mollet, from 1 February 1956 to 12 June 1957.

One must not exaggerate the degree of instability. Of the twenty-three Premiers under the Fourth Republic, sixteen had served in the previous Cabinet, and twelve were to serve in the next one. Especially in areas less central to political disagreement, ministers could occupy a single post for many successive years, as did Bidault and Schumann in the Quai d'Orsay (Foreign Affairs).[4] While the country grew strong economically, this was mainly due to continuity in the administration provided by civil servants.

The lack of political leadership was most pronounced in more contentious areas, such as the levying of taxes to meet expenditure and, in the later years of the Republic, to deal with the colonial crises.

Military Failures Like Britain, France did not leave its colonies without putting up significant military resistance. The military activities of the period were marked by notable defeats, leading to recriminations between the military leadership and the politicians at home. Vietnam was vacated in 1954 after a series of military defeats. Tunisia and Morocco followed suit in 1955. In addition, there was the Suez débâcle of 1956. Uneasy relationships existed between the military and politicians. Military repression was inadequately controlled in advance, as in February 1958, when the Algerian military command ordered the bombing of a civilian village at Sakhiet in Tunisia, close to a supposed base for Algerian liberation movements. Although politicians often felt obliged to defend military actions after the event, they preferred other policies that were not appreciated by the military. For example, the Government put an end to martial law in Algeria in March 1958, against the advice of military commanders and just at the time when the latter claimed that this policy was beginning to reduce the number of terrorist incidents.

Domestically, there were also problems with the police, who in March 1958 staged an anti-parliamentary, anti-Semitic demonstration outside the National Assembly.

The Transition to the Fifth Republic These problems came to a head on 13 May 1958, when the army occupied Government House in Algiers and set up a Committee of Public Safety, warning the President not to sell out to the liberation movements on Algerian independence.[5] This occurred the day before Pflimlin was to be confirmed by Parliament in Paris as Prime Minister, an event that the settlers took to be a sign of imminent capitulation to the independence forces. Some Gaullist

[4] See generally M. Larkin, *France since the Popular Front: Government and People, 1936-1986* (Oxford, 1988), 145-6.

[5] See Williams, *Crisis and Compromise*, 51 ff.; Larkin, *France since the Popular Front*, 261 ff.

supporters were already seeking ways of securing the return of the General, by illegal means if necessary, and they tried to take charge of events. The leaders of the rebellion called for de Gaulle to take charge, and de Gaulle declared his willingness to form a Government (contrary to what he had told the President a mere ten days before). Algerian-based troops occupied Corsica on 24 May, declaring for de Gaulle, and the CRS (Compagnies républicaines de sécurité) forces sent to counter them merely capitulated without a fight. There was clearly the danger of a military coup that would spread to the mainland. The rebels threatened that 'Operation Resurrection' would be launched unless de Gaulle formed a Government. The centre–right coalition collapsed on 28 May, and the President warned that he would resign unless Parliament accepted de Gaulle, which many parliamentary leaders were prepared to do. On 1 June the President invited de Gaulle to head a Government. The General agreed, on condition that he could draw up a new Constitution. A *loi* of 3 June 1958 gave his Government powers to make decrees on whatever matters it thought necessary for six months, while it prepared a draft of a Constitution to be put to the people by referendum.

The crisis atmosphere, and the limited time for the exercise, ensured that the writing of the Constitution was not going to be a slow process of reflection and discussion. The Government began work on a draft in July. This was submitted to the Comité consultatif constitutionnel, which met for eighteen sessions between 29 July and 14 August. The draft was then presented to the Conseil d'État at the end of August, and was voted by referendum on 28 September 1958. The majority in favour was substantial (80.1 per cent of those voting, constituting 66.4 per cent of the electorate), even though some Socialists and the Communists were against it.

Clearly, given the time-scale, there was no room for enormous originality. In its delegation of full powers to de Gaulle, Parliament had isolated five principles that the new Constitution had to reflect: (i) universal suffrage as the sole source of political power; (ii) the separation of legislative and executive powers so that each was assured the full exercise of its proper functions; (iii) the answerability of Government to Parliament; (iv) the independence of the judiciary; and (v) a relationship between the Republic and associated peoples. There were, in addition, several abortive proposals for reform of the Constitution that had been made in recent years. Furthermore, many of those involved in the drafting had been leading figures in previous administrations, which was especially the case with the Comité consultatif constitutionnel. Continuity with established principles and practices was to be expected.

Duvergier isolates novel inputs from three sources: de Gaulle, Debré, and ministers.[6] De Gaulle proposed that the President of the Republic should be elected by a wider constituency than just the members of Parliament, that he should have powers to act in an emergency (such as that of 1940), and that ministerial and parliamentary functions should be mutually incompatible. Debré concurred in this last idea, and also wanted a 'rationalized Parliament'. Ministers wanted some way of dealing with issues of confidence that would avoid governmental instability, put an end to the annual debate approving Government policy, and do away with the honorific role of the Senate. Beyond these general outlines, work was supervised by Debré but carried out by legal specialists. The 1958 text is, in consequence, technical, and focuses on issues of the operation of institutions rather than on grand principles either of government or of fundamental rights.

1.2. Features of the Fifth Republic Constitution

The principal novel features of the Fifth Republic Constitution are institutional in nature: presidentialism, rationalized parliamentarianism, and the Conseil constitutionnel.

Presidentialism

The role of the President in the Third and Fourth Republics was essentially that of a figure-head, smoothing the way for the creation of coalitions, but having no part in active politics. Experiences with Louis Napoleon Bonaparte, who organized a coup to stay in office and became Emperor in 1852, with Marshal MacMahon, who tried to stage a royalist coup in 1877, and with Pétain during the Vichy period were sufficient to encourage such limitations on the President's role. Since he was elected by members of Parliament, his independence of action was thus limited. As Weiss commented in 1885: 'The fundamental principle of the Constitution is, or ought to be, that the President hunts rabbits and does not govern.'[7] All the same, the President did chair the Council of Ministers, though not the Cabinet, which included the junior ministers. Given governmental instability, his function was bound to be more than honorific, and both Auriol and Coty in the Fourth Republic played significant roles in the development of political events, but this owed much to their personal authority and to the tolerance shown by parliamentarians rather than to any constitutional mandate.

The original conception of the presidential role is contained in article

[6] M. Duvergier, *La Cinquième République* (Paris, 1959), 12 ff.

5 of the Constitution, whereby the holder has the function of seeing that the Constitution is respected, and exercising arbitrament in securing the proper operation of government. In discussion before the Comité consultatif constitutionnel, de Gaulle stated that the President's function was that of an arbiter who had to secure the proper functioning of public powers, whatever happened, but whose influence did not extend to day-to-day policy.[8] He was to act when there was a gap in the institutions, such as in the crisis situation of 1940 or when there was no effective Government. All the same, the notion of 'arbitrament' (*arbitrage*) means that, even in normal times, he was to be no mere figure-head who appoints the Prime Minister and ministers, is head of the army, and signs treaties. Debré described his role as 'superior judge of the national interest', with the authority to require Parliament to reconsider bills, to refer them to the Conseil constitutionnel, and even to dissolve the National Assembly (a power that had only once been used in the previous eighty years, but has been used four times in the Fifth Republic). As such, he was to be a 'head of State worthy of the name'.

The original procedure for electing the President involved an electoral college composed of some 80,000 national and local representatives. Direct elections were thought impossible while there were so many French nationals in the African colonies, and, in any case, a popular election might be partisan in character, thus dividing the nation.[9] Following his success in ending the Algerian war, and the popular support he received after an assassination attempt against him at Petit-Clamart on 22 August 1962, de Gaulle managed to have a constitutional amendment passed by referendum instituting direct elections for the presidency. This created the kind of dialogue between President and people that de Gaulle had envisaged in his Bayeux declaration of 1946. It provided the presidency with an independent mandate that both increased its authority and later helped Mitterrand to remain in office between 1986 and 1988, when the parliamentary elections produced a majority for the opposition. This new situation was already evident from the statement of Pompidou on assuming office in 1969, that the President is 'both head of the executive and guardian and guarantor of the Constitution. In this double role, he is charged with giving the fundamental impetus, defining the essential directions, and assuring and controlling the proper functioning of public authorities: an arbiter with primary national responsibility at the same time'.[10] The President is the man in charge. Although

[7] Cited in Williams, *Crisis and Compromise*, 185.
[8] *Avis et débats*, 118-19.
[9] M. Debré, Speech to the Conseil d'État, 27 Aug. 1958, in *GT* 2-3.
[10] Press conference, 10 July 1969, ibid., doc. 5-101.

successive Presidents have denied that they are electoral agents for their party, they do in fact have an important role in parliamentary election campaigns, and build up a parliamentary majority for their policies.

According to article 8, the President appoints the Prime Minister and, on his recommendation, he appoints and dismisses other ministers. In practice, this has meant that the Prime Minister holds office at the President's good pleasure, and Presidents have changed Prime Ministers when they have wanted a new direction, as in 1972 with the dismissal of Jacques Chaban-Delmas by Pompidou, or in 1984 with the resignation of Pierre Mauroy after Mitterrand had simply announced the withdrawal of the controversial education reform bill from the parliamentary agenda, or in 1991 when Michel Rocard was forced to resign after a long period in which it was obvious that he no longer enjoyed the President's support. Presidents have equally appointed ministers of whom the Prime Minister does not approve.[11]

The capacity of the President to take emergency powers under article 16 was used at the height of the Algerian crisis, from 16 April to 30 September 1961, when a military rebellion in Algeria threatened the Republic. This power is limited, in that the Prime Minister and the Conseil constitutionnel must be consulted before a declaration is made, and Parliament meets automatically and cannot be dissolved while the emergency powers are in force. The Conseil constitutionnel offers advice on the measures to be taken. Deciding when to end the period of emergency powers is up to the President, and despite informal suggestions from the Conseil, de Gaulle continued them for longer in 1961 than was really necessary.

Overall, the President does meet Debré's idea of a 'republican monarch', though one with considerably more power than any unelected monarch in Europe.

Separation of Executive and Legislature

One of the principal criticisms made by the authors of the Constitution was that Parliament had, in the recent past, been too keen to intervene, making the task of governing difficult, if not impossible. The Fifth Republic set out to remedy this. Both de Gaulle and Debré considered that a clearer division of labour between Government and Parliament was necessary. This is carried out by a separation of both functions and of personnel.

Establishing the Government's capacity to govern involved setting limits on the scope of parliamentary legislation (*lois*) in article 34, in

[11] See Wright, *The Government and Politics of France*, 28-30, 87; *GT*, docs. 8-102-5.

order to free the Government to take charge of the direction of the nation without being bound by unacceptably detailed instructions from Parliament. Equally important was the curtailment of the power of Parliament to dismiss a Government. The Government was to be appointed by the President, not by the National Assembly, to whom it is responsible only in limited ways. A Government must resign if it is defeated in a vote on its general programme, but it is under no obligation to present one to Parliament at all. Thus, in June 1988 the minority Government of Rocard did not present a programme to Parliament, and thus did not court defeat. In addition, censure motions are restricted. The proposers of a motion must secure an absolute majority of the members of the Assembly—in other words, they must be more or less in a position to offer an alternative Government. Counting only the votes in favour of the censure motion could have curious effects: for example, in the United Kingdom the Callaghan Government was defeated on a censure motion in March 1979 by 311 votes to 310; the absolute majority at the time would have been 315. On the Fifth Republic's rules, Callaghan would have won the vote of censure. As Debré put it: 'the responsibility of the Government is established according to procedures that avoid the risk of instability.'[12]

The separation of personnel was simply part of this idea of the independence of the executive. Article 23 creates an incompatibility between being a minister and being a member of Parliament, with the result that ministers need have no parliamentary experience. This rule has some origins in the experience of de Gaulle as head of the Liberation Government of 1944–6, when he considered that ministers owed too much allegiance to their parties and not enough to him. Ministers have the freedom from party responsibility, with only limited control by Parliament. Parliament authorizes policies and votes *lois*, but it does not have the power to challenge ministerial appointments, as it had done under the Third and Fourth Republics. The executive was to have, in effect, a more managerial role.

Rationalized Parliament

The new role of the executive necessarily diminished the functions of Parliament, at least in a formal sense. It was no longer a regime dominated by the elected Assemblies. While formally weaker, Parliament came in practice to enjoy real powers that restored its importance. The rationalized system streamlined what Parliament was effectively able to achieve within a system with a strong executive.

[12] Speech of 27 Aug. 1958, in *GT* 7.

Though apparently weaker than the United Kingdom Parliament, the real position of the two institutions is actually much closer.

Formally at least, Parliament no longer enjoys legislative sovereignty. As will be seen in Chapter 3, article 34 enumerates a long list of matters on which Parliament enacts *lois;* in some areas it enacts the rules, while in others it merely sets out the general principles. For the rest, article 37 confers legislative power on the executive to make *règlements* (decrees). In other words, formal residuary legislative power lies with the executive, not with Parliament. In addition, since implementation of *lois* depends on the Government passing decrees, the Government is effectively able to block parliamentary legislation without any obstacle from Parliament. For instance, the *loi* on contraception passed in December 1967 was not implemented for seven years, and other legislation has met a similar fate.[13]

The power of Parliament is further reduced by the fact that its time for sitting is restricted. Article 28 establishes the *maximum* for parliamentary sessions, not the *minimum* set in previous Constitutions. It meets in two sessions, from 2 October for eighty days (including weekends), and from 2 April for ninety days. Since the budget occupies much of the autumn session, the time left for ordinary activities is limited. Even the Government has to resort to extraordinary sessions to pass its own legislation. Parliament may be convened at other times either by a majority of members of the National Assembly or, more usually, by the Prime Minister. In either case, there must be a specific agenda for the extraordinary session. Although this is more limited than the United Kingdom Parliament, it is less restricted than many other similar bodies.

What inhibits Parliament more is the Government's control over the agenda. Article 48 gives priority to the discussion of Government bills or private members' bills that it endorses. Effectively, this can mean that serious matters do not get discussed by Parliament at all, as happened, for example, with the events of May 1968.

Questions by members of Parliament to the Government have had a chequered history. Limited initially to one session a week (and that on a Friday, when many deputies were away in their constituencies), oral questions received scanty responses from Government; before the National Assembly in the Pompidou era, for example, the reply rate fell to below 20 per cent. In the period from 1970 to 1974 only 14 per cent of questions were answered within the proper time-limit, and often junior ministers were sent to reply on behalf of the whole Government. Since 1974, the number of sessions has increased, and so has Government diligence in replies, but, even then, 15 per cent of

[13] See Wright, *The Government and Politics of France*, 146-7.

questions from 1981 to 1984 remained unanswered. There is certainly not the same atmosphere as Question Time in the United Kingdom Parliament.

The Government possesses significant powers to push through its legislation. As will be seen in Chapter 4, under article 44 §3, the Government can require Parliament to vote on the whole or part of a bill as a block, with such amendments as the Government is prepared to accept. This procedure is often combined with making a vote an issue of confidence, thereby guaranteeing the bill's passage. The Government may also hasten the passage of a bill by making it a matter of urgency under article 45. And, furthermore, the Government always has up its sleeve the possibility of securing a delegation of power under article 38 to pass legislation by *ordonnance*, thus avoiding further parliamentary battles and saving time.

The financial powers of Parliament are closely controlled. As will be seen in Chapter 4, ordinary members of Parliament cannot propose increases in public expenditure or taxation. In addition, Parliament must deliberate on the annual budget within a very tight timetable.

From this it might appear that Parliament is an institution of limited value. While this could be said of the early years of the Fifth Republic, it is less true today. Governments of recent years have been more parliamentarian in outlook, and Parliament has been able to exert influence, even if it has not been able to stop the Government doing much of what it wished to do.[14] The extent of Parliament's influence depends in part on how far Government supporters are ready to vote in whatever is presented, and on the stability of the majority coalition. While the Constitution was drafted against the background of a number of small parties that existed in the Fourth Republic, the realities of the Fifth have seen a move to a semblance of two-party politics, particularly in the 1980s. Nowadays, there is a real situation of Government facing the opposition in a similar way to the United Kingdom, though with the major difference that the Prime Minister and other ministers are not members of the National Assembly. For much of the Fifth Republic the Government has had a working majority in the National Assembly, and has thus had a compliant Parliament, rather than needing to resort to its constitutional powers to push through its measures. In such a situation, the Government is much less likely to be aloof from Parliament, and the active co-operation of the two is more likely.

1.3. Constitutional Review

The creation of the Conseil constitutionnel was originally intended as an additional mechanism to ensure a strong executive by keeping

[14] See Wright, *The Government and Politics of France*, 149-55.

Parliament within its constitutional role. As will be seen later, this original intention has been departed from to a significant degree in subsequent years to create what is effectively a constitutional court.[15]

Constitutional Review in French History

In presenting the Conseil constitutionnel to the Conseil d'État in 1958, Debré stated that 'It is neither in the spirit of a parliamentary regime, nor in the French tradition, to give to the courts, that is to say, to each litigant, the right to examine the validity of a *loi*.' Until then, although the rule of law quite quickly involved the subordination of the executive to the law and to bodies that could be called courts, Parliament, as the lawmaker and representative of the nation, was in a different position. This is reflected both in the history of institutions and in currents of ideas.

Institutions In the *ancien régime* there was nothing quite like a modern national Parliament. The Estates General, composed of representatives of the three orders of society (nobility, clergy, and commoners), was the nearest equivalent to one, but it did not meet from 1614 until 1789. The regional *parlements* were lawcourts, membership of which was an office of profit to be bought and sold, and representatives were drawn from the nobility and the clergy. In order to be valid, a law made by the King had to be registered with the local *parlement*, from which grew up the view that it had the power to refuse to register laws made by the King. Although the King could impose his will by holding a *lit de justice*, remonstrances made by the *parlement* against a particular measure could ensure that it was altered. This power had been used in the years leading up to the Revolution of 1789 in order to block reforms introduced by Louis XVI. The clergy and nobility were thus able to resist change, and the *parlements* gained the reputation of being reactionary.

The separation of powers introduced by the revolutionary Constitutions sought to prevent the judiciary being able to obstruct Parliament. Hence the *loi* of 16–24 August 1790 insisted in article 10 that the judiciary was not 'to take part directly or indirectly in the exercise of legislative power', or to 'obstruct or suspend the execution of the decrees of the legislative body'. Article 127 of the Criminal Code backed this up by making it an offence for a judge to interfere with the legislative power. The Tribunal de cassation (as the highest ordinary

[15] See generally J. Beardsley, 'Constitutional Review in France' [1975] *Supreme Court Review* 189 at 191-212; F. Luchaire, *Le Conseil constitutionnel* (Paris 1980), ch. 1; L. Hamon, *Les Juges de la loi* (Paris, 1987), ch. 2.

court was known) decided as early as 1797 that 'the absolute terms in which the prohibition on the courts to stop or suspend the implementation of *lois* is drafted can admit of no exception or excuse'.[16] Apart from two possibly aberrant decisions of the Cour de cassation (as the Tribunal de cassation had then become) in 1851, the ordinary judiciary has consistently refused to challenge the validity of *lois*.

This reaction to the *parlements* did not mean that the French were not aware of the need to keep Government and Parliament within the limits of the Constitution. The excesses of the Terror brought home the necessity for such control. In the debates on the *Directoire* Constitution of 1795 Sieyès suggested that there should be a *jurie constitutionnaire* that would ensure that the Constitution was obeyed by annulling acts of the legislature and the executive that were contrary to it. Rejected then, the idea was taken up at the adoption of the next Constitution in 1799. The Constitution of year VIII established a Sénat conservateur that had the function of considering the constitutionality of provisions, and even annulling decisions referred to it by the Tribune (Assembly) or by the Government (article 21); this included annulling *lois*, though only before their promulgation (article 37). In fact, this body was totally ineffective. Composed of persons appointed by the consuls, who were irremovable, it did nothing to prevent the excesses of the Napoleonic period. It only decided to quash legislative acts two days before the capitulation of Paris in 1814. The institution was revived under Louis Napoleon (soon to become Napoleon III) in the Constitution of 14 January 1852. Again composed of persons appointed with security of tenure, it was intended as the guardian of liberties and other basic values. It could pronounce on the constitutionality of *lois* before promulgation, but it could also pronounce on any decision referred to it by the Government or on a petition of citizens. This was thought to open the way to review of *lois* even after they had been promulgated. But the provision was never put to the test, and the Sénat of the Second Empire was as ineffective as that of the First.

The contrary, and typically republican, tradition exhibited in other Constitutions before 1946 was that Parliament itself was the guardian of constitutionality. In the very first Constitution of 1791 the National Assembly was enjoined to refuse all proposals that infringed the Constitution. Self-limitation was the preferred institutional device for ensuring that the Constitution was respected, with an ultimate control exercised by the electorate.

It was the collapse of the Third Republic, and the actions of the Vichy regime that encouraged further consideration of institutional safeguards. The First Constituent Assembly of 1945–6 adhered to the

[16] Decision of 18 fructidor, an V.

predominant republican tradition, and did not even have a second chamber as a check on the National Assembly. Following the Second Constituent Assembly, the Constitution of 1946 established a Comité constitutionnel composed of the Presidents of the Republic, the National Assembly, and the Council of the Republic (as the Senate was called), and then seven persons nominated by the National Assembly and three by the Council of the Republic. The nominees appointed by the two Assemblies were to come from outside their membership. This body was designed to make prompt decisions on a limited range of matters concerned exclusively with the institutional provisions of the Constitution, and deliberately excluding the declarations of rights. It could examine *lois* before they were promulgated to see if a constitutional amendment was necessary. Since the procedures for such an amendment were cumbersome, this was, in theory, a serious obstacle to unconstitutional legislation. It was concerned merely to resolve conflicts between the two chambers, and was designed to redress the power of the National Assembly to override the Council of the Republic. It was called upon to make only one decision: in 1948, on the question of the time-limit within which the Council to the Republic had to vote on a bill when the National Assembly had classified it a matter of urgency.[17] Other attempts to have matters referred were not successful. Notably, in 1957 the Gaullists tried to have the bill ratifying the Treaty of Rome referred to the Comité on the ground that it was incompatible with national sovereignty. Since there was no conflict between the two Assemblies on the issue, the Comité had no jurisdiction to consider the matter.

The self-limitation of Parliament was not very effective, in that blatantly abusive *lois* were passed both in terms of procedure—for example, the delegation of blanket legislative powers to the executive—and substance—the anticlerical legislation of the turn of the century. Even the Fourth Republic offered no check where both chambers were agreed on a measure.

Constitutional review before 1958, where it was permitted, involved three main features. First, examination of bills before promulgation was typically the only way envisaged for dealing with questions of constitutionality. There was no sense in which there could be a challenge in the ordinary courts by way of defence (the so-called *exception d'inconstitutionnalité*). Secondly, this was reinforced by the persons who could refer matters to the relevant organ. Apart from the reference to the 'petition of the people' in the Second Empire, only the legislature and the Government were empowered to make such an application.

[17] The Constitution resolved the question by reference to the provisions of a standing order of the National Assembly that had never been made.

Thirdly, the organ of constitutional review was a political body, not a judicial institution, though it was sufficiently distinct to have some claim to be an independent guarantor of the Constitution. In many ways, the Conseil constitutionnel of the Fifth Republic is the heir to, and an improvement on, these institutional mechanisms of the past, rather than being designed as a constitutional court in its own right.

The Doctrinal Debate In the realm of ideas, constitutional review did not loom large as an issue in the public imagination. During the Third Republic, chiefly in the 1920s, the issue did excite constitutional writers, much as the Bill of Rights debate engages academic attention in the United Kingdom today. Indeed, the arguments deployed in that period bear striking similarity to those currently voiced on this side of the Channel. The specific problem that caused the controversy was that the Constitution of the Third Republic was made up of three *lois* of 1875 that briefly outlined the major organs of government, but little more. Even as institutional rules, they required much elaboration, and there was no declaration of rights, as there has been in every other Constitution before and since. Additional rules on both institutions and civil liberties were passed by ordinary *lois* over succeeding years. The Constitution did have a procedure for amendment to the constitutional texts of 1875, which involved both chambers of Parliament meeting as a single body. But this was rarely needed. With no formal limits on the legislative power of Parliament, and no institutional checks other than the agreement between the two chambers, there seemed little scope for constitutional review. All the same, measures of social reform and of emergency powers passed in the early years of the twentieth century generated discussion about constitutional controls over Parliament. This was fuelled after the First World War by the creation of constitutional courts in Europe and by consideration of the activity of the US Supreme Court.

The principal argument in favour of constitutional review was that the Constitution provided a higher norm, which bound the legislature just as much as any other organ of government. As Maurice Hauriou put it: 'Under the rule of a national constitution, no public authority is sovereign in the sense that it cannot be controlled in the exercise of its power or in the performance of its function. . . . Uncontrollable sovereignty lies only in the nation and is not delegated at all. All delegated sovereignty is controllable.'[18] For him and for others such as Duguit and Jèze, the need for control arose from the very nature of the legal authority granted to the legislature by the Constitution. Drawing on Kelsen, Eisenmann supported this view of the hierarchy of

[18] *Précis de droit constitutionnel* (2nd edn., Paris 1929), 266.

legal norms, and considered, as did Kelsen, that this in no way compromised legislative sovereignty: 'Any supposed *loi* which, from whatever perspective, infringes one of these provisions [of the Constitution] is in reality only a pseudo-norm. And . . . in refusing to take notice of it, the judge does not refuse to apply a legislative rule, but a rule that, improperly, claims to be legislative.'[19]

That remark, however, was written in a note on a case in which the Conseil d'État had stated that 'in the current state of French public law' an argument that a *loi* was unconstitutional could not be discussed by it. Both Jèze and Eisenmann were clear in recognizing that the argument from the hierarchy of norms, though logically impeccable, enjoyed no judicial support, and could not be said to be part of positive law.[20]

Duguit and Hauriou argued that declarations of rights and other values could be included among constitutional norms, despite their absence from the enacted texts of the 1875 Constitution. Duguit argued that the State was not absolutely sovereign, but existed to promote social solidarity, a public service in the interests of society. The State's authority was derived from a higher, pre-political norm, and this was binding on any constituent assembly drawing up a constitution: 'The system of the declarations of rights serves to define the limits that are imposed on the State, and, for that, higher principles are formulated, which the constituent legislator must respect just as much as the ordinary legislator. Declarations do not create these higher principles; they identify them and proclaim them solemnly.'[21] Hauriou wanted such declarations to protect the citizen from the arbitrariness of the State.[22] But such arguments ran contrary to the legal positivism that predominated in French legal thought. As Henkin points out, the French have typically believed in rights through law, rather than in some pre-political rights that define the legitimacy of the law.[23] Even when declarations were put into the Fourth Republic's Constitution, some leading writers argued that many of their provisions did not have constitutional force.[24]

Those authors who adopted one of these points of view argued that the task of identifying properly enacted *lois* falls within the competence

[19] Note on CE, 6 Nov. 1936, *Arrighi*, DP 1938.3.1.

[20] A similar view was taken under the Fourth Republic: see M. Mignon, 'Le Contrôle juridictionnel de la constitutionnalité des lois' D. 1952 Chr. 45.

[21] Léon Duguit, *Traité de droit constitutionnel* (2nd edn., Paris, 1923), iii. 560: see K. H. F. Dyson, *The State Tradition in Western Europe* (Oxford, 1980), 146-9.

[22] *Précis de droit constitutionnel*, 731-5; Dyson, *The State Tradition*, 162-3.

[23] L. Henkin, 'Revolutions and Constitutions' (1989) 49 *Louisiana Law Review* 1023 at 1044-5.

[24] F. Gény, 'De l'inconstitutionnalité des lois ou des autres actes de l'autorité publique et des sanctions qu'elle emporte dans le droit nouveau de la Quatrième République française', *JCP* 1947, I. 613.

of the ordinary judge; it does not involve trespassing on the legislative function, since he is merely following the declared will of the higher (constituent) legislator. Most rejected the idea of a special court to review the constitutionality of legislation, on the ground that it was unnecessary to create a specific institution for such a small amount of litigation, and its special status might seem more of a threat to Parliament. In any case, the issue of the constitutionality of legislation would inevitably arise before the ordinary courts after the enactment of the legislation, and would occur when political passions were less heated.[25] On the other hand, Eisenmann, whose thesis was on the Austrian Constitutional Court, saw merit in having a specialist tribunal as a channel for constitutional issues, and from which a single, definitive answer could be obtained.[26]

The main arguments used against constitutional review focused on the status of *loi* and the separation of powers.

Article 6 of the Declaration of 1789 stated that *loi* is the supreme expression of the *volonté générale*. This was interpreted as meaning that Parliament was the representative of the general will of the nation, and that its enactments thus enjoyed the status appropriate to the expression of the will of the sovereign. On this view, the State was the voice of the nation, and its authority was a pre-condition for liberty.[27] In no sense was it argued that constitutional review was incompatible with the idea of sovereignty. The people could easily agree to restrict that authority, and Carré de Malberg considered that such limitations were very much expressions of sovereignty. All the same, the introduction of constitutional review would be a change in the way in which it was to be exercised.[28]

The argument from the separation of powers has two elements. On the one hand there is the statement of the appropriate function of each organ; on the other, there is the appropriate deference that must be paid by other organs of government.

The function of the legislature was, according to Carré de Malberg, to be the continuing representative of the will of the nation, and, as such, to complete the task of making the Constitution.[29] The French situation under the Third Republic was different from that in other countries, since Parliament had the power to alter the Constitution, and was not subordinated in this task to a prior or external institution or mechanism. In addition, the legislature was to be the interpreter of

[25] Hauriou, *Précis de droit constitutionnel*, 267, 270-1.
[26] *La Justice constitutionnelle et la Haute Cour constitutionnelle d'Autriche* (Paris, 1928), 292.
[27] Carré de Malberg, *La Loi: Expression de la volonté générale* (Paris, 1931), 115 ff.: see Dyson, *The State Tradition*, 162-3.
[28] Ibid. 125.
[29] Ibid. 120.

the Constitution; this was appropriate, since it was representative of the people who made it. Parliament was to interpret the Constitution when it came to passing legislation, and if it decided that its interpretation was compatible with the Constitution, this could not be gainsaid by any other organ:

> Nothing is more natural than to make interpretation an act of the very person who made the text . . . In other words, it is for the legislature, at the very moment of making laws, to examine if the *loi* being considered is consistent with the Constitution, and to resolve the problems that may arise on this point. The legislature interprets in this way by virtue of its popular representation.[30]

The appropriate role of the judiciary here was to defer to the decision of the legislature. The demarcation of powers was to be understood not as a separation of functions, as in the United States, but as a separation of organs of government. In making constitutional amendments, Parliament had to meet in a different way from when it passed ordinary legislation, but it was still Parliament that was acting. For both ordinary and constitutional *lois*, the separation of powers required the judiciary to respect the actions of Parliament. To try to impose the will of Parliament constituted as constitutional legislator over Parliament constituted as ordinary legislator did not make sense.[31] The revolutionary texts, attacked as outdated by the proponents of constitutional review, were seen as merely expressive of the appropriate separation of powers in a democracy, since judges could not stand in the way of the will of the people.[32]

A significant concern of the writers was the conservative effect of providing constitutional review based on any declarations of rights. The authors like Duguit and Hauriou who proposed some form of substantive review envisaged that this would strike down measures such as the anticlerical laws of 1905, the provisions on secrecy of tax returns (which enabled private settlements between the revenue and the taxpayer), the granting of judicial powers to parliamentary committees, and attacks on property.[33] To Jèze, this would simply make the judiciary a block to social progress: 'Against a democratic Parliament, product of universal suffrage, and against its possible will for reform, it is desired in reality to set up bourgeois judges for the defence and irreducible preservation of the possessing classes characterized as élites.'[34] This concern seemed to be supported by the highly

[30] Ibid. 131.
[31] Ibid. 115.
[32] A. Esmein, *Éléments de droit constitutionnel français et comparé* (8th edn., Paris, 1927), 645.
[33] See Hauriou, *Précis de droit constitutionnel*, 289–90.
[34] G. Jèze, cited in L. Hamon, *Les Juges de la loi* (Paris, 1987), 75; id., *Les Principes généraux du droit administratif* (3rd end., Paris, 1925), 368.

influential study of the US Supreme Court published by Édouard Lambert in 1921. He described it as 'doubtless the most perfected tool of social inertia to which one can currently resort to restrain workers' agitations and to hold back the legislator from the slippery slope of economic interventionism'.[35] Time and again the opponents of constitutional review would point to the experience of the United States in the 1910s as an illustration of what could happen in France.

Other arguments centred on the effect that this would have on the judiciary. Currently, this career civil service was strongly influenced in its appointments by the Government, and there was no strong, fearless independence equivalent to the Supreme Court of the United States. In any case, judges would inevitably be drawn into the political forum by constitutional review, and this would lead to attacks on them for their political views that would further reduce the reputation of the French judiciary.[36]

In any case, if Jèze and Eisenmann were right about the nature of the Constitution of 1875, reduced to mere rules of procedure without any declarations of rights, an institution of constitutional review would have very little impact and importance, and would thus not be worth while.

The Intentions of the Drafters of the Constitution As originally conceived, the Conseil constitutionnel was not to be a radical departure from what had gone before in terms of its institutional competence. The essential difference was the new view of parliamentary sovereignty as limited by the role accorded to the executive. The Conseil was merely one institutional mechanism to ensure that this new function of Parliament was adhered to. The Comité consultatif constitutionnel saw the Conseil as 'a corset', 'an essential element for the harmonious operation of public authorities',[37] a body to keep Parliament and the executive within their proper limits. It was not to be some form of supreme court in constitutional matters, along the lines of the German Constitutional Court, a specialist body to which references can be made from all courts during litigation. As the *commissaire du gouvernement* (the civil servant representing the Government), Janot, stated:

> Such a system would be tempting intellectually, but it seemed to us that constitutional review through an action in the courts would conflict too much with the traditions of French public life. To give the members of the Conseil constitutionnel the power to oppose the promulgation of unconstitutional texts appeared sufficient to us. To go further would risk leading us into a kind of

[35] *Le Gouvernement des juges et la lutte contre la législation sociale aux États-Unis* (Paris, 1921), 224.

[36] Ibid. 235.

[37] *Avis et débats*, 205.

government by judges, would reduce the legislative role of Parliament, and would hamper governmental action in a harmful way.[38]

The nature of the body was shown by its composition. The debate in the Comité consultatif constitutionnel was on whether its membership would be incompatible with elective office. Like the Comité constitutionnel of the Fourth Republic, its structure was essentially political, and the mode of designation reinforced this, given that three politicians—the Presidents of the Republic, the National Assembly, and the Senate—nominated members.

Its functions were a bundle of tasks that were being withdrawn from Parliament. Debré wanted electoral litigation to be judged independently. In the past, Parliament itself decided on the lawfulness of parliamentary elections, and this gave rise to a number of partisan decisions. The Conseil would have more authority here, especially if its members were not currently active in political life. The reason for giving the competence over the constitutionality of treaties to the Conseil was likewise to avoid past difficulties, particularly the recent experience over the Treaty of Rome, which Debré had argued went contrary to national sovereignty, but which could never be put to the test before the Comité constitutionnel.

The new institutional arrangements justified further competences. The competence of the Conseil over parliamentary standing orders was justified differently, namely that, in the new system, Parliament could not be left to decide its own rules, since these might interfere with the proper function of the executive. Giving powers to the President to act in a state of emergency required some safeguards. Charles X had tried to use such powers against Parliament in 1830, and could only be prevented by revolution. The Conseil was empowered to provide a more practicable check on that power.

The primary function was, however, to check the constitutionality of legislation produced by Parliament, to make sure that it did not overstep the competence given in article 34 of the Constitution, discussed below in Chapter 3. Proposals that the Conseil should judge the constitutionality of executive *règlements* made under article 37 were quickly crushed by Debré. Likewise, proposals that the ground of review should include the Preamble to the Constitution and the declarations of rights to which it referred were rejected by the Government and by other members of the Comité. For instance, Teitgen argued: 'In empowering the Conseil to check whether a *loi* voted by Parliament is consistent with the Preamble to the Constitution, you fall back into government by judges, each one assessing subjectively the

[38] Ibid. 57.

express or implied meaning of this text.'[39] All the same, some did find it strange for the *commissaire du gouvernement* to argue that these principles would be applied by the Conseil d'État, but not by the Conseil constitutionnel.[40]

The longest debate in the Comité was on the question of who should be able to refer matters to the Conseil. The draft—as, indeed, the final text put to the people—only enabled the President of the Republic, and the Presidents of the Senate and of the National Assembly, and the Prime Minister to do so, representing the Parliament and the Government. Triboulet successfully moved an amendment in the Comité consultatif constitutionnel to allow one-third of the members of either Assembly to refer a *loi* to the Conseil. The argument was one of protecting the position of minorities within Parliament. Since the Presidents of the National Assembly and the Senate were, on past experience, likely to belong to the majority, the opposition had no way of having a matter raised before the Conseil.[41] Debré argued that this would be incompatible with parliamentary government, and that the constant participation in political life that such a proposal would entail could not be accepted. Similarly, Teitgen argued, 'Every time that a *loi* has given rise to an impassioned debate, the opposition will not fail to refer it to the Conseil constitutionnel, and in the end effective government will be in the hands of the pensioners who will sit in the Conseil.'[42]

The Conseil constitutionnel

The institution of constitutional review in France has been described as belonging to the 'European model', in that it is a specialist tribunal, independent of Parliament, making decisions binding on other bodies and courts, with constitutional questions concentrated before it.[43] That description hardly fits the original intention of the drafters of the Constitution, and demonstrates how far ideas have had to be revised since 1958. Indeed, in 1987 Raymond Barre commented:

> In France we have too often held the view that there was nothing but parliamentary law. Now parliamentary law can change according to majorities. But we have understood in the course of recent years that there is a constitutional law that is stronger than parliamentary law. Let us not remove this fundamental acquisition, because the guarantee of democracy and of citizens rests

[39] Ebid. 77.
[40] See Coste-Floret (later a member of the Conseil constitutionnel), ibid. 102.
[41] Triboulet, ibid. 76. See also the views of Malterre, ibid.
[42] Ibid.
[43] L. Favoreu, *Les Cours constitutionnelles* (Paris, 1986), 14–15.

on this recognition of a legal order emanating from the Conseil constitutionnel that is higher than the parliamentary order.[44]

That a potential presidential candidate could speak in this way shows the extent of the change.

The role of the Conseil constitutionnel has increased significantly since 1974, when members of Parliament were permitted to make references to it, and since 1981 it has considered most major pieces of legislation.[45] Developments have not been linear, and they have owed much to particular individuals within the Conseil. These individual contributions have left a legacy to the institution. In addition, the rolling renewal of membership every three years, and the collegiality of the Conseil's operation do justify the assertion of an institutional ethos when talking of the Conseil's role and performance.

The Functions of the Conseil constitutionnel The Conseil has five broad heads of jurisdiction, which are not necessarily related.

First, the Conseil is an election court and returning officer. It determines the existence of a presidential vacancy or incapacity, oversees the election process, and announces the results. It has a similar supervisory function in relation to referendums. With regard to parliamentary elections, it rules on disputed elections.[46] It also rules on the ineligibility of members of Parliament. The case-load is quite considerable. As a result of the parliamentary elections of June 1988, some eighty-five decision on electoral matters are reported in the annual *Recueil* of decisions of the Conseil constitutionnel.

In the case of parliamentary elections, the Conseil will judge after the event, though it has recognized that it may be appropriate to rule on an issue before elections taken place, where this affects a large number of constituencies. Thus in *Delmas*[47] it ruled on whether the duration of the election campaign was not too short, in breach of the Electoral Code. This affected all elections. In this area the Conseil is like any judge, seeing that the provisions of the Electoral Code have been obeyed. As an election court, it does not have jurisdiction to challenge the validity of the *lois* that set out the rules for elections.[48]

[44] *Le Monde*, 27 Jan. 1987: 8.

[45] L. Favoreu notes that the period from Feb. 1980 to Feb. 1989 accounts for 61.1% of the Conseil's decisions on the constitutionality of *lois*, 77.9% of the annulments of such texts, and 87.45% of the paragraphs setting out the reasons for decisions (called *considérants* because they always begin 'considering that . . .'): 'Le Droit constitutionnel jurisprudentiel', *RDP* 1989, 399 at 408–9.

[46] See arts. 7, 58, 59, 60 of the constitution (below, Part Two, Section I).

[47] CC decision of 11 June 1981, *Delmas*, D. 1981, 589, note Luchaire.

[48] CC decision of 5 May 1959, *A. N. Algérie, circ.* 15, *Rec.* 215. Equally, since the administrative orders and decrees governing elections are administrative decisions, they come within the competence of the administrative courts alone: Genevois, §§49–50, 59–62.

Secondly, the Conseil also advises the President both when he seeks to use emergency powers under article 16 and on the rules made thereunder. Such advice is not binding, but it is of considerable authority all the same. The practice of 1961 would suggest that the Conseil's formal advice is preceded by informal advice. This may, however, be due to the particular personalities involved in the 1961 crisis, and might not be so easily repeated.[49]

Thirdly, the Conseil may also be asked to rule on the constitutionality of treaties. Treaties are signed by the President, but require parliamentary legislation in most cases before they can be ratified. Once ratified, they have a status superior to *lois* (article 55). Although the Conseil constitutionnel will not strike down a *loi* for incompatibility with a treaty, other courts may refuse to apply it in such a case.[50] Prior examination of the compatibility of a treaty and the Constitution is thus desirable.

The Presidents of the Republic, the National Assembly, and the Senate, or the Prime Minister or 60 deputies or senators may refer a treaty for consideration by the Conseil to determine whether it is contrary to the Constitution. If it is, then it can only be ratified after a constitutional amendment has been passed (article 54). This procedure is merely an extension of the competence of the Comité constitutionnel under the Fourth Republic, about which there was controversy when the EEC Treaty was ratified. The President of the Republic has been the only one to make use of it in relation to EEC taxation (1970), European elections (1976), the additional Protocol to the European Declaration on Human Rights on the death penalty (1985) and the Maastricht Treaty (1992).

Fourthly, the Conseil also examines the constitutionality of organic laws and parliamentary standing orders. Both are subject to compulsory review by the Conseil before they are promulgated (article 61 §1).

Organic laws are required in a number of areas, such as on the judiciary, on the composition of Parliament, on finance laws, and on the procedure of the Conseil constitutionnel. The process for passing them is stricter than for ordinary *lois*, requiring the agreement of the Senate or an absolute majority of members of the National Assembly (article 46). Since these organic laws may be used subsequently as a basis for judging the constitutionality of *lois*, and may extend the body of constitutional rules, it is appropriate that the Conseil should review them before enactment.

The scrutiny of *parliamentary standing orders* is justified by the desire to ensure that Parliament does not overstep the boundaries set out for

[49] J. Boudéant, 'Le Président du Conseil constitutionnel', *RDP* 1987, 589 at 628.

[50] For a recent reaffirmation of this, see CC decision no. 91-293 DC of 23 July 1991, noted in *AJDA* 1991, 631. Treaties figure among the 'infraconstitutional values' discussed in ch. 2.

it in the Constitution. If Parliament were to adopt procedures that blocked the dominance of the executive, this could clearly upset the new arrangements of 1958. It may have been all right to leave such matters to the sole judgment of Parliament in an era of parliamentary sovereignty, but this could no longer be the case in the 1958 regime. This area will be examined further in Chapter 4.

Fifthly, the Conseil had, as its primary original function, to police the boundaries of the legislative competences of Parliament and of the executive. This is performed in any of three ways.

(1) Under article 37, the Government can only amend or repeal provisions in *lois* passed after 1958 by way of *règlement* if the Conseil constitutionnel has first declassified them, in other words, if it has ruled that the provision does fall within the domain of executive legislative competence. In this way, it ensures that the Government does not overstep its competence. The Government must take the initiative, and refer provisions of *lois* to the Conseil if it wishes to have them declassified.

(2) When private members' bills or amendments are proposed in Parliament that stray into the area of the executive's legislative competence, the Government may seek to have the proposed provisions ruled out of order. Where the President of the relevant chamber of Parliament disputes the claim of the Government, either he or the Prime Minister may refer the dispute to the Conseil, which has to give a ruling within eight days (article 41). Since 1979, this procedure has rarely been used.

(3) Once a *loi* has been passed by Parliament, the Conseil has jurisdiction to rule on its constitutionality if a reference is made to it by the President of the Republic, the President of either the National Assembly or the Senate, the Prime Minister, or (since 1974) sixty members of either Assembly (article 61 §2). The reform of 1974 effectively gave the opposition a chance to challenge legislation, and it has become almost the only challenger to *lois*.

Although originally designed to keep Parliament within the competences set out in article 34, the reference of enacted *lois* to the Conseil has become a procedure for challenging them on wider, substantive grounds, particularly for breach of fundamental rights. The importance of this procedure can be seen from Table 1.1.[51]

The Conseil only has jurisdiction to challenge a *loi* before it has been promulgated. In early decisions it stated that references cannot be used to challenge the validity of previously promulgated *lois*.[52] But in 1985

[51] Reproduced from J. Fournier, *Le Travail gouvernemental* (Paris, 1987), 93 (updated).

[52] CC decision no. 78–96 DC of 27 July 1978, *Monopoly on Radio and Television*, *Rec.* 29; see generally J.-Y. Cherot, 'L'Exception d'inconstitutionnalité devant le Conseil constitutionnel', *AJDA* 1982, 59.

TABLE 1.1. *References to the* Conseil constitutional *under article 61 § 2 of the Constitution*

Period	References under art. 61 § 2	Number not consistent	Proportion not consistent (%)
1959–September 1974	9	7	77.0
October 1974–April 1981	47	13	28.0
May 1981–March 1986	66	33	50.0
April 1986–January 1989	32	19	64.6

the Conseil declared *obiter*: 'though the validity with respect to the Constitution of a promulgated *loi* may properly be contested on the occasion of an examination of legislative provisions that amend it, complement it, or affect its scope, this is not the case when it is a matter of simply applying such a *loi*.'[53] More recently, the issue has arisen indirectly in relation to tax law. Procedures under article L96 of the Code of Tax Procedures for recovering penalties for infringements of stamp-duty legislation enabled the Fisc to impose a penalty without hearing the taxpayer beforehand. In the *Finance Law for 1990*[54] this procedure was extended to breach the new rule that non-businessmen should not use cash to pay for goods or services above a value of 150,000 F. The Conseil held that the application of article L96 to this case violated constitutional rights of due process, because, since no tax was involved, the normal safeguards inherent in tax procedures did not apply. Again, in the *Finance Law for 1991*,[55] an additional levy was imposed on betting tickets, and article L96 was to apply to failure to pay the levy. The Conseil noted that these provisions in no way obliged the tax authorities to observe due process before imposing a penalty, and accordingly struck down the provision. In these decisions the Conseil was effectively casting doubt on the validity of the provisions of article L96. Since the administration is bound by a general principle of law to withdraw illegal measures,[56] the decisions of the Conseil constitutionnel provide strong, but not binding, encouragement to the administration at least to reform its procedures, if not to encourage Parliament to repeal the dubious provision.

[53] CC decision no. 85-187 DC of 25 Jan. 1985, *Urgency in New Caledonia*, D. 1985, 361, note Luchaire.

[54] CC decision no. 89-268 DC of 29 Dec. 1989, *RFDA* 1990, 143, note Genevois: see p. 218 below.

[55] CC decision no. 90-285 DC of 28 Dec. 1990, *RFDC* 1991, 136.

[56] CE Ass., 3 Feb. 1989, *Cie Alitalia*, *RFDA* 1989, 391, *conclusions* Chahid-Nourai; *GA*, no. 116.

Composition of the Conseil constitutionnel Unlike the ordinary courts, there are no criteria for membership of the Conseil other than having been nominated by either the President of the Republic, the President of the Senate, or the President of the National Assembly. One member is appointed by each of these three every three years for a non-renewable term of nine years. Should a vacancy occur in the meantime, the office-holder who nominated the previous member nominates his replacement for the remainder of the term.[57] In addition, all past Presidents of the Republic are members for life. In practice, only the former Presidents of the Fourth Republic, Auriol and Coty, have ever taken their seats, and no past President has taken part in a decision since 6 November 1962.

The absence of specific qualifications for the post means that there are few limits on choice. The *ordonnance* of 7 November 1958 does set out certain functions that are incompatible with membership of the Conseil, and this does somewhat restrict the field of those willing to accept nomination. Under article 4, a member cannot also hold a position in the Government, in Parliament, or on the advisory Economic and Social Council. Thus Pompidou had to resign in 1962 to become Prime Minister, and Michelet resigned in 1967 to become a member of the National Assembly. Giscard d'Estaing ceased to be eligible to sit when he entered the National Assembly in 1984. During membership of the Conseil a person may not be nominated for any public employment or be promoted within the public service, except by reason of advancing age. Members are prohibited from taking a public position on questions that might be the subject-matter of a decision by the Conseil. In practice, this involves retirement from active political life, which few are willing to accept until the end of their careers. Despite suggestions to the contrary in the Comité consultatif constitutionnel, the Conseil constitutionnel is a committee of retired persons, with an average appointment age of 65 (the 1989 Conseil had an average age of 69, and that of 1992 has an average age of 64). It may well be that the absence of any political future gives the members the luxury of independence, and cushions them against adverse reactions from erstwhile colleagues.

The pattern of appointment is similar to the independent Comité constitutionnel of the Fourth Republic. Together with the designation of the Conseil as a 'Council' rather than a 'Court', this provides a clue to what was originally anticipated of its members. They are to be independent guardians of the republican constitutional tradition, freed

[57] Since that person is then usually nominated for a 9-year term in his own right, a member may serve for longer than 9 years. For instance, Louis Joxe served from Oct. 1977 until Feb. 1989.

from the vagaries of partisan politics, but also to typify the various strands within the mainstream of political life. Appointed by the three figure-heads of the State, they are meant to be more than mere party representatives. At the same time, the three nominators are unlikely to be from the same party. The nomination procedure provides for some form of participation by the expected protagonists—the Government and the Parliament. The sharing of power also confers legitimacy in the eyes of each.[58]

The most difficult feature of the nomination process concerns the dimension of competence. What characteristics make someone a suitable spokesperson for the republican constitutional tradition? First, since that tradition has, before 1985, been mainly a political one, experience in politics might be thought necessary. Two forms of experience might be distinguished here. On the one hand, there is party political, and in particular parliamentary experience, which lays stress on the democratic and parliamentary traditions of government. On the other, there is experience in government, related to the traditions of governmental effectiveness, public service, and the like. Both features are political, but in different ways. The distinction between the two has become more marked in the Fifth Republic, but it could be said to have existed before. In addition, since the contexts for decision-making will sometimes be politically charged, and since the Conseil cannot rely on long-standing authority for its judgments, some political sensitivity will be necessary to ensure that there is an awareness of the context of decisions and the limits of the Conseil's powers therein. Secondly, many of the constitutional values also have a legal source and force. Since the members of the Conseil have to interpret texts and to draft decisions that will have legal application and will be analysed by lawyers, some legal experience might also seem relevant. Thirdly, the persons appointed should command authority for their statements of the tradition. If the Conseil consisted of political hacks, its decisions would be effective, but not authoritative; every effort would be made to reverse them as soon as was opportune. If the Conseil is to help provide a stable political framework, which was certainly desired in 1958, then its decisions must carry some authority. The nomination process effectively secures a significant degree of independence from momentary political pressures, so that members can feel free to decide on the basis of tradition rather than on political expediency, but the character of the nominees is also important.

When one looks at the forty-eight appointments to the Conseil since

[58] On the characteristics of judicial selection processes, see J. Bell, 'Principles and Methods of Judicial Selection in France' (1988) 61 *Southern California Law Review* 1757 at 1769–79.

1959,[59] the features of parliamentary, governmental, and legal experience, as well as personal authoritativeness, can be seen to have shaped nominations. Of the forty-eight, fourteen had been Government ministers, and twenty-two had been members of Parliament at some time. Only eight had no known political affiliations, and ten had governmental experience as holders of a ministerial office. As far as legal experience is concerned, only eleven had no legal training. Of the rest, eleven were law professors, eleven were members of the Conseil d'État, twelve had been in private practice, and four were judges of the Cour de cassation; a further four had law degrees. The high proportion of law professors is explained by the fact that this is one of the few careers in the public service that can be pursued at the same time as membership of Parliament. It can also be continued during membership of the Conseil constitutionnel, and this has typically been done, with two members becoming Presidents of their university while they held office. The current President of the Conseil, Robert Badinter, continued to teach at the University of Paris 1 during his political career, and carries on while a member of the Conseil constitutionnel. A parallel explanation in the private sector can be offered for the number of practitioners. These two groups constitute the most likely source of lawyers with political or governmental experience. The Conseil d'État offers a career that enables members to move from judicial and legal advisory activity to work in a government department, or even to hold office as a minister.[60] Personal authoritativeness and distinction are evident among the lawyers, who include a President of the Criminal Chamber of the Cour de cassation, one past and one future Vice-President of the Conseil d'État, a former judge of the European Court of Justice, a First Advocate-General of the Cour de cassation, and five of the leading constitutional lawyers in recent times. Among the politicians, there have been a former President of the Senate, three Ministers of Justice, a former *médiateur* (Ombudsman), two ambassadors, and three leading figures in the human rights movement.

Despite the absence of any vetting procedure, few nominations have caused much criticism. The appointment of Robert Badinter, the Minister of Justice, as President of the Conseil in February 1986, just one month before the Socialists were almost certain to lose office, did give rise to disapproval. Since the existing President had 'resigned' after only three years, but was remaining on the Conseil, it appeared

[59] For a full list of the qualifications of members of the Conseil constitutionnel, see L. Favoreu, *RFDC* 1990, 604–5, supplemented by *Le Monde*, 27 Feb. 1992. See also below, Table 1.2.

[60] 5 Prime Ministers have come from this corps this century, including 3 in the Fifth Republic—Debré, Pompidou, and Fabius.

that the President of the Republic was trying to rig the composition of the Conseil in anticipation of a difficult period during which he would have to preside over a Government composed of his political opponents.[61]

The Conseil in office from March 1989 consisted of its President, Badinter (aged 61), a former Minister of Justice, criminal advocate and law professor; Mayer (79), a former minister and President of the League of Human Rights; and Faure (67), a former Minister of Transport, head of a political party, and doctor in geography and law, all nominated by the President of the Republic. The President of the Senate nominated Jozeau-Marigné (81), a former senator and honorary *avoué* with a doctorate in law; Latscha (62), a company director, company legal adviser, and (briefly) a professor of constitutional law; and Cabannes (64), a former First Advocate-General at the Cour de cassation, and head of the Minister of Justice's private office (*directeur de cabinet*). The President of the National Assembly appointed Fabre (74), a former Socialist deputy and *médiateur*; Mollet-Viéville (71), a distinguished advocate; and Robert (61), a leading professor of constitutional and civil liberties law. In February 1992 the first woman was appointed, Mme Noëlle Lenoir (aged 43). She was a member of the Conseil d'État, and an experienced administrator in government departments and in committees on civil liberties and bio-ethics. The other nominations were Rudloff (68), a leading senator and a practising *avocat*, and Abadie (63), an experienced prefect and administrator in a government department. All have law degrees. Of the eight appointees since March 1986, only two (Faure and Rudloff) have held office in Parliament, in Government, or in a political party. Most of the rest (Mollet-Viéville, Latscha, Cabannes, and Robert) have predominatly legal distinction, while the 1992 nominees, Lenoir and Abadie, are principally administrators. Indeed, the Conseil in March 1989 was the first to have a majority who had not been ministers or parliamentarians. This may be a reaction to the Badinter controversy of February 1986, but it also reflects the fact that nominees have not always been of the same political persuasion as the nominator, and quality has often prevailed over political allegiance.[62]

Even if nominees have been chosen with their political affiliations firmly in mind, their perfomance has not always conformed to the views of their nominators. The first Conseil was strongly marked by allegiance to de Gaulle, and was subservient to his ideas for the new

[61] See *Le Monde*, 21–2 Feb. 1986: 7; *Libération*, 20 Feb. 1986: 9–10.

[62] Even in an analysis of the early membership of the Conseil, it was noted that attention to political allegiance and experience did not diminish the legal and other distinctions of nominees: L. Favoreu, 'Le Conseil constitutionnel: Régulateur de l'activité normative des pouvoirs publics', *RDP* 1967, 5 at 73–88.

TABLE 1.2. *Membership of the* Conseil constitutionnel

Term of office	Name of member	Political experience	Legal qualification	Other experience
1959–62	POMPIDOU	activist	CE	adviser to President
1959–62	DELEPINE	—	*avocat*/CE	—
1959–62	CHATENEY	deputy	law degree	—
1959–62	PATIN	activist	Cass.	—
1959–65	NOEL	activist	law degree	ambassador/ CCC/*cabinet*
1959–65	LECOQ DE KERLAND	—	*avocat*	—
1959–65	PASTEUR VALLERY-RADOT	deputy	—	doctor
1959–68	GILBERT JULES	senator	*avocat*	CCC
1959–68	MICHARD-PELLISIER	deputy	*avocat*	—
1962–4	CHENOT	activist	CE	—
1962–7	MICHELET	minister/ deputy	—	—
1962–71	WALINE	activist	professor	CCC
1962–71	CASSIN	—	CE/*avocat*/ professor	ECHR
1964–8	DESCHAMPS	activist	CE	—
1965–74	PALEWSKI	minister/ deputy	—	diplomat
1965–74	LUCHAIRE	activist	professor	*cabinet*/adviser to CCC
1965–74	MONNET	activist	*conseiller juridique*	—
1967–74	ANTONINI	—	law degree	*cabinet*
1968–77	SAINTENY	minister/ deputy	—	colonial governor
1968–77	DUBOIS	—	Cass.	—
1968–77	CHATENET	minister	CE	—
1971–7	REY	minister/ deputy	—	—
1971–9	COSTE-FLORET	senator	professor	CCC
1971–80	GOGUEL	activist	professor	S-G Senate
1974–83	FREY	minister/ deputy	—	CCC
1974–83	MONNERVILLE	minister/ senator	*avocat*	President of Senate
1974–83	BROUILLET	activist	—	*cabinet*/ Comptes

TABLE 1.2. (*cont.*)

Term of office	Name of member	Political experience	Legal qualification	Other experience
1977–83	PERETTI	deputy	*avocat*	prefect
1977–84	GROS	senator	*avocat*	—
1977–86	SEGALAT	—	CE	S-G Govt
1977–89	JOXE	minister/ deputy	—	S-G Govt
1980–9	LECOURT	minister/ deputy	*avocat*/ECJ/ Cass.	—
1980–9	VEDEL	activist	professor	ESC
1983–6	LEGATTE	senator	CE	*cabinet*
1983–7	MARCILHACY	senator	*avocat*	CCC
1983–92	MAYER	minister	—	HR
1983–92	JOZEAU-MARIGNE	senator	law degree	—
1984–8	SIMMONET	minister/ senator	professor	—
1986–	BADINTER	minister/ deputy	professor/ *avocat*	HR
1986–	FABRE	deputy	—	*médiateur*
1987–92	MOLLET-VIEVILLE	activist	*avocat*	—
1988–	LATSCHA	—	professor	businessman
1989–	FAURE	minister/ deputy	—	—
1989–	ROBERT	—	professor	HR
1989–	CABANNES	—	Cass.	—
1922–	LENOIR	—	CE	*cabinet*
1992–	RUDLOFF	senator	*avocat*	—
1992–	ABADIE	—	law degree	*cabinet*/prefect

Note: *cabinet* = civil servant in ministerial office; Cass. = judge of Cour de cassation; CCC = Comité consultatif constitutionnel (member, unless otherwise indicated); CE = member of Conseil d'État; Comptes = member of Cour des comptes; ECHR = judge of European Court of Human Rights; ECJ = judge of European Court of Justice; ESC = member of Economic and Social Council; Govt. = Government; HR = human rights activist; S-G = Secretary-General.

Republic. Its President, Noël, was a close confident of de Gaulle and was consulted on numerous political matters, tending to defer to his interest in his decisions.[63] His successor, Palewski, admitted that the Conseil had acted as 'yes-men' to the General in the 1960s, but argued

[63] See Boudéant, 'Le Président du Conseil constitutionnel', 627–32.

that it was absurd to contradict the author of the Constitution as to its interpretation.[64] This is most clearly illustrated by the 1962 decision on the *Referendum Law*.[65] Nevertheless, it was the same, previously subservient Conseil under Palewski that decided that it could review legislation on grounds of conformity to fundamental rights in the *Associations Law* decision of 1971. Again, as Favoreu and Philip remark, 'In reality, the Conseil of 1977–1980, which one might consider one of the most political (Frey, Monnerville, Peretti, Joxe, Coste-Floret), was one of the most active, least servile, most protective of freedoms and most juridical.'[66]

Despite the absence of any member appointed by a Socialist, the Conseil of 1982 was not manifestly obstructionist to the Socialist change of direction, and, as has been noted, placed less constitutional objections in the way of the reforms than the Conseil d'État would have done. It has been argued that 'experience shows that nomination transforms the politician into a constitutional judge, i.e. that it substitutes for the passion, the commitment, and the partiality of the former, the serenity, wisdom, and loftiness of perception necessary to accomplish the tasks of the second.'[67] Shorn of its Gallic floridness, the comment rightly underlines the different kind of role that the member of the Conseil sees himself as undertaking. The appointment of distinguished individuals at the end of their careers enables the 'nine wise men of the rue de Montpensier' to stand somewhat aloof from political influence and to give effect instead to their own perceptions of the Constitution and the republican constitutional tradition. In this, they are more likely to have views in common with others who have been engaged in the various facets of that tradition, than over more clearly partisan questions. Although standing within the political process, the Conseil aims to impose stable and fundamental values in a situation where opportunism and short-term considerations have a significant place in the motives of the principal actors. The particular character of the political decisions to be taken by the Conseil may mean that members will frequently unite despite their party allegiances. But there will still be issues on which the Conseil divides along lines that correspond to party opinion. Voting figures are not known, but it has been widely stated that the decision of 23 January 1987 on the *Séguin Amendment* was adopted on the casting vote of the President, Badinter, after the four Socialists (Badinter, Fabre, Marcilhacy, and

[64] Ibid. 637.
[65] Ibid. 631–3, and below, p. 133.
[66] *GD* 378.
[67] C. Debbasch, J. Bourdon, J.-M. Pontier, and J.-C. Ricci, *Droit constitutionnel et institutions politiques* (Paris, 1984), 492.

Mayer) tied with the more right-wing members (Jozeau-Marigné, Lecourt, Simmonet, and Vedel) in the absence of Joxe.[68]

Apart from the lack of partisan pressure, the collegial character of the institution will tend to reinforce an independent line. Since the members have to work with each other over a significant period, they will tend to adjust to these internal constraints and expectations as well as to external pressures and expectations. The internal dynamics of the Conseil are not well known, but there are sufficient indications that a collegial spirit operates to produce unanimity in most cases. The importance of this will depend on the character of the persons in question—whether they operate entirely on their own or not—and on the animation that the President gives to team-work. Experience of working in such an environment may be more important than the particular origins of members.

The complexity of the factors that influence decision-making by any judge suggests that there is no straight correlation between any one element, such as political opinions or experience, and votes on particular decisions. Since the assumption is that several factors are relevant to appointments, and since internal institutional ethos may moderate the influence of external expectations about an individual's performance, the significance of any one feature of the members may not be great in determining the outcome of cases. It is the fact that certain experiences give a general sensitivity to problems, and confer a kind of *ex ante* authority on the members of the Conseil that is the most important element of the nomination process. The effect of all this on decisions is rather contingent, especially as there is no mechanism of control or formal criticism of the decisions made (and it is hard to know how individuals voted in any case).

Procedure and Methods of Working in the Conseil constitutionnel Since the Conseil is not formally a court, its procedures are not fully judicial, and in many ways they resemble those of an administrative inquiry. There is no set of procedural rules, other than the exiguous organic law contained in the *ordonnance* of 7 November 1958 (extracts from which are reproduced in Part Two, 'Constitutional Texts', Section IV). The fact that members of the Conseil are bound to secrecy about their deliberations has led to an air of secrecy about the way in which it functions. The following description of the procedure draws on a number of disparate remarks in the literature, and on discussions with members of the Conseil and especially of its legal service.

The procedure before the Conseil varies significantly depending

[68] CC decision no. 86-225 DC of 23 Jan. 1987 (DECISION 16); Boudéant, 'Le Président du Conseil constitutionnel', 620.

on the task in hand. It is most judicial when dealing with electoral disputes, and least judicial when dealing with the declassification of *lois* under article 37 §2.

Acting as an election court for parliamentary elections, the Conseil operates very much like the Conseil d'État when it deals with disputed local elections. Here the *ordonnance* of 7 November 1958 is at its fullest, providing fourteen detailed articles on the procedure to be adopted.

Within ten days of the election result, any candidate or elector for the constituency concerned may present a petition by way of a letter to the Conseil, or to the local prefect, or to his or her equivalent, setting out the grounds of complaint. For election matters, the Conseil then divides into sections of three members, chosen by lot, who are assisted by associate reporters selected from the Conseil d'État or the Cour des comptes on an annual basis. One of the associate reporters is appointed to act as reporter (*rapporteur*) on the particular petition. His task is to provide the analysis of facts and rules on which the section can prepare a report for discussion by the whole Conseil. This process of *instruction* is closely modelled on the procedure of the Conseil d'État.[69] Petitions that are inadmissible for some reason—because the petitioner is not an elector of that constituency, for example—or that clearly cannot affect the result of the election can be rejected without *instruction*. In other cases the reporter will collate the allegations and any observations from the elected deputy or senator. The section can conduct hearings under oath, and can require the communication of any official document relating to the election. A member of the Conseil or the reporter may be sent to conduct a site inspection. The results of these enquiries are provided to the parties concerned, who then make their observations (with or without the help of lawyers). Once the *instruction* is complete, the report is presented to the Conseil. Only members of the Conseil have a vote, though the reporter may present cases to the meeting.

It is clear that here the Conseil is operating much as any administrative court, with all the safeguards of hearing both sides.

The rules on the procedure for referring legislative texts are more exiguous. The *ordonnance* of 7 November 1958 merely states who should transmit the text, who should be notified in the case of a reference made by members of Parliament, and that decisions should be reasoned. For the rest, it is a matter of practice and personalities, both of which have changed over time.

Where there is a compulsory reference, the text is transmitted by the Prime Minister in the case of an organic law, or by the President of the relevant Assembly in the case of parliamentary standing orders. Since

[69] See Brown and Bell, *French Administrative Law*, ch. 5.

both have been the subject of a public vote in Parliament or in one of its Assemblies, there is no need to warn anyone that a reference has taken place, nor are there specific grounds for it.

Where the reference is optional, as in the case of ordinary *lois* and treaties, then notification of the fact that this has been done might appear more necessary. All the same, it is only formally required where a reference is made by sixty senators or deputies, so that the office-holders—the President of the Republic, the Prime Minister, the President of the National Assembly, and the President of the Senate—are aware of this and can make observations.

The Conseil must reach its decision within a short period of time. This is normally one month, though it can be reduced to eight days where the Government claims that it is a matter of urgency. The issue of who is to judge 'urgency' has never been put to the test, and is left obscure by both the Constitution and by the *ordonnance*. In practice, there is a *gentleman's agreement* not to use this procedure, and it was only officially requested in one case between 1981 and 1986 (the second decision on *nationalizations* in 1982).[70] Usually, the Secretary-General of the Conseil and the Secretary-General of the Government come to an arrangement about the time-scale for decision. It is notable that the urgency procedure was not invoked in the case of the *Urgency Law for New Caledonia* in January 1985, when the Conseil took barely twenty-four hours to make its decision. Equally, at the Government's request, the Conseil published its first decision on *Nationalizations* on a Saturday, when the Stock Exchange was closed, since the decision was bound to affect dealings in the companies in question.

There is no requirement that the President should wait to see whether there will be a reference before promulgating the *loi*. This can lead to nervousness on the part of members of Parliament that the text will be promulgated before a reference can be made, and this results in a number of stratagems. The most extreme came with the *Urgency Law for New Caledonia*, where a courier was stationed outside the Conseil with a reference already signed. When the final vote on the text had been taken in Parliament, he was contacted on citizens' band radio and promptly marched in to present the reference. It is more usual for the text of the reference to be discussed in the parliamentary group of the party wishing to make a reference, and then signed by the requisite number of deputies or senators, before the final vote is taken on the text.[71] Because the reference has been drawn up before the bill has been approved by Parliament, it can happen that it complains about articles that are not in the final text of the *loi*.

[70] CC decision no. 82-139 DC of 11 Feb. 1982, *Nationalizations II*, Rec. 31. See Fournier, *Le Travail gouvernemental*, 97.

[71] See M. Charasse, 'Saisir le Conseil constitutionnel' (1986) 13 *Pouvoirs* 81.

The reference can take the form of a single letter or several letters from individual deputies or senators; the Conseil only counts the first sixty, to reach the requisite number. It is usually reasoned, setting out grounds for challenging particular articles of the *loi*, but, as in the case of the *Associations Law* of 1971, the letter may simply request the Conseil to examine the constitutionality of the whole text.[72] In addition, the party leader may submit a memorandum containing more detailed arguments; this occurred in the *Vehicle Searches* case of 1977, for instance. These days the references are usually well reasoned, and are often based on legal arguments drawn up by consultants (frequently professors of constitutional law). They may, at times, be excessively inventive in argumentation. Since 1983, the text of the reference has often been published in the *Journal officiel*. (An illustration of a reference is provided in Part Two, DECISION 12.)

The Conseil considers itself free to examine any part of the text, not merely the specific articles contained in the reference. Since the decision certifies the constitutionality of the *loi*, the Conseil considers that it has the right to raise issues ex officio. This was done, for example, in the *Feminine Quotas* case of 1982,[73] where the article of the electoral law providing for a minimum of 25 per cent of candidates of each sex on the lists for local elections was struck down as unconstitutional, even though this point was not discussed in the reference. On the whole, issues are only raised ex officio where there is an obvious and serious question of unconstitionality.[74] This practice has become more frequent in recent years, rising from one decision in twenty between 1974 and 1981, to one in ten between 1981 and 1986, one in three in 1986–7, and one in five from 1987 to 1989, and there were twelve such arguments raised by the Conseil ex officio in 1989–90.[75]

The short time-scale for deliberation, and the limited scope of the argumentation in the reference may require the Conseil to do a lot of work very quickly. Quite sensibly, the Conseil has developed a practice of jumping the gun. When a bill is presented to Parliament, opponents will usually move a motion that it or some of its articles are unconstitutional, and are thus out of order, and will set out the reasons for this view. The Government will provide a reasoned reply to the motion, thereby making sure that the basic issues have been aired. Such motions put the Conseil on alert, and the President may well

[72] DECISION 1: there was a subsequent supplementary reasoned memorandum in this case, but this is not a requirement of a reference.

[73] DECISION 34.

[74] B. Genevois, *RFDA* 1990, 406 at 408.

[75] Favoreu, 'Le Droit constitutionnel', 426–7, and L. Favoreu and T. Renoux, *RFDC* 1990, 780.

then decide to appoint a reporter to start work, building up the file on the potential reference. The choice of reporter is at the discretion of the President. Attention will be paid to a reasonable distribution of the work-load, but also to the expertise of particular members. Although the identity of the reporter is not officially made public, it is known that Chatenet and Lecourt acted as reporters on European Community matters, Segalat on finance bills, Gros on broadcasting laws, and Vedel on nationalizations (1982), university professors (1984), privatizations (1986), the Competition Council (1987), and the press (1984 and 1986).[76]

The file will be built up by the reporter, acting usually with the help of the Secretary-General (an expert lawyer) and the small legal service under him.[77] This will consist initially of three elements. First, there are the legal texts (constitutional and otherwise) on this particular area, to show the context and scope of the contested provisions. Secondly, there are the parliamentary debates, with special attention to the arguments on constitutionality developed at different stages in the parliamentary proceedings (the Conseil is equipped to listen into parliamentary debates at any time). Thirdly, there will be other materials, such as case-law in the public or private law courts, doctrinal legal writings, or memorandums produced by the legal service to help the reporter (which can be more or less detailed). The advice that the Conseil d'État gives on a Government bill is not published, and there is no formal mechanism for it to be transmitted to the Conseil constitutionnel. Nevertheless, since a member of the Conseil d'État is likely to be on either the Conseil constitutionnel or its legal staff, no difficulty is usually encountered in obtaining the text of the advice. Depending on the personality and legal expertise of the reporter, the Secretary-General and the legal service may well have an important role in the preparation of the file and in the drafting of the text of the decision presented to the Conseil. (It is not, however, unknown for a reporter to seek legal advice from outside the Conseil's staff.)

Once a formal reference has been made, the reporter will have a precise set of grounds to work from, though they will be almost identical to points made earlier. The text of the reference is sent to the Secretary-General of the Government, who will usually provide observations on it or on the *loi* in general. These observations do not represent a full defence. As a recent Secretary-General has written:

[76] Boudéant, 'Le Président du Conseil constitutionnel', 615 nn. 107, 108; Favoreu, 'Le Droit constitutionnel', 411.

[77] The legal service is small; in 1986 it consisted of 3 people, with an additional consultant and 2 researchers. The President may well have his own advisers; for instance, the present President is advised by Luchaire, a former member of the Conseil and a leading constitutional expert.

'The written observations presented by the Secretary-General of the Government in no way constitute a memorandum in defence. Most often, they remain limited to replying to questions asked by the member of the Conseil constitutionnel designated as reporter during a working meeting held at the Conseil constitutionnel.'[78] Sine 1986, this memorandum has been sent to the authors of the reference, so that they can make comments. All the same, there is not the same kind of hearing of each side that occurs in disputes over electoral matters.

The reporter remains master of the procedure. He may consult or listen to whomsoever he likes, and take note of whatever he wishes. It is up to him how much of what he learns in this way is communicated to his colleagues when the Conseil meets. He may often confer with the reporters of the parliamentary committees that examined the bill. On 3 June 1986 Badinter suggested that this might be formalized, and that the reporters might be consulted officially. This proposal was rejected by the Presidents of both chambers on the ground that the committee reporters had no standing to speak on behalf of the whole chamber. All the same, the practice is of some importance.[79] The authors of the reference and the Secretary-General of the Government may well be called to meetings with the reporter, so that he can clarify issues. In addition, others may be invited to attend or may seek an audience. Pressure groups may telephone the reporter at home or invite him out to lunch. For example, in the *Nationalizations* case some of the directors of the affected companies were seen by the reporter. Interested persons may write letters to the reporter, and he makes such use of them as he considers fit. These are not necessarily referred to in the *visas* (the introductory phrases beginning 'Vu . . .') of the decision.[80] The point is that, like an administrator compiling a dossier, the reporter follows up all interesting leads until he considers that he has seen all sides of the question.

Reporters work in different ways: some very much on their own, others discussing matters with the legal service, and others discussing with colleagues. Since not all of them are in Paris, this last may be difficult to arrange much before the decision-making meetings. The

[78] R. Denoix de Saint-Marc, in *Conseil constitutionnel et Conseil d'État* (Paris, 1988), 108.

[79] See L. Favoreu (ed.), *Nationalisations et Constitution* (Paris, 1982), 29; Boudéant, 'Le Président du Conseil constitutionnel', 618.

[80] See the letter of the Green party (*Les Verts*) to the Conseil constitutionnel concerning a provision in a *loi* submitted to it that permitted new tourist developments in mountain regions: *Le Monde*, 17 July 1990. Although the Conseil constitutionnel annulled the provision criticized by *Les Verts* (art. 16) ex officio, and even though it was not challenged by the authors of the formal reference, commentators are sceptical of any link between the letter of *Les Verts* and the willingness of the Conseil to challenge the provision of its own motion: see Favoreu and Renoux, *RFDC* 1990, 730, and J.-C. Douence, *RFDA* 1991, 346. For the text of the decision, see CC decision no. 90-277 of 25 July 1990, *RFDA* 1991, 354.

President of the Conseil may play an important part. He may keep in touch with a reporter to see how things are going, and may suggest meetings of a few members from time to time. (Unlike members of the Conseil d'Etat, members of the Conseil constitutionnel have their own individual offices, and may well thus work on site.) The President may prefer just to have an occasional lunch with the reporter or with other members of the Conseil. Other Presidents have left reporters very much on their own. Although he does not act as a reporter himself, the President may well secure a good sense of how the draft judgment is going to look, and may seek to influence its content. But he is not typically a dominant figure among so many authoritative individuals. One former member of the Conseil wrote:

> In truth, important though the role of its President is in the functioning of the Conseil constitutionnel, this is only through the climate that he creates between its members, by the tone that he contributes to deliberations, and by the way in which he conducts these. This role includes no interference with the judgment of each of the Conseil's members. It would be altogether wrong to contrast the Gaston Palewski case-law or the Roger Frey case-law with the case-law of Léon Noël: there is only a case-law of the Conseil constitutionnel, the development of which is explained essentially by the widening of the grounds on which references are made.[81]

The reporter produces a draft judgment, which is circulated to all members of the Conseil at least one day before the decision-making meeting. These other members may also receive a general file from the Secretary-General setting out the legal texts, the parliamentary debates, and other matters that the reporter considers would be useful.

The actual decision-making occurs in the meeting-room of the Conseil, part of the Palais-Royal fitted out for Napoleon III's sister. Only members of the Conseil attend, and no minutes are kept. The report is discussed, and votes, if necessary, are taken after that. No time-limit is set on discussion, and some cases take more than one day (hence they have two dates in their official reference). It is not generally known how people vote. It is known that the Conseil split 6 to 4 on the *Referendum Law* (DECISION 14) decision, and that it split 4 to 4 on the *Séguin Amendment* (DECISION 16). However, it is reliably said that unanimous decisions are common, and that members do not necessarily vote in the way that their party allegiances might suggest. Once they are in the Conseil, with no political future ahead of them, they can act as free agents. In any case, the dynamics of collective decision-making, free from the public gaze, may produce different pressures from those in a politicized forum.

When the Government wishes to legislate by way of decree, using its

[81] F. Goguel, 'Le Conseil constitutionnel', *RDP* 1979, 5 at 24.

powers under article 37, it may need to ask the Conseil to declassify a provision contained in a *loi* enacted after 1958. (It does not have to follow the declassification procedure for earlier *lois*.) The procedure is entirely *ex parte*, in that nobody other than the Government and the Conseil is involved. The Government will submit a list of texts that it wishes to amend or repeal by decree, together with drafts of the provisions that it proposes to enact. This last part of the procedure is not necessary, but it helps the Government to obtain a useful ruling from the Conseil, in that the latter will try to frame its decision in such a way as to provide guidance to the Government on what it has specifically in mind.

Authority of Decisions of the Conseil constitutionnel Article 62 states that a provision that the Conseil has declared unconstitutional cannot be promulgated or implemented. That aspect of the decision effectively binds the President on *lois* and treaties, the Government on proposed *règlements*, or Parliament in respect of proposed bills, amendments, or standing orders. More widely, the same article states that its decisions 'are binding . . . on all administrative and judicial authorities'. In the case of the administration, a circular of the Prime Minister of 25 May 1988 reminded civil servants of the need not only to respect decisions of the Conseil constitutionnel, but also to anticipate potential breaches of the Constitution.[82] If the purpose of this kind of constitutional review is to obtain early and authoritative rulings on all aspects of a *loi*, then it is important that the decisions are adhered to by the courts. But French courts do not have a formal doctrine of *stare decisis*, and were initially reluctant to treat rulings on abstract points of law as authoritative when they come from what is, formally at least, a non-judicial body, though attitudes have changed in more recent years.

The normal policy for French courts, set out by article 5 of the Civil Code, is that they cannot lay down general rules for the future. The formal authority of the decision is thus confined to the case itself, and appeal to previous judgments is not, as such, a sufficient reason for a judicial decision. All the same, in practice, courts will follow earlier judicial decisions, especially those of the highest courts. *La jurisprudence* has thus a real authority, even if there is no rigid, formal rule of *stare decisis*.

Unlike the ordinary courts, the Conseil is not solving particular disputes between parties, but ruling in abstract on the validity of a *loi* that will affect a variety of future cases. It is said to judge a text, not litigants. In addition, the Conseil recognizes its responsibility for creating constitutional doctrine, a doctrine far more unsettled than

[82] See *RDP* 1989, 436.

private, criminal, or administrative law. In its decisions the Conseil has tried, therefore, to set out general principles of constitutional law, rather than simply to make specific rulings relating only to the particular *loi* under discussion. In this way general guidance can be offered to the Government, the Parliament, and the courts.

The Conseil takes a wide view of the binding force of its decisions. In the *Agricultural Orientation Law* of 1962.[83] the Conseil stated 'that the authority of the decisions [of the Conseil constitutionnel] mentioned [by article 62 of the Constitution] attaches not only to their result (*dispositif*), but also to the reasons that are its necessary support and constitute its very foundation'. The Conseil had already ruled in two decisions of 1961[84] that certain parliamentary amendments to an agricultural bill that sought to fix prices fell within the legislative competence of the Government under article 37. When asked the same question by the Government in relation to a *loi* of 1960, the Conseil merely replied that it did not require an answer since it had already ruled on that matter.

The scope of such binding authority extends to any provision with the same effect as one on which the Conseil has already ruled. Thus, in the case of the *Amnesty Law of 1989*[85] a provision was introduced both to amnesty and to make eligible for reinstatement those who had been guilty of serious fault during industrial disputes. A clause to this effect had already been struck down in the case of the *Amnesty Law of 1988* because it would impose an excessive burden on the victims of the fault.[86] Although the legislature had tried to modify the 1988 provision by making an exception for employers thus affected, on the Conseil did not consider that this had cured the problem, particularly in relation to the burdens that reinstatement would place on fellow employees. Relying on its previous decision, the Conseil struck down the new provision because it 'violates the authority that attaches, by virtue of article 62 of the Constitution, to the decision of the Conseil constitutionnel of 20 July 1988'.

The private and administrative courts have used techniques similar to those known in common law to distinguish decisions of the Conseil constitutionnel that they have not wished to follow. The first technique is to confine the decision to the text that was before the Conseil. This is often combined with a second technique of restrictive interpretation of the reasoning. Thus, in a decision of 1977 the Conseil held as unconstitutional a *loi* that intended to confer on the police an unlimited

[83] CC decision no. 62-18 L of 16 Jan. 1962, *Rec.* 31.
[84] CC decision nos. 61-3 FNR of 8 Sept. 1961, *Rec.* 48, and 61-4 FNR of 18 Oct. 1961, *Rec.* 50.
[85] DECISION 4.
[86] DECISION 3b.

power to search vehicles on the highway even where no crime had been committed and there was no threat to public order, on the ground that the imprecise nature of the grounds of intervention by the police threatened individual liberty. In 1979 the Chambre criminelle of the Cour de cassation held that the police could search vehicles belonging to any person under the general provisions relating to the investigation of 'flagrant offences'.[87] While formally consistent, the latter decision did much to undermine the effect of the decision of the Conseil constitutionnel.

The third technique is to draw a distinction between the necessary reasons for the decision and other points that may be raised (in other words, between the *ratio decidendi* and mere *obiter dicta*). This again arose with regard to criminal penalties. These are divided in French law between *crimes*, *délits*, and *contraventions*. Article 34 of the Constitution states that *loi* should lay down the rules for 'the determination of *crimes* and *délits* and the penalties applicable thereto'. A provision of the Rural Code established the fines for illegal joinder of agricultural properties (an offence in the nature of a *contravention*). In declaring this to be within the competence of the executive to amend by way of *règlement*, the Conseil constitutionnel ruled that 'the determination of *contraventions* and the penalties applicable to them falls within the province of *règlement* when those penalties do not include measures depriving a person of their liberty'.[88] The *procureur général*, Touffait, sought to argue in a subsequent case that the proviso relating to measures for the deprivation of liberty did not bind the criminal courts, because 'the Conseil constitutionnel stressed its reasoning by inserting a general principle, which in this case was incidental, not to say superfluous'.[89] The necessary reasons were only those most directly concerned with disposing of the case before the Conseil, namely, the classification of those provisions submitted to it, and Touffait was followed by the Cour de cassation in thinking that the proviso did not fall within that specific category.

Although the formal *ratio* of the Conseil's decisions may be understood narrowly, their practical importance reflects its function of giving authoritative rulings on the meaning of the Constitution upon which a variety of public authorities can rely.

Article 62 states that the decisions of the Conseil constitutionnel bind public powers and administrative and judicial authorities. In the first group are included the President, the Government, and the

[87] Cass. crim., 8 Nov. 1979, *Triganol*, D. 1980 Chr. 102; cf. CC decision no. 76-75 DC of 12 Jan. 1977, *Vehicle Searches*, DECISION 17.

[88] CC decision no. 73-80 L of 23 Nov. 1973, *Criminal Penalties (Rural Code)*, DECISION 10.

[89] *Conclusions* to Cass. crim., 26 Feb. 1974, *Schiavon*, D. 1974, 273.

Parliament. The President is not permitted to promulgate a *loi* or any provisions of it that have been declared unconstitutional by the Conseil, nor can he promulgate a text where the unconstitutional provisions have not been declared to be severable from the *loi* as a whole. It is up to the President to decide whether to promulgate a text without the severable, unconstitutional provisions, to require a new deliberation by Parliament on the whole text submitted to the Conseil, or merely on those articles that were declared unconstitutional,[90] or to require a new *loi* to be presented to Parliament. Decisions on declassification restrict matters on which the Government can legislate by way of decree. Similarly, decisions on the admissibility of amendments bind the parliamentary chambers as to the proposals that can be discussed or adopted. In each case, the binding effect of the decision is very firmly limited to the text considered by the Conseil.

As far as the Conseil itself is concerned, it is in no way bound by its previous decisions. All the same, to ensure its authority and effectiveness, these 'must be marked with the seal of continuity and coherence'.[91] From the earliest days, the Conseil has sometimes referred to its previous decisions either in the *visas* or in its actual reasons. But such citations are confined to instances where these do have binding force, as in the *Amnesty Law of 1989*, and the Conseil does not explicitly discuss how the current decision fits into the pattern of previous case-law, even though the precedents are frequently cited in the letter of reference to it. Where it intends to follow earlier rulings, it adopts the practice of the higher French courts of repeating the wording of the previously declared principle verbatim, but without attribution. As with these courts also, changes in the case-law can be noticed by attending to the formulations adopted. In some areas the Conseil has departed from its previous decisions. For example, on the matter of whether a decree can alter the constituent elements of a criminal offence, the Conseil constitutionnel has aligned itself with the Conseil d'État.[92]

The possibility that the Conseil will overrule its previous decisions to some extent justifies the narrow view of binding judgments taken by the ordinary courts. All the same, although these courts may have been wary of, or even hostile to, the Conseil in the 1960s and the early part of the 1970s, more recently they have accorded more authority to the Conseil's decisions.[93] This happens in two ways, reflecting the different levels of authority that these enjoy.

[90] Art. 10 §2 of the Constitution, a practice declared licit by the CC decision no. 85-197 DC of 23 Aug. 1985, *Elections in New Caledonia*, *Rec.* 70.

[91] D. Labetoulle, 'Les Méthodes du travail au Conseil d'État et au Conseil constitutionnel', in *Conseil constitutionnel et Conseil d'État*, 249 at 255.

[92] See below, Ch. 3.

[93] See generally Genevois, §§107–9. H. Donentwille identifies 3 stages in the attitude

Where a decision of the Conseil constitutionnel has binding force, this will provide a sufficient reason for the judgment of a subsequent court. This has been clearly recognized by the Conseil d'État, which applied a decision of the Conseil constitutionnel that certain 'pollution payments' made to water authorities constituted taxes, thereby reversing its own previous decisions on the matter.[94] Similarly, the Tribunal des conflits has applied a decision of the Conseil constitutionnel on the classification of sewage payments.[95] In both cases the decision of the Conseil constitutionnel was cited as the reason in the judgment itself. As yet, no decision of the Cour de cassation has formally been based on the findings of the Conseil constitutionnel, though some cases do refer to its rulings.

Beyond this, the Conseil constitutionnel frequently provides the inspiration for decisions of the courts. For example, the Cour de cassation held that criminal judges were competent to judge the legality of identity checks made by the administrative police. The justification for the decision was that article 66 of the Constitution confers the protection of civil liberties on (private law) judges. As the conclusions of the *avocat général* make clear, this was taken directly from the case-law of the Conseil constitutionnel.[96] Currently, the civil and criminal courts do pay attention to developments in the case-law of the Conseil, but this merely aids the discovery of principles leading to a solution, rather than providing the solution itself.[97] Like lower courts faced with rulings by the Cour de cassation or the Conseil d'État, they prefer to use their own judgment in deciding what the Constitution requires, while according great respect to the Conseil constitutionnel.

All the same, there have been significant divergences of opinion. In later chapters reference will be made to differences on the status of

of the courts to the Conseil constitutionnel: indifference, distance or resistance, and emergence: 'De l'effet des décisions des juridictions constitutionnelles à l'égard des juridictions ordinaires en droit pénal français', *Journées de la Société de législation comparée*, 1987, 431 at 435.

[94] CE Ass., 20 Dec. 1985, *Établissements Outters*, *RFDA* 1986, 513, *conclusions* Martin, applying CC decision no. 82-124 L of 23 June 1982, and reversing both an *avis* of 1967 and a decision of CE 21 Nov. 1973, *Société des papeteries de Gascogne*, *AJDA* 1974, 489. See generally B. Genevois, 'Continuité et convergence des jurisprudences constitutionnelle et administrative', *RFDA* 1990, 143; L. Favoreu, in *Conseil constitutionnel et Conseil d'État*, 178–81, 185–9.

[95] TC, 12 Jan. 1987, *Cie des Eaux et de l'Ozone* c. *SA Établissments Vetillard*, *RFDA* 1987, 284, *concl.* Massot. See generally L. Favoreu, 'Le Juge constitutionnel, le juge administratif et le juge des conflits: Vers une harmonisation des jurisprudences', *RFDA* 1987, 264; and id., *RFDA* 1989, 142.

[96] Cass. crim., 25 Apr. 1985, *Bogdan* and *Vuckovic*, D. 1985, 329, *concl.* Donentwille. See also 'La Cour de cassation, le Conseil constitutionnel et l'article 66 de la Constitution', D. 1986 Chr. 169.

[97] G. Rouhette, 'L'Effect des décisions du conseil constitutionnel à l'égard des juridictions civiles', *Journées de la Société de législation comparée*, 1987, 399 at 407.

treaties and criminal penalties but these are not the only ones.[98] Such a situation is not at all unusual for a supreme court in France. The highest courts of each judicial system meet resistance from below, and this may well cause them to reverse their original opinions. Uniformity is not as highly valued as correctness in the legal system as a whole, and the Conseil constitutionnel cannot expect any special treatment if it is to act as a court.

It must not be forgotten that the Conseil d'État also acts as adviser to the Government, vetting all bills before they are presented to Parliament. A significant part of its advice consists of deciding whether provisions, as drafted, are constitutional or not. In doing this, the Conseil d'État is inevitably driven to study carefully the case-law of the Conseil constitutionnel and to predict its likely reactions.[99] This advice will, as we have seen, find its way unofficially to the Conseil constitutionnel, so that there is an indirect dialogue between the two institutions over the scope and content of legislation. All the same, there may be differences of opinion. The most famous one was over the law on nationalizations passed in 1982. The Conseil d'État considered that certain changes had to be made to the indemnity provisions proposed by the Government in their draft bill. The Government followed this advice, only to find that the changes were condemned as unconstitutional by the Conseil constitutionnel.[100] While such a pre-emptive control may not be infallible, it does reduce the litigation before the Conseil constitutionnel and increase its influence over the whole legislative process.

Techniques of the Conseil constitutionnel The influence of the Conseil is also extended by the character of its judgments. It is not content simply to answer the straightforward question of whether a particular provision is constitutional or not. Because it is deciding the issue once and for all and in abstract, it tries to anticipate the various situations that may arise and to provide guidance as to the *manner* in which the provision can be constitutionally valid. The guidance comes in the form of 'reservations of interpretation' that condition the constitutionality of the clauses considered, so that the judgment may state that 'subject strictly to the reservations of interpretation' set out in the decision, the

[98] See notably Cass. Ass. plén., 19 May 1978, *Dame Roy*, D. 1978, 541, *concl.* Schmelck; *JCP* 1978, II. 19009, *rapport* Sauvageot: a Catholic school could legitimately dismiss a teacher who remarried following a divorce, despite a Conseil constitutionnel decision of 1977 recognizing the freedom of conscience of teachers.

[99] See Y. Gaudemet, 'Le Conseil constitutionnel et le Conseil d'État dans le processus législatif', in *Conseil constitutionnel et Conseil d'État*, 87.

[100] See *GD* 500–1, and see DECISION 29.

loi is not contrary to the Constitution. This technique dates from one of the earliest decisions of the Conseil.[101]

Three different forms of such reservations can be identified, and they can be illustrated from the important decision on *Security and Liberty* of 1981.[102] The first is *interpretation*, whereby the Conseil offers a reading of the text that will be consistent with the Constitution. Thus, in that case a provision made it an offence, *inter alia*, to use any means to hinder or obstruct the passage of vehicles on the highway. It was objected that this might interfere with picketing or demonstrations connected with the right to strike. The Conseil simply stated in paragraph 14 of its decision that 'there is no possibility that the application of these provisions might, in whatever way, prevent or interfere with the lawful exercise of the right to strike or union action.' The text was 'emptied of its venom' so that it could not be applied in an unconstitutional way. The second technique is that of *addition*, whereby the provision is filled out in such a way as to make it constitutional. Thus, article 39 of the *loi* simply provided that an extension of detention before being charged could be authorized by the investigating magistrate or by the President of the local criminal court. It was objected that, in the latter case, the detention would be extended without the judge having read the file. The Conseil replied in paragraph 19 that the judge authorizing the extension 'will necessarily have to examine the file to authorize the extension of the detention before charge'. A third technique is to address *injunctions* to the administration about how the law should be administered. Thus, in relation to identity checks, the Conseil remarked in paragraph 64 that:

> with a view to preventing abuses, the legislature has surrounded the procedure for controlling and checking identity that it has created with numerous precautions; that it is up to the judicial and administrative authorities to ensure that they are fully respected, as well as to the competent courts to punish, where necessary, illegalities that are committed and to provide compensation for their harmful consequences.

This was a strong encouragement to the courts, who are normally loath to provide damages for illegal acts by the police.

The use of such techniques depends on both the nature of the *loi* in question and on the body charged with its implementation. Where the *loi* is very general and really is no more than a framework for future Government discretion or for legislation that will not be subject to control by the Conseil, then there is good reason for the Conseil

[101] CC decision no. 59-2 DC of 17, 18, and 24 June 1959, *Standing Orders of the National Assembly*, DECISION 13.

[102] CC decision no. 80-127 DC of 19, 20 Jan. 1981, DECISION 17. On the interpretative techniques, see *GD* 452–4.

constitutionnel to be expansive in the reservations of interpretation that it lays down. This was noticeable in the range of *ordonnances* issued in 1986. In the context of privatization, there were detailed reservations on the way in which the price was to be calculated and on the protection of national independence.[103] On more specific *lois*, the extent of reservations may be more limited.

The other consideration is the body that has to implement the *loi*. In *Security and Liberty* the body in question was the police. The guidance provided in the decision was adopted by the *Garde des Sceaux* (Minister of Justice) and incorporated in a circular on the application of the *loi*. In many cases judges will control the implementation, and it will be relevant to consider their normal principles of interpretation (and, indeed, their attitude to the decisions of the Conseil). The more the interpretation put forward departs from established principles within the relevant jurisdiction, or the more there is resistance to the Conseil, the more the reservations will have to be set out. In an extreme case the Conseil may prefer to quash the *loi*, if it is not likely that these reservations will be adopted.

On the whole, however, the Conseil will prefer to uphold a text as constitutional rather than strike it down. The inconvenience of a ruling of unconstitutionality against a text is significant. The Conseil's decision will usually come after the end of the parliamentary session, with perhaps three months until the next session is due to commence. Unless a provision is minor and severable, the Government may be forced to convene an extraordinary session of Parliament just to get the bill passed. Interpretation may well be a kindness in preference to nullity. Total unconstitutionalities are very rare. In part this is because severance is used to a significant extent, though this itself will require some interpretation of the text.

The scope and procedure of the constitutional review now operated by the Conseil constitutionnel is very much like that of a constitutional court, but with substantial differences from the kinds of court that exist in the United States and Germany. The Conseil constitutionnel is a court in all but name, though its procedure for reviewing legislation lacks significant attributes of a judicial process, even when compared just to ordinary French courts. Its jurisdiction is limited to reviewing *lois* before they are promulgated; once promulgated, a *loi* becomes immune from challenge in the ordinary courts.[104] At the time of the bicentenary, proposals were made by the President of the Republic and the President of the Conseil constitutionnel that the ordinary courts

[103] See DECISION 30, and O. Beaud and O. Cayla, 'Les Nouvelles méthodes du Conseil constitutionnel', *RDP* 1987, 677 at 682.

[104] See recent reaffirmations of this by Cass. 1ère civ., 1 Oct. 1986, *Bulletin de la Cour de cassation*, I. 232; and CE Ass., 21 Dec. 1990, *AJDA* 1991, 158 (1st case).

should be allowed to refer to the Conseil constitutionnel the issue of the constitutionality of *lois* as they affected the fundamental rights of the citizen, even after they had been promulgated. But the proposed constitutional amendment to this effect met with resistance and was blocked in Parliament in autumn 1990.[105] The system depends heavily on the willingness of the opposition to refer *lois* to Parliament. For example, the penalty imposed by the *loi* of 13 July 1990 on racism, anti-Semitism, and xenophobia was to deprive convicted persons of their civic rights. Although this was of obvious constitutional significance, no political party with sufficient members was willing to refer the matter to the Conseil constitutionnel before it was promulgated.[106]

Within the balance of political institutions, the development of constitutional review places obvious limits on the sovereignty of Parliament, already reduced by the 'rationalized parliamentarianism' of the Fifth Republic. But since in political reality Parliament is now dominated by the Government (or even, arguably, by the President of the Republic), the growth of constitutional review may be seen as part of the process of redressing the balance between political institutions within the Fifth Republic. All the same, its emergence is a significant departure in terms of constitutional principle from previous Constitutions, especially in relation to conceptions of the separation of powers and parliamentary sovereignty. On the other hand, there is substantial continuity with concepts of the rule of law and with elements in the French State tradition. If Parliament and the executive are merely organs of the nation, then they must be kept within the powers that they have been given by the nation in the Constitution.

The significance of such changes will be seen in subsequent chapters both in relation to the operation of institutions of government in law-making and in parliamentary procedure, and in substantive terms in relation to fundamental freedoms and equality.

[105] See L. Favoreu, 'L'Élargissement de la saisine du Conseil constitutionnel aux juridictions administratives et judiciaires', *RFDC* 1990, 581. President Mitterrand repeated his proposal in Nov. 1991, and suggested that it should be put to a referendum, but (at the time of writing) this has not yet been done.

[106] Indeed, Favoreu states that, even after the reforms of 1974, only 10.4% of *lois* are submitted to the Conseil: ibid. 587.

2

SOURCES OF THE CONSTITUTION

It has been suggested by René Capitant that the 1958 Constitution of the Fifth Republic was one of the worst written of the many French Constitutions over the last two hundred years. Although this may be unfair on the style of drafting, there is much truth in the remark in so far as the sources of the Constitution are concerned. Nothing illustrates this more than the way in which fundamental rights are set out. Constitutional rights are found in any of four places: the Constitution of 1958, the Declaration of the Rights of Man and of the Citizen (DDHC) of 1789, the Preamble to the 1946 Constitution of the Fourth Republic, and in the 'fundamental principles recognized by the laws of the Republic'. Other constitutional writers suggest that this picture must be further complicated by the existence of infraconstitutional norms, breach of which does not directly make something unconstitutional, but they are important for the way in which the Constitution is interpreted, and thus they have indirect constitutional importance.

The provisions on institutions are better collated in the 1958 text itself, although they have been filled out, notably by organic laws and parliamentary standing orders. All in all, the search for 'the Constitution' is not as easy as one might first imagine, and this provides significant scope for the Conseil constitutionnel to define its content as well as to interpret its meaning.

This chapter is about the 'formal sources' (to use the French term) of the Constitution, in the sense of the authoritative standards by reference to which constitutional norms are identified. Necessarily, this also leads to a discussion of the 'material sources' of the Constitution, namely, the particular texts from which constitutional norms are quarried. This chapter is not directly concerned with the sources for the political legitimacy of these constitutional norms—why they have come to be treated as authoritative—but such issues will arise tangentially from time to time.

2.1. LAW AND CONVENTION

J. Boulouis[1] has written that 'French constitutional law, written though it may be, draws from the history of its development as many rules based on convention as other constitutions that are commonly regarded as customary.' This arises because no constitution can provide in advance for all eventualities. Conventions also enable a constitution to adapt and be flexible as its practice develops.

Conventions may be defined as rules of constitutional morality that establish the rights and duties of political actors. Since they give rise to obligations, they are to be distinguished from mere practices or habits.[2] If the constitution is a declaratory text in the sense outlined in the Introduction, and essentially reflects a tradition of constitutional morality, then it may be almost accidental whether a particular rule is laid down in law or convention. Indeed, both will have their foundations in the institutional morality of the constitutional order—that morality which underlies and justifies the institutions, practices, rules, and principles of the constitutional system.[3]

The importance of conventions can be illustrated through the development of the Constitution of the Fifth Republic. In the area of parliamentary practice, Pierre Avril and Jean Gicquel identify three kinds of convention: interpretative conventions, which interpret a text; creative conventions, which add new rules; and *contra legem* conventions, which nullify the effect of a constitutional rule.[4] Such a classification gives the impression that conventions function to adapt or fill out the written text. However, it would be right to suggest that the French Constitution does not contain *all* the rules and principles of constitutional morality, and that some conventions almost exist independently of the texts. For instance, the accidental exclusion of certain established practices from the Constitution can be seen in the area of ministerial responsibility. Article 20 of the 1958 Constitution establishes a *collective* ministerial responsibility to Parliament, but the text says nothing about the *individual* responsibility of ministers for their own mistakes or for those of their departments. Yet such a convention does exist, as was shown in 1985 when Charles Hernu, Minister of Defence, resigned after it was revealed that secret service agents acting on his orders had blown up the Greenpeace ship, the

[1] J. Boulouis, in *Le Domaine de la loi*, 206; also J. E. S. Hayward, *Governing France: The One and Indivisible Republic* (2nd edn., London, 1983), 1, suggests that the Constitution has both an unwritten and a written character.

[2] G. Marshall, *Constitutional Conventions* (Oxford, 1984), 8–9, 11; C. Munro, *Studies in Constitutional Law* (London, 1987), 54–5.

[3] On this idea of institutional morality, see R. M. Dworkin, *Law's Empire* (London, 1986), esp. pp. 211, 225.

[4] Avril and Gicquel, 18–19. On conventions generally in the 5th Republic, see J. Gicquel, *Droit constitutionnel et institutions politiques* (9th edn., Paris, 1987), 211–13.

Rainbow Warrior. Given the nature of the 1958 Constitution, however, most conventions will operate as interpretations or modifications of texts.

The most important illustration of conventions came with 'cohabitation'. Between March 1986 any May 1988 the majority of the National Assembly was held by the centre–right coalition under Jacques Chirac, while the President was a Socialist. Mitterrand appointed Chirac as Prime Minister, but had to define the lines of competence. He followed previous suggestions by de Gaulle and Giscard d'Estaing that the President should work with a Prime Minister of a contrary political persuasion. Since the 1958 Constitution was silent about what should be done in such a situation, there was a need for an interpretative convention. The precise division of functions had to be settled by an agreement in such a form.[5] It is best to consider this a convention, in that the criteria of Jennings are met here: practice, a belief in its obligatory character, and a good constitutional reason for the practice.[6] In essence, the President retained control of national defence and foreign affairs, in keeping with his role under article 15 of the Constitution, while the Prime Minister was left free to run the activities of the Government under article 20. The President went further than this simple exegesis of the various texts, and suggested that to the functions of the President must be added 'his obligation to guarantee the independence of the administration of justice and to safeguard the rights and liberties defined by the Declaration of 1789 and the Preamble of the 1946 Constitution'.[7] There was agreement among the political actors that this division of functions was the appropriate way of interpreting the Constitution. It had the consequence that François Léotard, leader of the Union pour la démocratie française (UDF), was vetoed as Foreign Minister, A right-wing career diplomat being appointed instead.[8] In addition, France was represented at international summits by both President and Prime Minister performing a double act. Only in the claim by the President that the Government could not go back on 'social acquisitions' (*les acquis sociaux*) was there no consensus. The agreements can be taken as conventions, since their authority lay not simply in the mutual convenience of potential presidential candidates, but in the way in which these arrangements helped to fulfil the constitutional division of functions between President and Prime Minister,

[5] The creation of a convention by specific agreement is not unusual: see the agreement of the Dominion Prime Ministers that gave rise to the statute of Westminster, 1931.

[6] W. E. Jennings, *The Law of the Constitution* (5th edn., London, 1959), 136; Marshall, *Constitutional Conventions*, 8–9.

[7] MATERIAL 1.

[8] Gicquel, *Droit constitutionnel*, 581–3; P. Avril and J. Gicquel, 'Chronique constitutionnelle française' (1987) 38 *Pouvoirs* 174; V. Wright, *The Government and Politics of France* (3rd edn., London, 1989), 71.

and were, as such, obligations on each. In addition, the constitutional necessity for such a set of new obligations was obvious, in that it was necessary to give effect to the conflicting choices of the voters. Of course, this was not the only way in which the Constitution could have been interpreted, but it was one that seemed appropriate for more than conjunctural party political reasons.

The creative function of conventions can be seen in the fact that questions to the Government on Wednesdays, as well as Fridays, were instituted in the National Assembly in 1974, on the initiative of President Giscard d'Estaing, and in 1982 in the Senate. Article 48 of the Constitution had provided for only one session (fixed on Fridays) of questions to the Government in the National Assembly, and none in the Senate.[9]

The adapation of texts *contra legem* (against their express wording) can be seen in the practice employed for the President's choice of the Prime Minister under article 8 of the Constitution. The text simply states that 'The President of the Republic appoints the Prime Minister. He terminates his functions when the latter presents the resignation of the Government.' On a simple reading, the President would appear to be more of a figure-head, appointing a person thrown up by political circumstances for as long as that person commands a parliamentary majority. In practice, this has not been the case. The Prime Minister has presented the resignation of the Government at the request of the President, thus enabling the latter to be effectively in charge of policy. Indeed, failing to retain the confidence of the President is a major reason for Prime Ministers to resign.[10]

Chazelle suggests that the very institutions set out in the written texts may create the need for conventions. Unlike the situation in previous Republics, Parliament no longer has exclusive control over its own rules of procedure. They must be approved by the Conseil constitutionnel, which has been keen to keep to the letter of the 1958 arrangements. As a result, developments *contra legem* have evolved entirely by convention. For instance, article 27 says that voting in Parliament is strictly personal, yet it is a commonly accepted practice for the keys used to record votes to be given to colleagues so that one can be absent: indeed, very few need be present for large majorities to be recorded. The situation is described graphically by M. Larkin: 'In

[9] See Avril & Gicquel, 226–7; R. Chazelle, 'Continuité et tradition au sein de la seconde chambre: Le Sénat et le droit parlementaire coutumier', *RDP* 1987, 711, 731.

[10] See above, p. 16. Compare this practice to the statement of the *commissaire du gouvernement*, Janot, in the debates in the Comité consultatif constitutionnel: 'Messrs Dejean and de Bailliencourt asked how the President of the Republic could get rid of a Prime Minister whose policy no longer conformed to his views. The reply is very simple and very clear: he cannot do it.' (*Avis et débats*, 54.)

practice . . . deputies merely supplied colleagues with a duplicate key, and it was a not uncommon sight to see a handful of deputies in a semi-deserted Assembly, turning a succession of keys, like night-porters illuminating an office-block in preparation for the nocturnal incursion of cleaners.'[11] Parliament could not achieve this by formal changes to its procedure, for the Conseil constitutionnel would not approve them, but a convention does not have to be formally submitted to it, and it is thus easier for the Conseil to ignore such apparent breaches of the Constitution.[12]

The place of conventions is important, in that it brings into relief the issue of whether the Constitution is a legally binding text or a statement of political principle. Mitterrand's interpretation of the Constitution in 1986 was in no straightforward sense legally binding. His behaviour was constrained essentially by political opinion. But the character of enforcement does not deny the existence of a constitutional obligation; it is merely one factor in identifying the kind of obligation at issue. Many important constitutional obligations in France are conventional and are enforced by political rather than legal pressures. One might, of course, adopt a rather Diceyan line that the ultimate source and sanction are legal, and say that, if the President oversteps the mark, he will eventually run up against decisions of the Conseil constitutionnel or the Conseil d'État.[13] A more appropriate view would follow Jennings, suggesting that the President is not amenable to legal sanction for his interpretations of the Constitution, but that these therefore fall into the area of conventional constitutional obligations.[14]

It is quite clear that the provisions on fundamenteal rights mentioned in the Preamble to the 1958 Constitution were originally intended to be conventional, not legal. When asked specifically whether they were to be of constitutional value, the *commissaire du gouvernement*, Janot, replied: 'certainly not!' They were to be binding on the Government, but not on Parliament.[15] In other words, they would be legally enforceable on the Government by the administrative courts, but only

[11] *France since the Popular Front* (Oxford, 1988), 288. See the recent example of a request for a new vote on a measure because the Socialist deputies had not finished turning the keys for their absent colleagues before the vote was counted: *RFDC* 1990, 472–3. More generally, see J.-C. Nemery, 'Le Principe du vote personnel dans la Constitution du Ve République', *RDP* 1987, 995.

[12] The Conseil's earlier case-law was very strict on this issue, and an organic law was struck down for failing to make it sufficiently clear that the delegation of the right to vote in Parliament had to be exceptional: CC decision no. 61-16 DC of 22 Dec. 1961, *Delegation of Voting*, *Rec.* 24, but cf. *Séguin Amendment*, DECISION 16 §§2–5.

[13] Cf. A. V. Dicey, *An Introduction to the Law of the Constitution* (10th edn., by E. C. S. Wade, London, 1959), 445–6.

[14] See R. Romi, 'Le Président de la République, interprète de la Constitution', *RDP* 1987, 1265.

[15] *Avis et débats*, 101. See generally *Droits et libertés*, 14–16.

politically enforceable on Parliament, a solution that some found unacceptable.[16]

Even though conventions are an important source of constitutional rules, they are not directly enforceable at law, except as a background against which legally binding constitutional rules are to be interpreted. It is, of course, possible that not all the legal standards are to be found in written texts, and so one can talk of unwritten or 'customary' law. Although some French writers do talk of 'customary law' in this sense, particularly in relation to parliamentary procedure, the scope for constitutional rules of this kind is small, since all legally-enforceable rules have to be traced back to the text of the 1958 Constitution and the sources to which it refers, and none of these provides direct scope for customary law. Rules that political actors as a matter of practice consider to be binding are most likely to be conventions rather than customary law.

2.2. Institutional Rules

Those who drafted the 1958 Constitution were primarily concerned to have an effective set of governmental institutions. Therefore, the 1958 text is fairly full and complete in its provisions establishing the organs of government and their powers. Remaining rules are found in organic laws and parliamentary standing orders.

The 1958 Constitution

Duvergier remarked that 'this Constitution is the work of lawyers more attached to details than to the whole, more miniaturists than architects'.[17] As a results, the bulk of the text of the 1958 Constitution (reproduced in Part Two, Section I) concentrates on setting out rules for the operation of the various institutions of government. Title II (articles 5–19) deals with the election, functions, and powers of the President of the Republic; title III (articles 20–3) with the Government; title IV (articles 24–33) with the membership and convening of Parliament; title V (articles 34–51) with the legislative powers of Government and Parliament, and with the parliamentary procedure for passing legislation and censure motions; title VI (articles 52–5) with treaties; title VII (articles 56–63) with the Conseil constitutionnel; title VIII (articles 64–6) with the judiciary; title IX (articles 67–8) with the High Court of Justice, which impeaches the President or members of the Government; title X (articles 69–71) with the advisory Economic

[16] See Coste-Floret, in *Avis et débats*, 102.
[17] *La Cinquième République* (Paris, 1959), 12.

and Social Council; title XI (articles 72–6) with local government and overseas territories; title XII (articles 77–87) with the Community of former French colonies; title XIII (article 88) with agreements for association with the Community; and title XIV (article 89) with amendment. Title XV (articles 90–2) dealt with the transitional provisions. A large part of the text is on parliamentary procedure, setting out such details as the dates on which Parliament can convene, and how long each session lasts. Title V contains comprehensive rules on the division of legislative powers, which will be discussed in detail in Chapter 3.

Thus the basic institutional rules and principles are contained in the text of the 1958 Constitution. This text itself authorizes further institutional rules to be specified in organic laws and parliamentary standing orders, which will be discussed in Chapter 4.

Organic Laws

The 1958 Constitution provides that organic laws will complete the text in relation to elections and eligibility of members of Parliament (article 25), the proxy voting by members of Parliament (article 27), the scope of Parliament's legislative power (article 34 §7), finance law procedure (article 47), the organization and functioning of the Conseil constitutionnel (article 63), the judiciary (article 65), and the Economic and Social Council (article 71). All such laws have to be approved by the Conseil constitutionnel before they are promulgated. These provisions have given rise to organic laws, especially in relation to parliamentary elections and to finance laws, the latter of which will be discussed in Chapter 4. These organic laws have generated a significant case-law from the Conseil constitutionnel. Organic laws have also been passed on the other areas mentioned above. Although they were important for completing the Constitution in the early years of the Republic, they are rarer today, accounting for only 4.8 per cent of *lois* passed between 1981 and 1986.

As will be seen in Chapter 4 in relation to finance laws, failure to comply with the povisions of an organic law may lead to the invalidity of a *loi* passed by Parliament.

Standing Orders of the Parliamentary Assemblies

The procedures of Parliament are of constitutional importance, since they affect the effective control that the elected chambers exercise over the Government and over legislation. In previous Republics Parliament was left free to determine its own procedure as set out in its standing orders. Since the Fifth Republic introduced a 'rationalized Parliament', however, it was thought undesirable that Parliament should be able to undo this by altering its own standing orders. For that reason, article 61 requires that the Conseil constitutionnel approve any standing orders of the National Assembly or the Senate before they are put into effect. There has developed a body of law that determines what Parliament can or cannot do, which is more rigid than rules having a status akin to the conventions that govern parliamentary procedure in the United Kingdom, and, unlike the situation in the United Kingdom, this body of law is enforced by an external body. This area will be examined further in Chapter 4.

As will be seen in Chapter 4, breaching parliamentary standing orders does not make a *loi ipso facto* unconstitutional. This will only occur where constitutional rules are breached, and where the result might well have been substantially different had the procedure been followed.[18]

2.3. SOURCES OF FUNDAMENTAL VALUES

Constitutional Values

No significant effort was made in 1958 to enumerate fundamental rights. In part, this was due to constraints of time, but it was also considered that, because the Constitution was being elaborated by the Government, not by a Constituent Assembly, the task of setting out fundamental values was not appropriate. Moreover, the experience of the two Constituent Assemblies of 1946—which had successively struggled to formulate such values, and had ended up with a simple cobbling-together of texts—was enough to dissuade most people from embarking on such an exercise. The concern of the drafters of the Constitution was to achieve a workable system of governmental institutions, not to produce a complete set of provisions on all aspects of constitutional life. In any case, if fundamental values were not to be legally enforceable against Parliament, little specific attention need be paid to their formulation. The contemporary practice of the Conseil d'État was not to cite such texts as the Declaration of 1789 directly, but to make a judicious selection from its provisions and to enforce them as 'general principles of law'. So, even for enforcing fundamental values against the Government,[19] there was no need for any precise and exhaustive enumeration. As a result, the 1958 Constitution merely lists a hotchpot catalogue of sources from which fundamental rights can be defined. The list attempts to be declaratory, rather than adopting the

[18] See CC decision no. 86-225 DC of 23 Jan. 1987, *Séguin Amendment*, DECISION 16, §4.

[19] F. Batailler, *Le Conseil d'État: Juge constitutionnel* (Paris, 1966), 96: see generally below, ch. 3.

constitutive approach of the writers of the 1793 and 1795 Constitutions or of the abortive draft of May 1946. The declaratory approach was inherent in the *loi* of 3 June 1958 that empowered the drafting of the Constitution. In giving full powers to de Gaulle's Government, it restricted its authority to issue decrees where the 'republican constitutional tradition' reserves matters to *loi*. The tradition is stated to result 'especially from the Preamble to the 1946 Constitution and from the Declaration of the Rights of Man and of the Citizen'.

In recent years it has been common to talk of the 'bloc de constitutionnalité' (the block of constitutional norms) to designate the totality of constitutional provisions. This phrase rightly suggests that there is a degree of coherence and interrelationship between the texts in question. It also introduces a hint of indefiniteness into the notion of 'the Constitution'.[20] Since 1976, the Conseil constitutionnel has typically referred to 'principles having constitutional value' (*principes à valeur constitutionnelle*), frequently without mentioning a specific source. This phrase includes both the written texts and other materials, drawn from fundamental principles recognized by the laws of the Republic, some general principles of law, and some objectives of constitutional value.[21] It is still worth while, however, to identify the various possible sources of constitutional values.

The 1958 Constitution contains a few individual fundamental rights. These include the equality of all citizens before the law (article 2 §1), the right to vote (article 3 §4), the right to organize political parties (article 4), and the right to personal status (article 75). The collective rights of local government are set out in title XI. The 'indivisible, secular, democratic, and social' character of the Republic is proclaimed in article 2. This last principle gave rise to a strange annulment of a purely programmatic article of a *loi* of 1991 reforming the government of Corsica. Article 1 of the *loi* provided that: 'The French people guarantees to the historic and living cultural community that makes up the Corsican people, a constituent part of the French people, rights to the preservation of its cultural identity and to the protection of its specific economic and social interests.' The Conseil constitutionnel struck the provision down, on the ground that the legal concept of 'the French people' had constitutional status, and the Constitution only recognized that entity without distinctions. Thus, to endorse the existence of a separate people within the French people was contrary to the Constitution. All the same, it was not unconstitutional either for a particular regime of local government to be established for the special

[20] See S. Rials, 'Les Incertitudes de la notion de Constitution sous la Ve République', *RDP* 1984, 587.

[21] See J.-P. Costa, 'Principes fondamentaux, principes généraux, principes à valeur constitutionnelle', in *Conseil constitutionnel et Conseil d'État* (Paris, 1988), 133.

requirements of a specific area, or to include Corsican language and culture in the school curriculum.[22]

The Preamble to the 1958 Constitution announces a commitment to the rights of man and to principles of national sovereignty to be found in the Declaration of 1789 and the Preamble to the 1946 Constitution, together with the self-determination of peoples within the Community of former colonies. It may seem strange to differentiate the Preamble from the rest of the Constitution, but this was clearly the original intention. In essence, the Preamble was to be the expression of pious intentions, fine words, and possibly political obligations without any clear legal consequences. It was first treated as a legally binding constitutional text by the Conseil constitutionnel in its decision on *Changes to the Budgetary Provisions of the EC Treaties*,[23] but the most famous instance was the *Associations Law*[24] decision of 1971, which begins: 'In the light of the Constitution, and especially of its Preamble'. Since then, the Preamble has been treated as giving legally binding constitutional force to all the values that it mentions.

The Declaration of 1789 (the text of which is reproduced in Part Two, Section II) contains the bulk of specific texts on individual freedoms. It sets out the basic liberal philosophy of the 1789 Revolution, and constitutes the touchstone of the republican tradition in France. In article 2 it enunciates the principle that 'the final end of every political institution is the preservation of the natural and imprescriptible rights of man. These rights are those of liberty, property, security, and resistance to oppression.' It then sets out basic provisions on three of these: liberty, equality, and national sovereignty. The most elaborated principle is that of equality: article 6 covers equality before the law and equal access to public office; and article 13 covers equality before public burdens (including taxation), which is understood to incorporate, by analogy, equality before the public service.

The Declaration also enunciates a number of fundamental freedoms: freedom from arrest (article 7), proper treatment on arrest (article 9), freedom from unnecessary criminal penalties (article 8), free expression of religious and other opinions (article 10), freedom to publish (article 11), and freedom of property (article 17). In addition, there is the principle of residual liberty: the freedom to do anything that does not cause harm to another, and which is not prohibited by law (article 4).

According to article 3, sovereignty belongs to the nation and is exercised by the people. This is repeated in article 3 of the 1958 Con-

[22] CC decision no. 91-290 DC of 9 May 1991, *Statute of Corsica*, *RFDC* 1991, 305, note Favoreu. See also CC decisions nos. 82-137 DC, *Decentralization*, and 82-138 DC, *Corsica*, *GD*, no. 34; L. Favoreu, 'Décentralisation et Constitution', *RDP* 1982, 1259.

[23] CC decision no. 70-39 DC of 19 June 1970, *Rec*. 15.

[24] CC decision no. 71-44 DC of 16 July 1971, DECISION 1.

stitution, and, indeed, it forms its basis, for that text originates from a referendum. The supremacy of *loi* is confirmed by article 6, which notes that it is the expression of the general will (*la volonté générale*).

Even in recent years some have doubted whether all the provisions of the Declaration were of constitutional value, but such doubts are no longer justified.[25]

The Preamble of 1946 (the text of which is reproduced in Part Two, Section III) is more social in character. It was written after an attempt to draft a new Declaration of Rights had been rejected by the referendum of May 1946. Failing a radical revision, the Constituent Assembly preferred to list additional fundamental political, economic, and social rights that it felt were 'particularly necessary for our times'. It contains provisions on equality of the sexes (§3), equality before national calamities (§12), and equal access to education and training (§13). Fundamental rights are affirmed for workers (the right to work in §5, union rights in §6, the right to strike in §7, the right to collective bargaining and participation in the management of the company in (§8), as well as rights to free education (§13), asylum (§2), and health (§11). The State also has a duty to provide the conditions necessary for individual and family development (§10).

To see in this Preamble a set of specific rights is problematic. It is very much a set of pious hopes and objectives lacking specificity. Take paragraph 10: 'The nation shall assure to the individual and to the family the conditions necessary for their development.' What does that mean in concrete terms? Is there a constitutional right to the *revenu minimum d'insertion* (a social welfare benefit ensuring a minimum income) introduced in 1989? The same might be said about paragraph 5: 'Every individual has the duty to work and the right to employment.' Is unemployment unconstitutional? Considerable doubts have been expressed, even in recent years, about the constitutional value of all these provisions, but it now appears that (except for the now obsolete provisions on the Community of former colonies) they are to be treated as having such validity.[26]

Even where the Preamble has been accepted as containing constitutional values, the relationship with the Declaration of 1789 causes problems, since there is a clear clash of philosophies between the rather individualistic text of 1789 and the more social approach to

[25] See *Droits et libertés*, 21, and Genevois, 330. Cf. L. Philip, 'La Valeur juridique de la Déclaration des droits de l'homme et du citoyen du 26 août 1789 selon la jurisprudence du Conseil constitutionnel', in *Études offertes à Pierre Kayser* (Aix, 1979), ii. 317.

[26] See *Droits et libertés*, 27. Cf. M. Debene, 'Le Juge constitutionnel et "les principes particulièrement nécessaires à notre temps"', *AJDA* 1978, 531; L. Philip, 'La Valeur juridique du Préambule de la Constitution du 27 octobre 1946 selon la jurisprudence du Conseil constitutionnel', in *Mélanges R. Pelloux* (Lyons, 1980), 265.

rights of 1946. The clash is best illustrated by the *Nationalizations* case of 1982.[27] Here paragraph 9 of the 1946 Preamble clashed with article 17 of the 1789 Declaration. The former provided that 'Any property or business whose exploitation has or acquires the character of a national public service or a *de facto* monopoly should become the property of the community.' This seemed a positive encouragement to nationalization. By contrast, article 17 of the 1789 Declaration defended the right of individual ownership of property: 'Property, being an inviolable and sacred right, none can be deprived of it, except when public necessity, legally ascertained, evidently requires it, and on condition of a just and prior indemnity.' The Conseil constitutionnel used historical arguments to justify the priority of the 1789 text. Since any attempt to depart from the attachment to property in the 1789 Declaration had failed with the rejection of the first draft Constitution by referendum in May 1946, the reaffirmation of the 1789 text by the Preambles to the 1946 and 1958 Constitutions, duly approved by referendums, showed that article 17 retained its full force. The 1946 Preamble suggested that its own statements of rights were complementary to those of 1789, and this was the appropriate way to construe the provisions. The Conseil's decision on the legitimacy of the nationalizations proposed by the *loi* focused on whether the conditions laid down in article 17 had been met. While that of public necessity was not evidently violated, that of the 'just and prior indemnity' was.

This decision is important in two respects. First, the 1958 Constitution is seen as retaining the fundamental liberal values of the 1789 Revolution. As the Socialists had recognized in their *programme commun* of 1979, a more idealistic effort to nationalize for general policy reasons, rather than just for public necessity, would require a constitutional amendment. Secondly, the social provisions of the 1946 Preamble were of limited application, being concerned to develop, rather than to contain, the area of rights laid down in 1789.

Fundamental principles recognized by the laws of the Republic complete the picture. These are referred to in paragraph 1 of the 1946 Preamble as an 'in case we have forgotten anything' provision. In essence, the 'French republican constitutional tradition' contains values that are not expressed in the Declaration of 1789 or the Preamble of 1946, but which are, nevertheless, of constitutional value. The principal Republic in question is the Third Republic, from 1870 to 1940. It did not have any Declaration of Rights. Rather, there were a series of *lois* passed during

[27] DECISION 2. G. Drago, 'La Conciliation entre les principes constitutionnels', D. 1991 Chr. 265, argues that the assignment of priorities between constitutional norms is purely pragmatic. On the other hand, those inside the Conseil constitutionnel argue that it is principled: see R. Badinter and B. Genevois, 'Normes de valeur constitutionnelle et degré de protection des droits fondamentaux', *RFDA* 1990, 317.

its lifetime that set out basic values, such as freedom of the press (1881), freedom of association (1901), and freedom of religion (1905). The express purpose of this phrase, 'the fundamental principle recognized by the laws of the Republic', was to designate such principles without having to provide an exhaustive enumeration.[28] These have formed the basis of numerous decisions of the Conseil constitutionnel, following the pattern set at the end of the Fourth Republic by the Conseil d'État.

Since there is no official list of which provisions of which laws recognize such fundamental principles, it means that the category of fundamental rights is open-ended, and has to be settled by decisions of the Conseil constitutionnel. In making these, the Conseil has to sift through provisions found in 'republican legislation passed before the coming into force of the Preamble of the 1946 Constitution'[29] to establish what is fundamental. There are therefore three criteria: the principle must be in a *loi*, of a Republic before 1946, and it must be fundamental. The first condition means that *decrees* are not an adequate source. Thus, in deriving the principle of the independence of administrative judges, the Conseil could not refer to the decree of 31 July 1945 on the Conseil d'État, but had to claim that it was attached to the *loi* of 24 May 1872 which that decree had repealed.[30] (Since one is only looking for principles evidenced in *lois*, the current validity of those *lois* is not important.) The second condition means that some non-republican texts must be ruled out. Thus, in deriving the principle of the separation of public and private law courts, the Conseil could not rely on the classical and clearest statement of this principle in the *loi* of 16–24 August 1790, since this was a *loi* of a monarchy.[31] The Civil Code, promulgated finally on 29 March 1804, would qualify, since the Empire was not proclaimed until 18 May 1804. The third condition requires that the principle expressed in a *loi* is constitutionally fundamental. Many provisions of a *loi* may be fundamental to a particular branch of law, and may thus be 'fundamental principles' within article 34. As will be seen in Chapter 3, these latter principles define the legislative competence of Parliament, but they are not constitutionally fundamental in any other sense. For example, the *loi* of 29 March 1880 decided that there were to be no more tolls on national or departmental highways, thus establishing the principle of the gratuity of traffic on

[28] See Genevois, §333. Its unstated purpose was also to smuggle in certain values, notably freedom of education, about which there was considerable disagreement in the Second Constituent Assembly of 1946: ibid.

[29] See DECISION 3a, §3.

[30] CC decision no. 80-119 DC of 22 July 1980, *Validation of Administrative Decisions*, *Rec.* 46.

[31] See CC decision no. 86-224 DC of 23 Jan. 1987, *Competition Law*, *Rec.* 8.

these roads. The Conseil did not consider this to be sufficiently fundamental to prevent the legislature reintroducing tolls on some roads.[32]

The essential character of the exercise of deriving a fundamental principle can be seen in DECISION 1, the *Associations Law* case of 1971. This concerned the freedom to form an association without prior approval by a public official. The French have long taken the view that, if you have a right to do something, then you should not need prior official permission to exercise it; the official should only intervene afterwards if you have acted illegally or abusively in the exercise of those rights. In this case the Government had banned a left-wing organization as subversive. Simone de Beauvoir and other left-wing intellectuals formed an association, the Amis de la cause du peuple, in support of the banned organization and in protest at the Government's action. The Prefect of Paris refused to register the association, claiming that it had an illegal purpose, but this decision was quashed by the Conseil d'État as an unlawful prior restraint. The Government then promoted a bill that empowered prefects to refuse to register an association whose objects appeared to be unlawful, pending a reference of the matter to the courts for a ruling. This was passed despite the opposition of the Senate, and was referred by the President of the Senate to the Conseil constitutionnel. Although the Conseil d'État had approved it, the Conseil constitutionnel declared the provision unconstitutional by reference to the freedom of association:

> Considering that, among the fundamental principles recognized by the laws of the Republic and solemnly reaffirmed by the Constitution is to be found the freedom of association; that this principle underlies the general provisions of the *loi* of 1 July 1901; that, by virtue of this principle, associations may be formed freely and can be rendered public simply on condition of the deposition of a prior declaration . . .'

Now, the reasoning here is in three stages. First, the Conseil identifies a text that talks about the value in question. In doing this, the Conseil needs to have a prior conception of what right is at issue and whether it is potentially fundamental. Typically, this will come from the claims of the authors of the reference. The text will either be a whole *loi*, or even a disparate provision, as will be seen in the case of the right to freedom of education that was 'discovered' in article 91 of the finance law of 31 March 1931.[33] Secondly, the Conseil looks not so much at the precise wording of the text as at the fundamental value that it expresses, a general principle underlying its specific provisions. Finally, having elicited a principle of general import, like the freedom of association, it has to produce a specific rule capable of resolving the

[32] CC decision no. 79-107 DC of 12 July 1979, *Toll Bridges*, *Rec*. 31.

[33] See DECISION 21 §3.

question before it, such as the rule against prior restraint. In performing this task, the Conseil is at its most creative. Apart from the freedom of association, the independence of administrative judges, and the separation of public and private law courts, the Conseil has also declared as fundamental the rights of due process in legal proceedings,[34] the freedom of education and the freedom of conscience,[35] the freedom of teachers in higher education,[36] the freedom of movement and the right of privacy,[37] the continuity of public services,[38] the freedom of commerce and enterprise,[39] and respect for all human beings from the beginning of life.[40]

Although the expression used in the 1946 Preamble was intended to designate values that had been regarded as fundamental by the French republican tradition, this is not to say that all the values of that tradition have been constitutionalized. The Conseil constitutionnel has had occasion recently to distinguish between the two ideas. In DECISION 3a, the *Amnesty Law of 1988*, the Conseil stated that breach of the republican tradition could only provide the basis for the argument that a provision was unconstitutional 'to the extent that this tradition has given rise to a fundamental principle recognized by the laws of the Republic'. Other values, such as Mitterrand's 'social acquisitions', may be contained in the republican tradition, but they merely give rise to political and not legal obligations.

Objectives of constitutional value are not laid down in any text. They are means for implementing constitutional values. As Genevois states: 'the objective of constitutional value appears as the necessary corollary of the implementation of a constitutionally recognized value.'[41] Since these are matters of means rather than ends, it might be thought that the legislature is free to identify and define these objectives, and to pursue their implementation. But the Conseil constitutionnel has been reluctant to leave Parliament with total liberty in this area. If an objective is identified as of 'constitutional value', it has a special status as a means by which the legislature must realize a fundamental constitutional value. The law cannot be changed in such a way as to weaken the constitutional protection afforded to individual rights. In this way, the objectives restrict the freedom of action of the legislature. Thus far, the Conseil has recognized the search for criminals and

[34] See DECISION 37 §35.
[35] Both in CC decision no. 77-87 DC of 23 Nov. 1977, DECISION 21 §3.
[36] CC decision no. 83-165 DC of 20 Jan. 1984, *University Professors, Rec*, 30.
[37] CC decision no. 76-75 DC of 12 Jan. 1977, DECISION 17.
[38] CC decision no. 79-105 DC of 25 July 1979, DECISION 23.
[39] See DECISION 2, and below, ch. 5.
[40] CC decision no. 74-54 DC of 15 Jan. 1975, *Abortion Law*, DECISION 20.
[41] Genevois, §342. He draws a parallel with the *Nold* v *Commission* [1974] ECR 491 (European Court of Justice).

the prevention of threats to public order, especially to persons and property,[42] the preservation of the pluralist character of socio-cultural expressions of opinion,[43] and the pluralism of daily newspapers of political and general information.[44]

The clearest illustration of the way in which these objectives work was in DECISION 26, the *Press Law*, in 1986. The Chirac Government sought to alter the restrictions on the ownership and management of the press that the Socialists had introduced in 1984. In particular, it wished to raise the ceiling on individual control of political and general dailies over the whole of the country from 15 per cent to 30 per cent. In its *Press Law* decision of 1984 the Conseil constitutionnel had stated that the pluralism of such dailies was 'in itself an objective of constitutional value', since it was a way of ensuring that article 11 of the Declaration of the Rights of Man and of the Citizen, which set out freedom of thought and opinion, could be realized. As will be seen in Chapter 5, the Conseil considered that, in repealing some of the 1984 provisions designed to achieve this objective, the new *loi* had reduced the legal protection afforded to the constitutional principle established in article 11, and so that part of the *loi* was struck down. In doing this, the Conseil stated:

> Considering that it is permissible at any moment for the legislature, deciding in the province reserved to it by article 34 of the Constitution, to amend previous texts, or to repeal them and substitute for them other provisions, as the situation requires; that, in order to achieve or reconcile objectives of constitutional value, it is no less permissible for it to adopt new methods, of whose appropriateness it is the judge, and these may involve the amendment or repeal of provisions that it considers excessive or unnecessary; that, however, the exercise of this power should not lead to the removal of legal safeguards for requirements having constitutional value.

By including objectives of a constitutional value within the class of legal provisions that the legislature can modify only in limited ways, the Conseil has restricted its freedom more severely than is evident from the written sources of the Constitution.

At the same time, objectives of constitutional value can justify the restriction of rights laid down in the written texts. Thus, in DECISION 18, *Security and Liberty*, it was held permissible to restrict the freedom of the individual by imposing identity checks in order to realize the constitutional objectives of the search for criminals and the prevention of threats to public order, especially to persons and property.

These various sources work together to build up a body of norms for

42 DECISION 18, §54.

43 CC decision no. 82-141 DC of 27 July 1982, *Audio-Visual Law*, *Rec*. 48.

44 See the *Press Law* decisions of 1984 and 1986.

the protection of fundamental freedoms in France. These will be discussed in Chapter 5, though the more specific issues relating to the doctrine of equality will be left to Chapter 6.

Infraconstitutional Values

Although it might appear strange to classify as 'fundamental' values that are not protected by the Constitution, it is important to isolate some infraconstitutional rights as fundamental. First, the line between constitutional and infraconstitutional rights has really been of major significance under the Fifth Republic. Before then, no judge could challenge the validity of *lois*, so a mix of constitutional and other fundamental rights was protected by interpretation in the ordinary courts. Secondly, this practice continues to some extent today. General principles of law and international treaties cannot be used to invalidate *lois* at a constitutional level, but they can be used by the ordinary courts to limit the effectiveness of *lois* and other legal rules to a very significant extent. This is particularly important in the area of *règlements*, where the competence of the executive to make legislation is restricted by reference to such values.[45]

General principles of law were elaborated by the Conseil d'État particularly during and after the Vichy period, to make concrete the 'republican constitutional tradition'. Even though the administrative and civil courts could not strike down legislation, they could limit executive acts and they could interpret legislation. A series of general principles were elaborated by the administrative courts, drawing on both the specific Declarations of Rights and on more comprehensive values. In 1951 Rivero identified four sources of general principles of law: (i) the traditional principles of 1789, such as equality, freedom of trade and conscience, and the secular character of the State; (ii) general principles derived by analogy with private law and private law procedure (the binding force of decisions, rights of due process); (iii) principles drawn from 'the nature of things' and the logic of institutions, such as the continuity of the public service; and (iv) necessary ethical principles, such as the administration seeking to serve the common good.[46] These therefore include consitutional values, such as freedom of education, religion, and commerce, as well as procedural safeguards, like the right to a hearing and the right to challenge decisions of the administration in the courts. Even rules of procedure, such as the right of appeal, might be included. The *commissaire du*

[45] See below, ch. 3.

[46] D. 1951 Chr. 21 at 22. See generally B. Jeanneau, *Les Principes généraux du droit dans la jurisprudence administrative* (Paris, 1954); and id., 'La Théorie des principes généraux du droit à l'épreuve du temps', *Études et documents du Conseil d'État*, 1981–2, 33.

gouvernement, Gentot, stated in *Dame David* that: 'If the general principles of law express—or reflect—commonly accepted ideas that are at the base of our legal system, they have to be consecrated by history and traditions, and be characterized by a certain permanence and a certain appeal to universality.'[47] Letourneur linked this with statutory interpretation:

> When the legislator of a specific nation votes a particular text, he does so within the framework of the political, social, and economic organization existing at the period in question, a framework determined by a certain number of principles that represent the state of the evolution and civilization which that nation has reached; when the judge does not find in a written text the solution to the litigation that is submitted to him, he is necessarily led to apply the same principles that the legislator is accustomed to take as his guide.[48]

Although the general principles have an important status, not all of them are of *constitutional* status. As Chapus suggested, some are binding on the legislature, while others merely bind the administration in its legislative and administrative functions.[49] Both have some importance in the interpretation of the Constitution. All general principles are of relevance in defining the scope of the executive's power to legislate under article 37 of the Constitution. As will be seen in Chapter 3, only Parliament may alter general principles of law. But some of these, though unwritten, may also bind the legislature, in that they constitute fundamental principles recognized by the laws of the Republic or objectives of constitutional value, such as the continuity of public services, and these Parliament cannot alter. The point is well made by the *commissaire du gouvernement*, Fournier, in *Syndicat général des ingénieurs conseils* in 1959:

> there are the general principles of law properly so-called, laid down by the declarations of rights, or deduced by judges from them. Among these fundamental principles, which are at the foundation of our political system, one must undoubtedly place the equality of citizens, the guarantee of essential freedoms, the separation of powers and the finality of judicial decisions, the non-retroactivity of the decisions of public authorities and the inviolability of acquired rights, the right of citizens to challenge administrative decisions, a right that has a passive form (the right to a hearing) and an active form (the right to bring an action for judicial review). Equally, this should include, as a

[47] CE, 4 Oct. 1974, *Leb*. 464; D. 1975, 369, note Auby; *JCP* 1975, II. 19967, note Drago. Although the case only concerned civil procedure, the *commissaire* took the view that the finding of a general principle of civil judicial procedure might lead to the procedural rules of administrative courts being called into question.

[48] 'Les Principes généraux du droit dans la jurisprudence du Conseil d'État', *Études et documents du Conseil d'État*, 1951, 19 at 20.

[49] R. Chapus, 'De la valeur des principes généraux du droit et des autres règles jurisprudentielles de droit administratif', D. 1966 Chr. 99 at 104.

counterbalance, the continuity of public services, essential to the life of the nation.[50]

Among those important general principles of law of infraconstitutional status are that administrative silence is tantamount to a decision to reject a request from a citizen,[51] that only laws and not administrative decrees or decisions can have retrospective effect,[52], and that of *audi alterem partem* (*le principle de la contradiction*).[53]

Treaties contain protections for fundamental rights and set out other basic values. Since France has signed the European Convention on Human Rights and the Geneva Convention on refugees, these have a special status in French law.[54] Of even greater significance are the treaties establishing the European Economic Community, which profoundly affect the institutional structures of French government and the rules of French law.

Article 55 of the Constitution states that treaties or agreements duly ratified or approved shall, upon their publication, have an authority superior to that of statutes, subject, for each separate agreement or treaty, to reciprocal application by the other party. Two problems have occurred here. First, there is the problem of how one can talk of 'reciprocal application' in the case of multilateral treaties. While the Conseil has not insisted that reciprocity is essential to the validity of a law ratifying a multilateral treaty, it has failed to endorse explicitly the view that the concept is unnecessary for priority to be given to treaty provisions over inconsistent *lois*.[55] Secondly, and more importantly, there is the question of the role of the courts in the face of conflicts between statutes and treaties. In its *Abortion Law* decision of 1975, and ever since,[56] the Conseil constitutionnel decided that it could not review the constitutionality of a proposed *loi* simply on the grounds of its incompatibility with a treaty. By contrast, the Cour de cassation has taken the view that it should give priority to a treaty over an incompatible *loi*.[57] This latter position has been recently adopted by the Conseil constitutionnel when sitting as an election court.[58] The Conseil

[50] CE, 26 June 1959, *Leb.* 394; *GD*, no. 96. See further G. Morange, 'Une catégorie juridique ambiguë: Les principes généraux du droit', *RDP* 1977, 761.

[51] See DECISION 7.

[52] CC decision no. 69-57 L of 24 Oct. 1969, *Course Fees at the École polytechnique*, *Rec.* 32.

[53] CC decision no. 72-75 L of 21 Dec. 1972, *Administrative Procedure*, *Rec.* 36.

[54] See H. Labayle, 'Le Contrôle contentieux des "expulsions dirigées" ou les prolongements de la jurisprudence *Bozano* au Palais-Royal', *RFDA* 1989, 3 (European Convention); and *concl.* van Ruymbeke, CE Sect., 27 May 1988, *Mujica Garmendia*, *RFDA* 1989, 46 (Geneva Convention).

[55] See P.-H. Teitgen, in Luchaire & Conac, 112–13.

[56] See above, n. 40; for a recent affirmation of this, see CC decision no. 91-293 DC of 23 July 1991, *Civil Service Statute*, noted in *AJDA* 1991, 631

[57] Ch. mixte, 24 May 1975, *Société Cafés Jacques Vabre*, D. 1975, 497, *concl.* Touffait.

[58] CC, 21 Oct. 1988, *5e circonscription du Val-d'Oise*, *AJDA* 1989, 128, note Wachsmann.

d'État finally came round to the same view in 1989, after initially insisting on giving effect to *lois* in such cases.[59] The approach adopted by the Conseil constitutionnel is formalistic. Both it and the ordinary courts recognize that a treaty is superior to a *loi* in the hierarchy of norms, but this does not suffice to make the *loi* incompatible with the Constitution.

Luchaire has suggested that there are what he calls *'paraconstitutional' principles* that are important in the decision-making of the Conseil, though it has never specifically used this term.[60] The principles are established values drawn from public and private law rather than from constitutional texts, which the legislature must apply to everyone, unless a good reason can be adduced for doing otherwise. Since they are principles of the ordinary law, 'the legislature is free to repeal them totally, but not to limit either their content nor their sphere of application without the agreement of the Conseil constitutionnel'.[61] As Genevois rightly notes, these principles form part of the doctrine of equality. Not merely must equality be respected in the application of certain constitutional maxims or in the provisions of a particular law, it must also be observed in the way in which certain long-established general principles of law are applied. It is this last category that Luchaire's 'paraconstitutional' principles are meant to designate.

Luchaire identifies three areas in which the equal application of these principles has been required. The first concerns the principles of liability. The Conseil has held that the public law principle of no-fault liability for public works, and the civil law principle of liability for fault cannot be derogated from without serious justification.[62] The second concerns the principles of judicial procedure, such as the right of appeal, the collegiality of a higher court, and a jury trial for serious criminal cases.[63] The third area is the application of the same law throughout the territory, so that, for example, the public freedom of education could not operate differently in different parts of France.[64] Luchaire's points relate exclusively to the doctrine of equality, and will be discussed further in Chapter 6.

The Constitution contains three levels of legal standard. First, there are specific rules, which provide clear instructions as to what should be done in particular instances—for example, on what date Parliament

[59] CE Ass., 20 Oct. 1989, *Nicolo*, *RFDA* 1989, 812; [1990] *PL* 134; *GA*, no. 117.

[60] See his note to CC decision no. 85-198 DC of 13 Dec. 1985, *Eiffel Tower Amendment*, D. 1986, 345, and *Droits et libertés*, 241–5.

[61] D. 1986 at 351.

[62] *Droits et libertés*, 241–3: see below, p. 223.

[63] *Droits et libertés*, 243–4: see below, ch. 6.3, and DECISION 19 and 36.

[64] *Droits et libertés*, 244–5: see DECISION 22.

must be convened for its spring session. At a higher level, there are constitutional principles, such as national sovereignty or the right to strike, which require greater specificity before they provide concrete directives about what should be done. For the most part, the fundamental values are of this kind. Finally, there are the objectives of constitutional value, which are merely means for achieving constitutional principles. Unlike rules, they have rather unspecific contents, yet, unlike principles, they have no ultimate constitutional value of their own. Although the institutional rules can be found exclusively in written texts, this is less true of constitutional principles, and not at all true of objectives. Whereas the rules concern institutions and are closely bound up with the 1958 text and its ancillary documents, principles and objectives concern the way in which those institutions should operate, and the ends that they should serve. Such ideas are difficult to specify and are more appropriately the product of experience, hence the amorphous nature of the various written and unwritten provisions on which the Conseil constitutionnel has drawn. Even the rules that define institutions are not easy to get right at one go. The authors of the 1958 Constitution may have drawn on some 170 years of experience with different kinds of institution, but even this has not proved sufficient, and constitutional conventions provide unwritten sources of rules in this area.

In the end, the decisions of the Conseil constitutionnel will provide a concrete source to which most French lawyers will turn. They do not constitute a formal source of the Constitution in the way that some English court decisions do—for example, those defining an Act of Parliament.[65] All the same, given the variety of possible sources, the Conseil's decisions provide an authoritative statement of what is constitutional and what is not, which is not easily available elsewhere. Despite the written texts, there is an important sense in which the French Constitution is 'judge-made', albeit in a different way from that intended by Dicey's similar epithet for the British Constitution.

[65] *Prince's Case* (1606) 8 Co. Rep. 1: see E. C. S. Wade and A. W. Bradley, *Constitutional and Administrative Law* (10th edn. London, 1985), 75–6.

3

THE DIVISION OF LAWMAKING POWERS: THE REVOLUTION THAT NEVER HAPPENED?

ANY constitutional organization of powers, at least in the modern world, has to come to terms with the fact that Parliament cannot make all the rules necessary for running the country. The executive must be permitted to make some rules on its own initiative. The issue for any constitutional settlement is rather more how this can be achieved within the framework of the sources of political legitimacy.

The Constitution of the Fifth Republic represents an attempt to work out a solution to this problem by resorting to a novel device. Instead of defining areas in which the executive is empowered to make *règlements* on its own initiative, it is the legislature that finds itself with a defined field of competence for its *lois*. Outside the field of Parliament's competence defined in article 34, the executive enjoys, by virtue of article 37 (the texts of which are reproduced in Part Two, Section I), the competence to legislate on all other matters. The constitutional innovation is thus that residuary legislative power lies in the executive, and not with Parliament. Parliamentary *lois*, once seen as the supreme expression of the will of the nation, would seem no longer to be sovereign. Although important matters are included in the defined field of Parliament's competence, the executive enjoys the power to govern effectively. During the deliberations of the Consultative Committee on the Constitution Marcel Waline, a leading professor of public law (later a member of the Conseil constitutionnel), remarked:

> [Articles 34 and 37] state a new principle, that of the superiority, or at the very least the equality, of *règlement* over *loi*. That is contrary to the fundamental principle of the superiority of *loi* over *règlement*, a principle on which the Conseil d'État . . . has for the last 150 years based a universally admired case-law. It cannot be denied that this is an innovation, perhaps necessary, but in any case revolutionary.[1]

[1] *Avis et débats*, 51 (4th Session, 31 July 1958).

Twenty years later, however, another eminent public law professor was able to remark that, in the event, the revolution had not occurred.[2] The apparent shift from parliamentary sovereignty to residuary legislative power in the executive had not produced a real change in the way in which the country was governed. Once a stable parliamentary majority could be secured for the Government, there was no need for recourse to executive decrees as a way of ensuring that legislation could be passed. In any case, the constitutional tradition was stronger than mere forms of words. It is not simply that old habits and ways of thinking die hard; the values and experience that the constitutional tradition represents do not lose their importance very easily.

In studying the allocation of legislative power within the 1958 Constitution, it is possible to see how far the French republican constitutional tradition provides the real point of reference for that document's interpretation. Jean Boulouis has remarked in this context that 'French constitutional law, written though it may be, draws from the history of its development as many rules based on convention as other constitutions that are commonly regarded as customary.'[3] In addition, one can also observe the way in which the Conseil constitutionnel and the Conseil d'État interact as interpreters of the Constitution.

3.1. The Constitutional Tradition before 1958

Constitutions Prior to the Third Republic

As was noted in Chapter 1, there is an important strand in the French political tradition that stresses that the State is the representative of the nation, and thus that Parliament should be a supreme authority since it incarnates that representative character. On the other hand, another strand holds that executive public authorities are not the creation of the nation, but are a necessity of the existence of the State, and should have an independent zone of authority in order to serve the national interest. Legislation may direct and even legitimate this action of the executive, but its authority stems primarily from its serving the national interest.[4] Against this, the experience and fear of an authoritarian executive give rise to the need to limit executive power. Such considerations have produced a variety of constitutional arrangements over the last two centuries, but are well illustrated in those of 1791.

The Constitution of 1791 provided for a clear superiority of *loi* as the expression of national sovereignty. Article 3 of title III, chapter ii,

[2] Jean Rivero, in *Le Domaine de la loi*, 261.

[3] Ibid. 206.

[4] See above, pp. 24–5, and K. F. H. Dyson, *The State Tradition in Western Europe* (Oxford, 1980), 162–3, 224–5.

provided that 'There is no authority in France higher than the *loi*.' As stated in article 6 of the Declaration of the Rights of Man and of the Citizen, statute (*loi*) was the expression of the general will, but the task of lawmaking was delegated exclusively to the National Assembly (title III, chapter iii, article 1). By contrast, the executive merely implemented such *lois*. As title III, chapter iv, article 6 put it: 'The executive power cannot make *lois*, even provisionally, but only proclamations consistent with *lois* for the purpose of ordering or encouraging their execution.' The limited role of the executive was emphasized in the *loi* of 27 April–25 May 1791 on the organization of ministries, which provided that: 'Proclamations of ministers relating to their respective departments shall be made in Council; that is to say, those which, in the form of an instruction, prescribe the details necessary either for implementing a *loi* or for the welfare and activity of the [public] service, and those which order or encourage the implementation of *lois* in case of ignorance or neglect.' As in other aspects of the first revolutionary Constitution, the principle of the separation of powers appeared strongly. The King was to be refused legislative power, though it was impossible to deprive him of all authority to make orders.[5]

Other early republican Constitutions left little place for lawmaking by the executive. It was only with the Consulate Constitution of year VIII (1799) that it was explicitly provided in article 44 that the Government could 'make *règlements* and *ordonnances* necessary for the implementation (*exécution*) of [*lois*]', and thus have some independent authority for complementary rule-making. Later Constitutions made similar provisions, but they diverged significantly in the extent to which executive rule-making was permitted in other cases.[6]

The pattern established at the beginning of the Third Republic, therefore, was that statutes made by Parliament had superiority, but that the executive had an inherent power to make complementary rules for the implementation of those statutes. What remained unclear was the basis of this tradition persisting through monarchies, republics, and empires. Three issues can be identified here: legitimacy, hierarchy, and the effectiveness of government.

The legitimacy of executive lawmaking seemed to conflict both with the separation of powers and the idea that *loi* was the expression of the general will—and, indeed, with prudence, given the way in which many Governments behaved. But in other conceptions of government the executive would draw its legitimacy simply from its service of the national interest, being, as de Gaulle later put it, 'a decisive,

[5] See L. Duguit, *Traité de droit constitutionnel* (2nd edn., Paris, 1923), iv. 671.

[6] See Charter of 1814, art. 14; Charter of 1830, art. 13; Constitution of 1848, arts. 49, 75; Constitution of 14 Jan. 1852, art. 6; Senatus consultus, 21 May 1870, art. 14.

ambitious, and active institution, expressing and serving only the national interest'.

The hierarchy of legislative norms was more easily established. Even if the executive was to have some independent legislative power, it was always limited by, and subordinated to, the *lois* made by Parliament. The principle enunciated in 1791 was maintained in subsequent Constitutions: *loi* was clearly superior to *règlement*. In this way, the executive power could (in theory) always be kept in check.

The effectiveness of government, on the other hand, meant that certain rules had to be made by the executive and could not await the eventual decision of Parliament. After all, it was not clear that Parliament would be able to act. The 1791 Constitution contained provisions in the event of insufficient elected representatives attending the National Assembly for it to be able to act; and other Constitutions made the sitting of Parliament dependent on convocation by the head of State. Furthermore, there were bound to be matters, such as the internal operations of government departments and public services, which, as the *loi* of 27 April–25 May 1791 noted, would fall into the province of the ministers.

The Third and Fourth Republics

Mainly because of the brevity of the constitutional texts of 1875, the separation of powers within the Third Republic really developed as a matter of constitutional custom rather than of explicit provision. The supremacy of *loi* was implicit in article 3 of the *loi* of 25 February 1875, which stated that the President of the Republic 'shall supervise and ensure the implementation of *lois*'. But that text was insufficiently clear about what executive legislation was to cover. Writing in 1923, Duguit identified four areas.[7]

(1) *Colonial laws* had been made by the executive since the Senatus consultus (a *loi*) of 3 May 1854.

(2) *Complementary règlements* had been authorized since 1799 not merely to encourage compliance with the provisions of *lois*, but also to ensure their effectiveness. The inherent power to make complementary *règlements* was confirmed by an opinion of the Conseil d'État on 18 August 1807 relating to the implementation of article 545 of the Civil Code, and was recognized by a general provision of article 3 of the *loi* of 25 February 1875. Such *règlements* could also be authorized expressly by the legislature (in which case they were called 'règlements de l'administration publique'). In either case, what was involved was essentially filling out the details of a *loi*, while not adding any new

[7] *Traité de droit constitutional*, 694.

obligations or extending or restricting its scope.[8] The power to make such complementary legislation was limited by the hierarchy of norms, in that it could not infringe provisions of the parent or any other statute.

(3) *Autonomous règlements* were not permitted explicitly in the constitutional laws of 1875. In part, this was a deliberate attempt not to give the President too much power. All the same, it became apparent that certain powers were essential to the operation of government, and constitutional custom, endorsed by decisions of the Conseil d'État, came to recognize them. Three areas for such *règlements* were identified: the internal functioning of the administration and public services; public order; and national emergency, each of which had been recognized as contexts for executive legislation under some previous Constitutions.

The *internal functioning of the public service* had been within the province for *règlement* by ministers since the Revolution. Like the head of any organization, the minister has to have the power to give day-to-day instructions on how the service operates, who should be its employees, and what they should do. Only in the absence of an executive, could this level of direction be taken on board by Parliament, and not even the Convention tried to do that.[9] Such powers are almost an inevitable consequence of the legal personality of the State as administration, and really need no specific authorization.

In the leading case of *Babin*[10] it was established that, even without such authorization, a minister could make a decree governing the disciplinary procedures applicable to a body of civil servants. As the *commissaire du gouvernement*, Romieu, remarked in his *conclusions*:

> it is, in principle, the executive power that regulates the internal organization of public services and the conditions for their operation, which do not affect the rights of third parties. In particular, it fixes the rules of the contract between the administration and its agents, their recruitment, promotion, discipline, dismissal, etc. . . . [T]he executive power has full authority to decide on, and can freely fix, the terms of this contract, except on those matters which the legislature has, by way of exception, made its own by regulating them itself.

In that case, the minister was able to apply the procedures laid down in legislation of 1834 to a newly created body of civil servants. Contractual and employment powers were an inevitable part of being head of an organization. The same was true of other directions related to the

[8] See CE, 19 Feb. 1904, *Chambre syndicale des fabricants constructeurs de matériel pour les chemins de fer et tramways*, S. 1906.3.75, *concl*. Romieu.

[9] See arts. 65, 73 of Constitution of year I.

[10] CE, 4 May 1906, *Babin*, MATERIAL 2.

functioning of the public service. Even without a legislative text, the Conseil d'État was able to rule that a minister could give such directions and make such decrees as were necessary for the functioning of the service, including instructions on who should be allowed to enter premises used for the service.[11] Indeed, such powers would stem not only from the operation of the public service, but also from the occupation of public land. Legislation of this kind only affected directly those who were part of the public service. The users of the service, the citizens, would be affected indirectly by the character of the service that was provided to them, but they could not have their rights (to civil liberties, property, or contract, etc.) altered by the *règlements* that the minister laid down. Only *loi* could alter those.

Public order is a traditional function of government, permitting restrictions on individual liberty to be imposed so as to 'maintain public security, tranquillity, and health.'[12] The Constitution of 1791 specifically granted such powers to the King; but, following the abuses of Charles X, later Constitutions did not explicity do so, mistrusting the way in which the executive might use such authority. The exclusion from the provisions on the President in the 1875 Constitution was, thus, not an oversight. All the same, it became established quite soon that the President did enjoy such a power, though in fact the decrees that he issued were actually made by ministers. For instance, the President promulgated a decree of 2 October 1888 on the control of foreigners residing in France, and a decree of 13 November 1896 on the control of vagabonds.

A good example of such public order measures were the decrees of 10 March 1899 and 10 September 1901 on the construction and circulation of motor cars. Without any legislative authorization, the President promulgated a Highway Code and instituted driving-licenses. The legality of these measures was confirmed by a decision of the Conseil d'État, *Labonne*.[13] The Conseil noted that public order powers were explicitly conferred on municipalities by the *loi* of 14 December 1789, and thus could be extended *a fortiori* to the national level to deal with such problems as motor traffic. All the same, it was clear that such public order decrees could not impose penalties (of imprisonment), since such serious interference with individual freedom could only be effected by the legislature.[14]

Public emergency règlements were more controversial. The taking of exceptional measures in times of emergency was precisely the kind of power likely to be abused, particularly if it enabled *lois* to be

[11] CE, 7 Feb. 1936, *Jamart*, *Leb.* 172; *GA*, no. 56.
[12] Duguit, *Traité de droit constitutional*, 728.
[13] CE, 8 Aug. 1919, MATERIAL 5.
[14] Duguit, *Traité de droit constitutional*, 737.

suspended, as Charles X tried to do. Apart from specific legislation on the state of siege that was passed in 1849, no inherent power was conferred on the executive in any of the post-1830 Constitutions. All the same, the exceptional circumstances of the First World War made some such authority imperative. In its *Heyriès* decision of 1918[15] the Conseil d'État recognized the inherent power of the executive to take such decisions as were necessary to deal with an emergency, including suspending the statutory rights of a civil servant against dismissal. Although initially controversial, this came to be applied in later situations, before becoming constitutionalized in part in the 1958 Constitution.

Express delegation by Parliament to the executive of the power to legislate was not included in the 1875 arrangements. All the same, delegations of two kinds did take place: the power to make decrees on specific aspects of a programme, and the power to make decrees on a broad field. Whereas the former could be seen as mere implementation by way of *règlements de l'administration publique*, the latter regime of so-called 'décrets-lois' (decree-laws) amounted to a derogation from what was properly the province of Parliament. The executive was called upon not simply to fill out the details of a statutory scheme, but to go beyond the principles set out in the enabling law and to create a novel framework, even if this meant repealing previous *lois*. Because of political instability caused by the lack of lasting parliamentary majorities after the First World War, very broad delegations of power (the so-called 'lois-cadres' (framework-laws)) to make *règlements* on topics was a feature of the last years of the Third Republic.[16] Furthermore, once made, they could be altered by the executive making further *règlements*, until such time as the legislative authorization expired or was revoked.[17] After much hesitation, the Conseil d'État decided that it could consider challenges to the legality of such *règlements*.[18]

It was the delegation of such wide legislative powers that the Constituent Assemblies founding the Fourth Republic sought to restrict. Article 13 of the Constitution of 27 October 1946 stated plainly that 'The National Assembly alone passes *lois*. It cannot delegate this right.'[19] All the same, the political instability that had precipitated recourse to framework-laws under the Third Republic quickly resurfaced to require some definition of the contexts in which the execu-

[15] CE, 28 June 1918, *Leb.* 651; *GD*, no. 35.

[16] See esp. the enabling laws of 22 Mar. 1924 (S. 1926 L 295), 3 Aug. 1926 (S. 1927 L 1039), 8 June 1935 (S. 1935 L 1494), and 30 June 1937 (S. 1947 L 465).

[17] CE, 6 Dec. 1907, *Chemins de fer de l'Est*, S. 1908.3.1, *concl.* Tardieu, note Hauriou, dealing with decrees made in 1901, altering an *ordonnance* of 1846 on the regulation of railways, made under art. 9 of the *loi* of 11 June 1846.

[18] Ibid.

[19] See the similar provision in art. 66 §1, of the draft Constitution of 19 Apr. 1946.

tive could act without resort to Parliament. Article 7 of the *loi* of 17 August 1948 (MATERIAL 3) sought to specify a number of areas that lay within the competence of the executive to make *règlements*. In brief, they included the operation and control of public services, public undertakings, and nationalized industries, social security and social assistance, public finance and currency, and the rationing of energy and industrial goods and materials. While the Prime Minister, A. Marie, saw this as a rational division of tasks, some members of Parliament, notably René Capitant and Alain Poher, saw it as the loss of major areas of social reform from Parliament.

The Prime Minister's view was upheld by an opinion of the Conseil d'État's administrative sections in 1953, when it was stated that Parliament could delegate competence to the executive except in certain specific areas reserved to Parliament by constitutional tradition, and provided that the delegation was not so vague and imprecise as to amount to the abandonment of national sovereignty (MATERIAL 4). As a result, the scope for *décrets-lois* remained almost as wide as under the Third Republic.

Summary: The Scope of loi *and* règlement *in the Republican Constitutional Tradition*

Thus, even before 1958, the republican constitutional tradition can be said to have recognized certain inherent powers of the executive to make legislation. Those powers consequent upon being a legal person included that of running the internal life of the State organization—employing personnel and directing their activity, making contracts, and deciding how public land was to be used. The very nature of the executive branch of government conferred authority to ensure the implementation of *lois*, though this was understood more broadly than in Britain, and included the ability to fill out the detail of legislation. As the executive, it also required powers to maintain public order and to deal with emergencies. Finally, it could, of course, receive specific delegation from Parliament to make decrees in other areas.

Such a pattern of power goes beyond the line drawn by Romieu in *Chambre syndicale* and *Babin*, discussed above. He had argued that the legislature was required to intervene whenever unilateral, authoritative power was involved, especially where it affected individual rights and liberties. For him, the role of the executive was confined to the implementation of *lois* and the organization of the public service, unless contract or specific legislative delegation permitted otherwise, and provided no rights were affected. Public order and emergency powers could seriously impair the rights of citizens, and not just public servants.

The criteria developed under the Fourth Republic show how assumptions were changing about what it was 'natural' for the executive to do. Social provision, especially by way of social security, and economic intervention in currency, prices, and regeneration were areas in which Governments after the First World War, and particularly after the Liberation, were expected to be involved. The boundaries of the executive's natural or inherent competence to legislate had accordingly to be shifted.

On the other hand, it was well established that the legislature alone could act in certain matters, namely amnesty,[20] the creation of criminal offences and their penalties,[21] the declaration of a state of siege,[22] the levying of taxation, the creation of new courts,[23] and the creation of new categories of public bodies or services.[24]

3.2. The Allocation of Legislative Power in the Fifth Republic

Attempts were being made in the last years of the Fourth Republic to make this division of labour more explicit.[25] Furthermore, the aim since 1948 of producing a rationalized division of functions had not been fulfilled. Parliament was still spending an inordinate amount of time on making rules on trivial matters. If legislative self-restraint had failed, some more radical solution was considered necessary, especially by de Gaulle's advisers.[26] In many ways, the Constitution of the Fifth Republic represented a development of trends over the previous ninety years, rather than a massive shift in political power. This was the view taken by the *commissaire du gouvernement*, Janot, when introducing the future articles 34 and 37 to the Consultative Committee on the Constitution: 'The principle itself seems to be incontrovertible, because, whatever may be said, it is not a question of a revolution, or at least there is only one at the level of legal theory, because, at the level of practical realities, it only sanctions practice.'[27] Indeed, the rationalization of Parliament's activity would ensure that it would not be forced to

[20] Constitutional law of 25 Feb. 1875, art. 3. On the whole issue, see M. Waline, 'Les Rapports entre la loi et le règlement avant et après la Constitution de 1958', *RDP* 1959, 699.

[21] DDHC 1789, art. 8: CE Ass., 18 June 1958, *Syndicat des grossistes en matériel électrique de la région de Provence*, D. 1958, 656, *concl*. Tricot.

[22] *Loi* of 3 Apr. 1878; *loi* of 7 Dec. 1954, revising art. 7 of 1946 Constitution.

[23] CE, 2 June 1911, *De Pressensé*, *Leb*. 665.

[24] *Loi* of 13 July 1925: CE, 14 May 1948, *Sargosse*, S. 1949.3.25; CE, 13 May 1960, *Diop*, *Leb*. 454.

[25] The proposals of Guy Petit (1954) and of Pflimlin (22 Mar. 1958), following the scheme proposed by the General Committee of the Resistance in 1944.

[26] See B. Janot, in *Le Domaine de la loi*, 62.

[27] *Avis et débats*, 45.

abdicate sovereignty by a scheme of *décrets-lois*. It would provide a practicable framework for an effective Parliament.[28]

All the same, Janot had to admit that it was a significant innovation to move from having a list of matters reserved to the legislature, to a category of regulatory matters that would be forbidden to it.[29] At the level of constitutional theory, Parliament was no longer to be sovereign over all potential areas of legislation.

Without plunging into the bulk of very detailed case-law of the Conseil constitutionnel and the Conseil d'État, a number of issues of general importance can be identified: (*a*) How far is article 34 exhaustive of Parliament's legislative competence? (*b*) What is the importance of the distinction between 'determining the rules' of some areas and 'laying down the principles' of others? (*c*) How far does the executive have real inherent power to legislate?

The Sources of Parliament's Competence

Although the list of matters in article 34 (see Part Two of this book) might appear to be exhaustive, there are, in fact, a number of other provisions of the Constitution that also require Parliament to legislate upon specific matters. In addition, the Conseil constitutionnel and the Conseil d'État have developed a case-law that requires the legislature alone to authorize departures from general principles of law. Finally, and contrary to Janot's supposition in 1958, Parliament has been recognized as empowered to legislate even in regulatory areas, provided the Government does not avail itself of the constitutional procedures to object to this. As a result, Parliament has a field of competence that is, in practice, as wide as that which it enjoyed before 1958.

Specific Constitutional Provisions *Article 34* is the principal text in the Constitution to detail Parliament's competence to legislate. In broad terms, the article identifies five areas where *lois* are required, though the matters within each category often have little in common. First, *loi* should establish the *rules* concerning the system of basic civil liberties and their protection. These cover civic rights and duties; fundamental safeguards of civil liberties; the law of persons (nationality, status, marital property, and succession); criminal law and procedure, as well as courts and the judiciary; and taxation. Secondly, certain basic *rules* of public law are also to be established by *loi*. These cover the electoral

[28] Janot, in *Le Domaine de la loi*, 77. This power might also make it easier to implement EEC legislation into French domestic law.

[29] *Avis et débats*, 58.

system; the creation of public undertakings, nationalizations, and privatizations; and the fundamental guarantees of civil and military employees of the State. Thirdly, *loi* specifies the *fundamental principles* of national defence; local government; education; labour and union law; social security; property rights; and civil and commercial obligations. Fourthly, finance laws determine the resources and charges on the State. Fifthly, programme laws (*lois de programme*) determine the economic and social objectives of the State (by defining 'Plans'). The first three categories are the most important. Certain areas, such as civil and administrative court procedure, the administration of the public service, public order, economic regulation, and the management of the economy, are not included, but most major sectors are embraced in one form or another.

A number of other texts in the Constitution refer to the need for intervention by *loi*. Thus, the legislature must determine the qualifications of voters (article 3 §4), and the conditions whereby the judiciary safeguard individual liberty (article 66), it must declare war (article 35), or continue a state of siege beyond twelve days (article 36). It must vote finance laws (article 47) and ratify a large number of treaties (article 53). In addition, articles 72 to 74 specify that *loi* must determine the creation, organization, free administration, and control of overseas departments and territories, as well as modify institutional arrangements for the specific situation of the territory in question. These provisions very much circumscribe the executive's power over dependent territories, treaties, and emergency powers. In addition, article 3 complements article 34 in requiring Parliament to determine most provisions of the electoral system. Furthermore, article 66 has been taken to complement article 34 in requiring that criminal penalties be fixed by *loi*, since they affect individual liberty.[30]

The Preamble to the 1958 Constitution refers to *two further specific texts* that have provisions involving legislative intervention. The Declaration of the Rights of Man of 1789 requires that *loi* determine the circumstances under which a person may be accused, arrested, or detained (article 7), and also the matters giving rise to a criminal offence and the penalties attached to it (article 8). The Preamble to the 1946 Constitution requires that *loi* should regulate the right to strike (paragraph 7). The same Preamble also refers to 'the fundamental principles recognized by the laws of the Republic'. No such principle has yet been identified that would add a further restriction on the executive power to legislate.[31]

As we noted in the previous chapter, even before the adoption of the

[30] CC decision no. 73-80 L of 28 Nov. 1973, *Criminal Penalties (Rural Code)*, DECISION 10.
[31] The conclusions of Fournier in *Syndicat général des ingénieurs conseils* (noted above, p. 74) would suggest that this is indeed possible.

1958 Constitution the Conseil d'État protected constitutional values through the notion of *general principles of law*. These are of two kinds: those of constitutional value, and those of merely legislative value. The former are drawn from the 1789 Declaration, the 1946 Preamble, and, thus, also from fundamental principles recognized by the laws of the Republic. They are constitutional, in that they circumscribe the activity of both Parliament and the executive. The latter, legislative principles, merely restrict the executive, but can be freely altered by the legislature. This ground was first clearly invoked by the Conseil constitutionnel in the *Protection of Beauty Spots and Monuments* decision of 1969.[32] The issue was whether a *règlement* could amend a provision in a *loi* of 1930 so that the owners of sites of natural beauty or historic monuments required explicit permission from the Ministry of Culture to alter them in any way. The Conseil replied that 'according to a general principle of our law, silence maintained by the administration is equivalent to a refusal, and, in the present case, can only be derogated from by a legislative decision.'

Since 'general principles of law' are nowhere mentioned in constitutional texts, such a decision by the Conseil constitutionnel, confirming the approach of the Conseil d'État, limits the scope for the drafting of *règlements* by reference to what is effectively a constitutional tradition. Indeed, such an idea appears explicitly in the *conclusions* of Gentot in *Dame David*, (cited in Chapter 2).

The importance of this category is that it brings into the province of *loi* matters that are not listed in article 34. The most obvious one is civil and administrative court procedure. Conseil d'État decisions illustrate a number of these: the right to a public hearing, the right of action against administrative decisions, the right to due process, the requirement that a judge be able to order an *astreinte* (a financial penalty for non-compliance with a court order), and respect for the rights of defence.[33] The Conseil constitutionnel has also used general principles to outlaw retroactive decrees and those altering the right to a hearing in administrative courts, as well as to affirm the right of the taxpayer to bring a court action.[34]

[32] CC decision no. 69-55 L of 26 June 1969, DECISION 7 §5. The principle relied on by the Conseil constitutionnel conflicts with the Conseil d'État's understanding of French public law: CE Ass., 27 Feb. 1970, *Commune de Bozas*, *AJDA* 1970, 232, which suggests that this is not a general principle of law, from which regulations may not derogate.

[33] See respectively CE, 4 Oct. 1974, *Dame David*, *Leb.* 464; CE Ass., 17 Feb. 1950, *Dame Lamotte*, *Leb.* 110, *RDP* 1951, 478, *concl.* Delvolvé, note Waline; CE, 13 Dec. 1968, *Association des propriétaires de Champigny-sur-Marne*, *Leb.* 645; CE Ass., 10 May 1974, *Barre and Honnet*, *Leb.* 276; CE Ass., 31 Oct. 1980, *F.E.N.*, *RDP* 1981, 499, *concl.* Franc, *JCP* 1983, II. 20003, note Auby.

[34] See respectively CC decision no. 69-57 L of 24 Oct. 1969, *Course Fees at the École polytechnique*, *Rec.* 32; CC decision no. 72-75 L of 21 Dec. 1972, *Administrative Court Procedure*, *Rec.* 36; CC decision no. 80-119 L, *Litigation Procedure in Tax Matters*, *Rec.* 74.

The Exclusiveness of Article 37 Janot's explanation of the scope of article 37 suggested that it was an area into which Parliament would be unable to tread, thereby preventing *lois* on many minor matters that so encumbered the parliamentary timetable during the Third and Fourth Republics. Two early decisions of the Conseil constitutionnel seemed to confirm this opinion. In the *Radiodiffusion-Télévision française* decision of 11 August 1960 the Conseil struck down a provision concerning the accounts of the nationalized broadcasting service as 'infringing the powers of the supervisory authority in this area'.[35] Similarly, in the *1965 Budget* case an extension of the powers of the auditing commission with regard to public undertakings was declared unconstitutional as infringing the authority of the regulatory authority.[36] Since both were decisions under article 61 of the Constitution, this seemed to suggest that Parliament was totally incompetent in these matters. Nevertheless, the Conseil constitutionnel changed its position in the *Freeze on Prices and Salaries* case of 1982. The *loi* in question sought to impose a fine on companies that distributed dividends above a certain level. In their reference, the deputies claimed that a *loi* could not determine fines for regulatory offences, since this belonged within the province of article 37. The Conseil replied that article 41 permitted the Government to object to such a trespass on the regulatory sphere, and article 37 §2 enabled it to restore the area to the sphere of *règlement* but both were merely optional for the Government. In consequence:

> it appears thus that the Constitution did not intend, through articles 34 and 37 §1, to make unconstitutional a provision of a regulatory character contained in a *loi*, but had wanted, alongside the province reserved to *loi*, to grant the regulatory power an inherent province, and to confer on the Government, by the implementation of the special procedures of articles 37 §2 and 41, the power to ensure that it is protected against possible encroachment by *loi*.[37]

Such a change of view merely accorded with Government practice. In 1967 Pompidou (then Prime Minister) stated in the National Assembly that the Government had the power 'to submit provisions of a regulatory character to the vote of Parliament whenever the completeness and exhaustiveness of the discussion of a *loi* of general import required it'.[38]

[35] CC decision no. 60-8 DC of 11 Aug. 1960, *Rectifying Finance Law for 1960*, *Rec.* 25. See also CC decision no. 78-95 DC of 27 July 1978, *Agricultural Education and Training*, *Rec.* 26, and the final *considérant* of CC decision no. 79-104 DC of 23 May 1979, *New Caledonia*, *Rec.* 27, which deal with trespassing by *loi* on the power to deal with the implementation of *lois*.

[36] CC decision no. 64-27 DC of 18 Dec. 1964, *Rec.* 29. See also the wording of CC decision nos. 82-139 DC of 11 Feb. 1982, *Nationalizations II*, *Rec.* 31, and 82-141 DC of 27 July 1982, *Audio-Visual Law*, *Rec.* 48, which talk of *loi* infringing the province of *règlement*.

[37] CC decision no. 82-143 DC of 30 July 1982, DECISION 6 §11. See also CC decision no. 82-140 DC of 28 June 1982, *Rectifying Finance Law for 1982*, *Rec.* 45, which applies the same principle to a *loi* levying a parafiscal charge.

The net result of this new approach is that, as under the Fourth Republic, Parliament can always legislate, even within areas reserved to the executive by article 37. The only difference is that the executive can object or can regain the initiative. In practice, once it has achieved a majority in Parliament, it may well be the executive that takes this view.

Once Parliament has intervened in an area, the executive is no longer able to make *règlements* until the parliamentary text has been declassified under article 37 §2.[39] As a result, any *règlement* that attempts to modify a *loi* that has infringed the sphere of regulatory competence will be invalid.

The Distinction betwen Rules and Principles

The major categories of legislative competence defined by article 34 can be divided into two groups: those in which the legislature has to fix the rules, and those in which it merely has to specify the fundamental principles. This distinction did not appear in the original draft of the Constitution submitted to the Consultative Committee on the Constitution, but it was added in response to pressure to include more matters specifically within the competence of Parliament. By conceding competence to Parliament over fundamental principles, the de Gaulle Government was able to limit the concessions it made to those wishing a wide scope for the sphere of *loi*.[40]

Although it may have been a clever political ploy, it lacks legal coherence. Given that the executive is specifically empowered to take measures to implement *lois* (articles 21 and 37), then it would appear always to have the power to complement legislative rules on matters of detailed application. In consequence, even where Parliament makes the rules, it is not expected to make all of them, especially the minor details. The Conseil constitutionnel pointed this out in a decision of 2 December 1976, which noted that 'even if the Constitution . . . reserves to the legislature the fixing of the rules [on voting] . . . it leaves to the regulatory power, by virtue of article 37, the task of laying down measures for their application.'[41] As a result, the difference in generality between the rules laid down by Parliament and fundamental principles is elusive. Both will be more general than the detailed rules that complementary decrees implementing the *loi* will provide. Indeed,

[38] Cited in *Le Domaine de la loi*, 102.

[39] CE, 13 July 1962, *Conseil national de l'Ordre des médecins*, *RDP* 1962, 739, *concl.* Braibant; CE Ass., 27 Feb. 1970, *Dautan*, *Leb.* 141. *Lois* passed before 4 Oct. 1958 may be amended or repealed by decree without any declassification procedure.

[40] See Janot, in *Le Domaine de la loi*, 69.

[41] CC decision no. 76-94 L of 2 Dec. 1976, *Proxy Voting*, *Rec.* 67.

most commentators now consider that the key distinction is simply that of the importance of the rules to be determined.[42] This view would seem to be confirmed by the wording of article 34, which in many of its provisions requiring rules to be laid down by the legislature merely talks of them concerning 'fundamental safeguards'.

Among the multitude of decisions on this matter, a few might usefully illustrate the similarity between rules and principles in this context. Most decisions on the meaning of fundamental principles have come in the area of social security, often because of the frequent need to change or codify them. Thus the Conseil constitutionnel decided that the existence of maternity allowances, as well as the nature of the conditions to be fulfilled by a mother (conditions as to her age, the duration of her marriage, and the intervals between births), were fundamental principles of social security, but *règlement* could determine the precise maximum age (here 25), the duration of the marriage (here two years), and the interval between births (here three years).[43] Similarly, the categories of social security payments, such as sickness payments, have to be laid down by *loi*, but the exact nature of the payments—for example, whether hospital charges include the cost of thermal or climatic cures—can be settled by *règlement*.[44] Such fundamental principles cover not only rules but institutional structures, so that the definition of the mission of social security organizations as promoting health care is legislative, but the precise scope of that mission is regulatory—for example, whether this includes thermal cures.[45] As long as the executive *règlement* does not undermine the basic values of the existing scheme, it can more or less do what it likes.[46]

The position where the legislature has to specify the rules seems little different, even if one takes a matter, like the status of persons, which is not couched in terms of 'fundamental guarantees', 'systems', or 'categories'. For instance, the requirement that the adoption of persons born abroad or whose place of birth is unknown be notified to a registration office affects the status of persons, and so is legislative. *Règlement*, however, may specify the particular register on which this must be recorded (that of the *1er arrondissement* of Paris), and the period within which registration may take place.[47]

This synthetic approach to rules and principles is compounded by the way in which provisions of the Constitution are combined to

[42] See *Le Domaine de la loi*, 33; Genevois, 151.
[43] CC decision no. 61-17 L, *Antenatal and Maternity Allowances*, DECISION 8.
[44] CC decision no. 60-5 L of 7 Apr. 1960, *Social Security*, *Rec.* 32. See also CC decision no. 90-163 L, *Old Age Allowances*, *RFDC* 1990, 338.
[45] Ibid.
[46] See DECISION 8 §3.
[47] CC decision no. 64-30 L of 17 Sept. 1964, *Rec.* 41.

require legislative action. Thus, in a decision on the State broadcasting network the Conseil constitutionnel linked the obligation to make rules on the fundamental guarantees of civil liberties (here the communication of ideas and information) with the obligation to lay down fundamental principles on the categories of public undertaking (in which the network was unique).[48] The same is true when different articles or texts are combined, as in the case of criminal penalties.[49]

The addition of general principles of law as a criterion fixing the competence of Parliament beyond the list of matters contained in article 34 or in other constitutional texts reinforces the view that, these days, essential matters are resolved by the legislature, while accessory matters are decided by the executive. Whatever may have been the early analysis of the Constitution,[50] it is not true that article 34 assigns subjects to Parliament irrespective of their importance.

Does the Executive Have Inherent Powers to Legislate?

The debate about whether the executive has autonomous power to legislate has attracted much academic comment. For Favoreu, there is no scope for really autonomous legislative power, because *règlements* are restricted to areas in which the legislature is not required to act by provisions of the Constitution, other constitutional texts, organic laws, the tradition of public law, and previous legislative provisions.[51] Others would argue that there is such a sphere of autonomous action, but that it is not very different from that which existed under previous Constitutions.[52] From what has already been seen, it is difficult to envisage that there are whole areas of the law totally unencumbered by any intervention by the legislature. It is thus more useful to consider the traditional areas of regulatory competence, and see how far they are still open to the executive, albeit providing that no constitutional or legislative provision is infringed.

The implementation of lois *by decree* now finds its authority, according to the Conseil constitutionnel, not in article 21, but in article 37.[53] The two articles form a unity whereby article 21 designates the person empowered to act under article 37. In early decisions the Conseil

[48] CC decision no. 64-27 L of 17, 19 Mar. 1964, *Radiodiffusion-Télévision française, Rec.* 33.

[49] DECISION 10 §2.

[50] See e.g. the *concl.* of Fournier in CE, 18 May 1960, *Karle, Leb.* 333.

[51] See L. Favoreu, 'Les Règlements autonomes, existent-ils?', in *Mélanges G. Burdeau* (Paris, 1977), 405; id., '"Les Règlements autonomes n'existent pas"', *RFDA* 1987, 781.

[52] See R. Chapus, *Droit administratif général* (3rd edn., Paris, 1987), i. 54.

[53] CC decision no. 76-94 L of 2 Dec. 1976, *Proxy Voting, RDP* 1977, 458, *observations* Favoreu and Philip.

constitutionnel was concerned to prevent parliamentary encroachment in the execution of *lois*.[54]

Nevertheless, more recent decisions question the exclusiveness of the executive sphere in implementing *lois*. There is nothing to prevent the legislature providing that legislation shall come into force immediately,[55] nor, as will be seen below, that a particular executive body shall have the task of implementing the *loi*. In a 1986 decision the Conseil constitutionnel held further that the legislature could not leave the commencement of a *loi* to a decree, without setting a framework for the conditions under which that decree had to be made. In that case a *loi* was to abolish a provision of the Tax Code under which the transporters of fruit and vegetables needed to issue a certificate of delivery that was open to inspection by the tax authorities. The date of abolition was to be fixed by decree. The Conseil constitutionnel stated:

> Whereas, within the areas of its competence, the legislature has the power to fix the terms for the coming into force of the rules that it enacts; even if it is permissible to leave to the Government the power to fix the date on which the repeal of a *loi* establishing the obligations of taxpayers will take effect, it cannot, without violating the competence it has under article 34 of the Constitution, confer on it in this matter a power that is not subjected to any limitation . . .[56]

This decision, and the requirement that lesser criminal penalties have immediate effect,[57] do significantly limit the autonomy of the regulatory power, and would prevent some of the delays that have occurred in the past over implementing *lois*.

Delegations of power to issue règlements *made before 1958* are still respected. The whole purpose of the 1958 Constitution was to strengthen the hand of the executive, so that it would be absurd if the system established by articles 34 and 37 ended up by making the executive more restricted than before. The Conseil d'État acknowledged this in 1960 by holding (albeit implicitly) that delegations of power to the executive made before 1958 could still confer authority to issue *règlements* even after the coming into force of the new Constitution.[58]

Furthermore, the legislature alone has the exclusive right to legislate on aspects of individual freedoms that were not already restricted when the Constitution came into force. Thus, in *Price of Agricultural*

[54] See CC decision nos. 70-41 DC of 30 Dec. 1970, *Rectifying Finance Law for 1970, Rec.* 29, and 78-95 DC of 27 July 1978, *Agricultural Education and Training, Rec.* 26, both of which imposed requirements of consultation before implementation decrees were made.

[55] CC decision no. 79-104 DC of 23 May 1979, *New Caledonia, Rec.* 27.

[56] CC decision no. 86-223 DC of 29 Dec. 1986, *Rectifying Finance Law for 1986, Rec.* 184, *considérant* 14.

[57] CC decision no. 80-127 DC of 19, 20 Jan. 1981, *Security and Liberty*, DECISION 18 §71.

[58] CE Ass, 27 May 1960, *Lagaillarde, Leb.* 369.

Leases,[59] the Conseil constitutionnel had to deal with an objection raised by the Prime Minister to a bill proposed by two senators that sought to stabilize the cost of agricultural leases by repealing a decree of 7 January 1959. In his reference the President of the Senate took the view that such restrictions on the leasing of land affected the fundamental principles of property law and of civil obligations, and so were within the province of *loi*. The Conseil, however, held that:

Considering that those of these principles which are involved here—that is to say, the free disposal by any owner of his property, the freedom of the will of contracting parties, and the unalterability of contracts—have to be understood within the framework of the generally applicable restrictions that have been imposed on them by previous legislation to permit certain interventions judged necessary for public authorities in the contractual relationships between individuals.

Since intervention by *règlement* in the area of agricultural leases had been authorized by the *loi* of 23 March 1953, no fundamental principles were being infringed.

This approach was quickly followed by the Conseil d'État in *Martial de Laboulaye*,[60] which concerned restrictions on the production and marketing of wine. Such restrictions had existed since 1931, and the basic regime had been laid down by a decree of 30 September 1953. Thus, the freedom of property had already been limited significantly before 1958. The Conseil d'État noted that, prior to 1958, decrees had been able to regulate this area of economic intervention, and it held that *règlements* were competent to do so in 1959.

The key determinant of whether a matter is a fundamental principle lies in the novelty of the change that it introduces into the previous state of the law. Thus, to set out conditions for access to the nursing profession, when this had been totally unrestricted before, is to alter the fundamental safeguards for the exercise of civil liberties, and requires intervention by Parliament.[61] Likewise, the extension of transport supplements as a compulsory part of a salary, then only available to those working in Paris, to the rest of the country was held to be the creation of a novel charge covering both new categories of employer and of employee, and so could only be decided by *loi* (though once a *loi* had settled the principle, a decree should fix the amount that employers had to pay).[62] By contrast, a mere alteration of the composition of a body making arrangements for social security

[59] CC decision no. 59-1 FNR of 27 Nov. 1959, DECISION 5.

[60] CE Sect., 28 Oct. 1960, *AJDA* 1961, 20, *concl*. Heumann.

[61] CE Ass., 22 June 1963, *Syndicat du personnel soignant de la Guadeloupe*, *AJDA* 1963, 483.

[62] CC decision no. 63-5 FNR of 11 June 1963, *Transport Supplements*, *Rec*. 37.

payments to doctors for medical care was held to be within the province of *règlement*.[63] Obviously, the degree of novelty involved depends on the interests affected. Thus, in that case, changing the composition of the body fixing the tariffs was less important than changing the composition of a disciplinary body. As de Villiers has suggested: 'It is not necessary that the new limitations are of the same kind as the previous ones, but simply that they are not more extensive. What seems decisive is the *importance* of the obligations imposed.[64]

Public order powers were recognized as part of the inherent competence of the executive under previous Constitutions, but they are not specifically authorized in the 1958 Constitution. All the same, they have been held to survive. The Conseil d'État had explicitly held that the powers had not been withdrawn in cases concerning the control of publicity along public streets.[65] The Conseil constitutionnel followed suit in 1987, but set limits:

> Considering that, even if article 34 of the Constitution has not withdrawn from the head of the Government the general policing functions that he used to exercise before, by virtue of his inherent powers and outside any specific legislative authorization, the creation of a special policing régime for hunting calls into question fundamental principles of property; that it follows that, to the extent that they confer the task of policing hunting on a State body, the provisions submitted for consideration by the Conseil d'État are within the province of *loi*.[66]

This would very much suggest that, whereas established areas of competence, such as traffic regulation or public health,[67] will continue to be earmarked for executive action, newer areas of regulation will typically fall within the province of Parliament.

Emergency powers can also be exercised as before 1958.[68] All the same, much of their importance has been lost because of article 16 of the Constitution. Only localized emergencies, or those of insufficient gravity to come under article 16 are likely now to be covered by inherent powers.

[63] CE, 13 July 1962, *Conseil national de l'Ordre des médecins, RDP* 1962, 739, *concl.* Braibant.

[64] M. de Villiers, 'La Jurisprudence de "l'état de la législation antérieure"', *AJDA* 1980, 387.

[65] CE, 17 Feb. 1978, *Association dite Comité pour léguer l'esprit de la Résistance, Leb.* 82; CE Sect., 22 Dec. 1978, *Union des chambres syndicales de l'affichage et de publicité extérieure, Leb.* 530.

[66] CC decision no. 87-149 L of 20 Feb. 1987, *Protection of the Countryside*, DECISION 9.

[67] CE Ass., 13 May 1960, *S.A.R.L. Restaurant Nicholas, Leb.* 323 (decree of 1957 controlling the sale of game as a measure to deal with myxomatosis).

[68] See CE, 18 May 1983, *Rodes, Leb.* 199 (measures taken by a prefect when the explosion of a volcano was expected).

The Reasons for These Developments Good reasons have been offered as to why the division between articles 34 and 37 has not produced the radical shift of power that was expected. The most fundamental reason is that the early commentators on the Constitution thought in terms of the antagonism between Parliament and Government that had characterized the Fourth Republic. In fact, within a few years the Government was able to count on a stable majority in the National Assembly, and this reduced the Government incentive to implement legislation by decree.[69] In addition, the complexity of the relevant case-law developed by the Conseil d'État and the Conseil constitutionnel made it difficult for the executive to be sure of the procedure for implementing measures. The saga of the Code of Civil Procedure, implemented by decree but annulled in a number of decisions spread over several years for infringing the competence of Parliament, was a salutary illustration of the difficulty of trying to rely on the regulatory process for implementing reforms. With a *loi,* one could be sure of dealing with everything at once, and if there were to be a challenge to its constitutionality, this would be settled before promulgation by the Conseil constitutionnel, and could not be challenged thereafter.

In essence, the Government had little to lose by using *lois* to implement its reforms. It could guarantee their passage (with, as will be seen in Chapter 4, if necessary, a little help from articles 44 §3 and 49 §3), and could secure their effectiveness and unavailability to challenge. On the positive side, a number of gains might be established. First, as has been noted, the inclusion of regulatory provisions relating to the implementation of a text would render it more intelligible, reassuring deputies and senators about what was planned.[70] Secondly, Latournerie[71] suggests that pressure groups behind a reform prefer a *loi,* as it carries a greater guarantee of permanence. Equally, he suggests that ministers may desire the additional solemnity that a *loi* would give to their reform. Thirdly, Latournerie also notes that most *lois* are prepared in government departments by non-lawyers who may well be oblivious of the subtleties of the separation of *loi* and *règlement* (until faced with them by the Conseil d'État when it vets the text before it goes to Parliament). Not merely ignorance, but caution in wishing to avoid litigation will make the passing of a *loi* the safest course. Fourthly, Latournerie notes the way in which widening the scope of *loi* improves relations with Parliament, and gives its members a sense of a more productive existence than voting on taxes and crimes. Others note

[69] J.-L. Pezant, 'Loi/règlement: La construction d'un nouvel équilibre', *RFSP* 1984, 922, 934.

[70] See the speech of Prime Minister Pompidou, *JO, Débats Ass. Nat.*, 1967, 1069, cited in *Le Domaine de la loi,* 100.

[71] *Le Domaine de la loi,* 217–18.

the way in which the Government would be unwilling to cause difficulties for the Presidents of the parliamentary chambers by pressing its prerogatives.[72]

There is also little evidence that Parliament wishes to overstep the mark, ensuring that a harmonious relationship is maintained. (In recent years the objections of inadmissibility raised in Parliament by the Government have become non-existent.[73]) On the other hand, no one wanted the reform of 1958 to make the position of the executive more limited than it had been under previous Constitutions. As a result, the twists and turns of case-law reflect a desire at least to maintain the status quo.[74] While there has been great concern to protect civil liberties in line with the spirit of 1789 and 1946, economic regulation within established fields has been subject to much less strict scrutiny.[75] This reflects a view that executive intervention is more suspect in some areas than in others, as well as the view that established practices should not be disturbed.

3.3. Areas of Difficulty

It has been claimed that 'The joint delimitation by the Conseils of the areas of *loi* and *règlement*, far from harming the coherence of case-law, has favoured the development of two complementary case-laws that render each other more complete and precise.'[76] This happy complementarity of the decisions of the Conseil constitutionnel and the Conseil d'État has ensured that the division between *loi* and *règlement* has been able to work smoothly. The advice of the Conseil d'État prior to the making of *lois* and the issuing of important decrees (*décrets en Conseil d'Etat*), and its judicial decisions on enacted decrees have fitted in well with the views of the Conseil constitutionnel on the appropriate line to be drawn. Boulouis argues that this reflects a political consensus in favour of the constitutional tradition.[77] There have, however, been divergences of opinion. In some cases the two Conseils seem to have exerted reciprocal influences to bring them to a common position; but in others these differences remain. The issue of criminal penalties illustrates this point.[78]

[72] Pezant, 'Loi/règlement', 934.
[73] Ibid. 938–9.
[74] See J. Chardeau, in *Le Domaine de la loi*, 165.
[75] Ibid.
[76] Franck, 291.
[77] In *Le Domaine de la loi*, 206–7.
[78] Other issues include the creation of public corporations, and legislation affecting the autonomy of local authorities: see Genevois, §§187–9, 193–4.

Criminal Offences and their Penalties

Perhaps the most famous confrontation between the Conseil constitutionnel and established courts has occurred over the competence of the executive in determining criminal penalties. Article 8 of the 1789 Declaration states that 'no one may be punished except according to a *loi* passed and promulgated prior to the offence'. The obvious reference is to the *lettres de cachet*, executive warrants prior to the Revolution that enabled detention without trial. The executive was to have no power to institute criminal offences and impose criminal penalties without the prior approval of Parliament. Article 34 of the 1958 Constitution merely confirms this established part of the French constitutional tradition by stating that *loi* should determine the rules specifying *crimes* and *délits*[79] as well as the penalties applicable to them. The difficulty with this apparently simple division of labour lies in the growth of other, regulatory offences (*contraventions*) to cover such matters as trading without a licence and the like, as well as in the more traditional areas of public order.

Two views could hold away in this area. The first suggests that, since article 34 only mentions two of the three classes of criminal offence, the third category, that of *contravention*, must fall within the province of *règlement* by virtue of article 37. This simple textual argument was the one first adopted by the Conseil constitutionnel in its decision of 19 February 1963, in which it stated: 'Considering that . . . if article 34 of the Constitution reserves to the legislature the task of fixing "the rules concerning . . . the determination of *crimes* and *délits* as well as the penalties applicable to them", the determination of *contraventions* and the penalties applicable to these belongs to the province of *règlement* . . .'[80] Thus, changes in fines on the police court scale, which are the penalty for a *contravention*, were a matter for *règlement*. Such a view coincided with that of the drafters of the Constitution,[81] since minor matters and details were left to the executive, but major offences and penalties were to be left to Parliament. The Conseil d'État had already adopted this stance in *Société Eky* in 1960, permitting the executive to decide what penalty was to be imposed for a regulatory

[79] The Penal Code distinguishes between 3 classes of offence. The most serious (such as murder) are *crimes*. The next most serious are *délits* (such as theft). The distinction between these 2 has much in common with the old common law distinction between felonies and misdemeanours. The 3rd class is that of *contraventions*, which are typically regulatory offences (failure to carry identity documents, failure to observe many traffic regulations, and so on).

[80] Decision no. 63-22 L of 19 Feb. 1963, *Police Fines*, *Rec.* 27. See also decision no. 64-28 L of 17 Mar. 1964, *Mutual Credit Funds*, *Rec.* 35; decision no. 65-35 L of 2 July 1965, *Financial Markets*, *Rec.* 79.

[81] See J. Foyer, in *Le Domaine de la loi*, 71.

offence, even if this involved imprisonment. Article 34 was thus seen as derogating from the 1789 Declaration.[82]

The difficulty with this view was that the line between regulatory and other offences did not correspond to the division between non-custodial and custodial penalties. From 1810 to 1945, when regulatory offences carried a maximum penalty of five days' imprisonment, this had more or less been the case. During the Fourth Republic the penalty was increased to ten days, but it was only after the Fifth Republic Constitution came into force that a regulatory offence could carry a maximum of two months in prison.[83] Thus, the established position before 1958 was that imprisonment could be imposed by decree for breach of a *contravention*, but that the period involved was, until then, minimal.

The second view was therefore that the spirit of the 1789 Declaration required that all imprisonment had to be sanctioned by *loi*, and could not be decided upon by the executive, even for a very short period. It was this position that the Conseil constitutionnel adopted, albeit *obiter*, in its decision of 28 November 1973.[84] There it stated: 'Considering that it follows from the provisions of paragraphs 3 and 5 of article 34 and of article 66 of the Constitution, combined with the Preamble, that the determination of *contraventions* and the penalties applicable to them are within the province of *règlement* when those penalties do not involve measures depriving liberty.' *A contrario*, it would follow that when imprisonment was involved, then this would have to be authorized by *loi*.

The process of reconciliation between these positions involved two issues: the definition of the constituent elements of an offence; and the specification of the penalties attached to it. As far as *the constituent elements of offences* were concerned, the problem was that criminal penalties were often imposed for failure to observe administrative formalities, such as obtaining planning permission, the non-observance of rules on matters such as maximum prices, and so on. The issue on which the Conseil d'État and the Conseil constitutionnel initially diverged was the extent to which the Government could use a *règlement* to amend these administrative formalities, the breach of which triggered a criminal offence. Early decisions of the Conseil constitutionnel seemed to suggest that *règlements* could determine such matters, even if failing to meet the terms of the *règlements* constituted a *délit*.[85] The Conseil d'État, however, retained its pre-1958 position that the con-

[82] CE, 12 Feb. 1960, *Société Eky*, MATERIAL 6. This view merely confirms the advice given in its administrative capacity by the Section de l'Intérieur of the Conseil d'État on 16 Oct. 1951 as to the position under the Fourth Republic.

[83] *Ordonnance* no. 58-1297 of 23 Dec. 1958; see Genevois, 98.

[84] Decision no. 73-80 L of 28 Nov. 1973, *Criminal Penalties (Rural Code)*, DECISION 10.

[85] See decision no. 64-28 L, *Mutual Credit Funds*, and decision no. 65-35 L, *Financial Markets*.

stituent elements of *délits* had to be determined by *loi*, not *règlement*.[86] The Conseil constitutionnel came to agree with this point. In 1969 it held that *loi* had to fix the period of notice that an owner of a site of natural beauty had to give to the prefect before commencing major works on it, since the breach of such a period of notice gave rise to a *délit*.[87] This position coincided with that adopted by Parliament, which, in 1965, had refused to delegate the power to fill in the details of serious crimes. As René Capitant had argued: 'Over and above even the terms of the Constitution, we consider that the creation of an obligation sanctioned by a penalty appropriate to a *délit* or a *crime* should be the work of *loi*. Therein lies an essential safeguard for individual liberty.'[88] The position thus agreed upon was that, if the breach of an administrative *règlement* constituted a serious crime, then the constituent elements of the offence had to be determined by *loi*, even if the matter, viewed in isolation from its penalty, would fall within the province of the executive alone.

Criminal penalties had been the subject of the 1973 decision of the Conseil constitutionnel (DECISION 10). It had held that, where the penalty involved imprisonment, even for the brief period appropriate for the breach of a *contravention*, it must be laid down by *loi*. The Criminal Chamber of the Cour de cassation and the Conseil d'État were quick to reject this view, and to reaffirm the traditional line that the executive was free to alter both the constituent elements and the penalty appropriate for a regulatory offence.[89] As the *procureur général*, Touffait, argued before the Cour de cassation, to rule otherwise would invalidate a large number of penalties imposed for various regulatory offences:

> immediately, the punishment of *contraventions* for a breach of the Traffic Code, in public health matters, on safety at work, on cheques for less than 1,000 francs without the funds to back them . . . would be seriously disrupted, thousands of challenges to rules would be made, the enforcement officers would be plunged into uncertainty, and the consciences of our courts would be troubled by a doubt as to the legality or constitutionality of established practice. A real legal anarchy would result from the yawning gap, which it would be necessary to fill promptly to avoid serious disorder.[90]

By distinguishing the decision of the Conseil constitutionnel on a technicality, the Cour de cassation and the Conseil d'État were able to

[86] CE, 18 June 1958, D. 1958, 656, *concl.* Tricot; CE Ass., 3 Feb. 1967, *Confédération générale des vignérons du Midi*, *AJDA* 1967, 159, *chronique* Lescaut and Massot, 164, *concl.* Galmot (the constituent elements of the *délit* of illegal mixing of wines from France and abroad had to be fixed by *loi*).

[87] CC decision no. 69-55 L of 26 June 1969, *Protection of Beauty Spots and Monuments*, DECISION 7.

[88] Cited in Franck, 231.

[89] Cass. crim., 26 Feb. 1974, *Schiavon*, D. 1974, 273, *concl.* Touffait, *SB* 106; CE, 3 Feb. 1978, *CFDT-CGT*, *AJDA* 1978, 388.

[90] *Schiavon*, D. 1974, 273.

evade a direct conflict. They pointed to the fact that penalties of imprisonment for *contraventions* were authorized by articles 464 and 465 of the Penal Code, which, having the status of *loi*, could not be challenged as unconstitutional by the ordinary private and public law courts.

As Touffait also suggested, the Conseil constitutionnel may have been more concerned to sound a warning note to the Government than actually to alter so much of established practice. Certainly, in a recent decision the Conseil constitutionnel seems to have reversed its view, again by way of *obiter dictum*. In a case concerning penalties for damaging public property (the *contraventions de grande voirie*) it stated:

> Considering that, by virtue of article 34 of the Constitution, *loi* fixes the rules concerning 'the determination of *crimes* and *délits* as well as the penalties applicable to them'; that, by these provisions, the drafters of the Constitution intended, in criminal matters, to confer on the legislature the competence to determine the most serious offences; that the scale of seriousness results from the distinction created by *loi* between *crimes* and *délits* on the one hand, and *contraventions de police* on the other hand, as well as the respective penalties that apply to them.[91]

In the end, the issue is really one of practicality rather than principle. In theory, imprisonment should be reserved for serious crimes, so that the penalties imposed for regulatory offences would not infringe the principle stated in the 1789 Declaration. But imprisonment for a short duration has been considered an appropriate penalty for such offences, and its maximum length was increased after the Constitution was adopted in 1958. In such a situation the executive has, within limits, to be given the power to determine the conditions under which people will be sent to prison and for how long, without reference to Parliament. As Genevois suggests, the real problem lies not in the difficulty of drawing the line between *loi* and *règlement*, but in the illegitimate expansion of imprisonment as a penalty for *contraventions*.[92]

3.4. Enabling Laws under Article 38

The Growing Importance of Procedure

If most attention in the literature has been paid to the division in principle of legislative power between the executive and Parliament, it

[91] CC decision no. 87-161 L of 23 Sept. 1987, *Post Office Code*, *AJDA* 1988, 62, *observations* Prétot. The case confirms, in identical wording, a decision of the Conseil d'État that the *contraventions de grande voirie* are not to be treated as criminal offences for the purposes of art. 34 (CE, 22 June 1987, *Rognant*, *AJDA* 1988, 62). Cf. §11 of CC decision no. 85-139 L of 8 Aug. 1985, *Social Security Code*, *Rec.* 94, and §8 of CC decision no. 85-142 L of 13 Nov. 1985, *Social Security Code*, *Rec.* 116, which both make a point of specifying that provisions are regulatory where they provide for fines for regulatory offences.

[92] Genevois, §168.

has to be recognized that the late 1970s and especially the 1980s saw the increasing use of special delegation by Parliament to the executive of the power to legislate on specific areas by means of article 38. (Such legislation takes the form of an *ordonnance*, an executive rule having the same authority as a statute, but which must eventually be submitted to Parliament for ratification.) That article was not intended to bring back the recourse to framework-laws of the kind seen in the Third Republic, but merely to enable a delegation of power if the inherent legislative powers of the executive proved to be too restrictive.[93] In so far as the scope of what is properly the province of *loi* remains unclear and expanding, such delegation is the surest method available to the executive of securing authority to take decisions. Whereas many of the instances in the 1960s and 1970s had concerned overseas dependencies, both the new Socialist administration in 1981 and the new right-wing administration of 1986 obtained wide-ranging powers to carry out major items of their programmes at rapid speed by means of delegation from Parliament. The decline in insistence on the executive's inherent right to legislate upon matters has led to the use of specific delegation to enable the executive to arrogate the power that it considers necessary to get on with the job. Contrary to the expectations of Boulouis in 1977, it is not the minority Governments that have made most use of article 38 delegations, but majority Governments, wanting to proceed faster than would be permitted by the ordinary course of parliamentary procedure and timing.[94] All the same, the nightmare envisaged by one member of the Consultative Committee on the Constitution has equally not materialized. Dejean feared that the Government would make use of article 38 to obtain the power to carry out its programme, and then send Parliament on holiday.[95] In part, the reason why this has not happened lies in the greater parliamentarian instincts of Governments, but it has also been the product of deliberate interpretations of the Constitution on the part of the Conseil constitutionnel.

Scope of the Procedure

Three of the Conseil's decisions are of particular importance in this context, in 1977, 1982, and 1986, representing the growth of the increasing use of article 38 delegations. In the first case the Conseil greatly restricted the kind of delegation that was possible.[96] The new Barre Government sought a delegation of authority to modify by *ordonnance* the constituencies for the election of representatives to the

[93] See Janot, in *Avis et débats*, 54, describing it as 'a safety-valve'.
[94] Cf. J. Boulouis, in *Le Domaine de la loi*, 201.
[95] *Avis et débats*, 48.
[96] CC decision no. 76-72 DC of 12 Jan. 1977, *Djibouti Elections*, DECISION 11.

Chamber of Deputies of the new independent state of Djibouti. The opposition made two objections. The first was that article 38 could not be used in this case, since the article referred to the use of *ordonnances* to carry out the Government's 'programme', but the matter in question had not been included in the programme approved by Parliament. The second objection was that the bill made no reference to implementing any part of such a programme. The Conseil preferred to interpret the word 'programme' in article 38 more narrowly than in article 49, where it referred to the Government's political programme, submitted to a confidence vote in Parliament. As a result, the authorization that it permitted was limited to a specific policy objective defined at the time of the request for a delegation of power. Article 38 required the Government 'to indicate with precision to Parliament, when the enabling bill is presented and in order to justify it, the objective of the measures that it proposes to take'. Sufficient detail had been given to Parliament in this case, but the observations of the Conseil demonstrated a deliberate attempt to prevent any return to the kinds of framework-laws of earlier Republics.

The 1982 decision[97] upheld a wide-ranging enabling law covering public service retirements and pensions, as well as introducing a more generally applicable earlier retirement age and a penalty for those who tried to combine a retirement pension with the salary from a job. Here the Conseil noted that it was valid in so far as it did not trespass on any area in which the Constitution required an organic law, and did not attempt to dispense the Government from conforming to constitutional requirements, notably with regard to the right to work.

The approach to the right-wing Government of Chirac in 1986 was similar. It came to power in March 1986, but, fearing an imminent end to 'cohabitation', and wishing to have tangible results from its policies before the end of 1986, it sought to obtain two enabling laws: the first to introduce a programme to 'liberalize' the economy, and the second to change the electoral system back to single-member constituencies and to bring an end to proportional representation.

The first *loi* empowered the Government to deal with competition, to end price and rent controls, and to privatize a list of sixty-five companies. The opposition claimed that the objectives of the delegations were insufficiently precise. For instance, article 1 of the *loi* provided that:

> In order to assure businesses a greater freedom of operation and to define a new competition law, the Government is authorized . . . to amend or repeal certain provisions of economic legislation concerning prices and competition,

[97] CC decision no. 81-134 DC of 5 Jan. 1982, *Enabling Law on Social Measures, AJDA* 1982, 85.

especially those of *ordonnances* no. 45-1483 of 30 June 1945 on prices, and no. 45-1484 of 30 June 1945 on the determination, prosecution, and punishment of economic offences. . . .

The opposition maintained that the idea of defining a new competition law was too broad. Furthermore, in this, as in other provisions of the *loi*, the terms were too loose. For instance, what was meant by 'freedom of operation'? The Conseil was not willing to be unduly restrictive, otherwise there would be little real difference between an enabling law and an ordinary *loi*.[98] All the same, the text was submitted to a careful process of 'interpretation'. First, the Conseil noted that the text did not authorize the modification or repeal of all the rules of civil and commercial, criminal and administrative law relating to the economy. Looking at the various preparatory materials (known as the *Travaux préparatoires*) that document the legislative history of a measure, especially the presentation that the Government made to Parliament, the Conseil noted that only certain texts were envisaged, and that 'within these limits' there was nothing unconstitutional about the provisions. Secondly, the Conseil added a couple of glosses by way of interpretation. The safeguards that the *loi* required the Government to provide were to be in addition to those required by constitutional values, such as the right to judicial review and to the defence of one's interests, as well as the rights of those who were not economic actors. Furthermore, nothing in the *loi* was to permit a derogation from France's international obligations. This approval 'with reservations' was applied to other provisions of the enabling law, notably those on privatization.

The same approach was adopted with regard to the enabling law on electoral reform. Given the precedent of 1977, the Government had been very careful in drafting the *loi*, giving little room for criticism about lack of precision. The opposition tried to claim first that an organic law was necessary, and secondly that it was an abuse of procedure. On the first, the text of article 34 was clear that the electoral system fell within the province of ordinary *lois*, and so could be delegated under article 38. The determination of electoral boundaries came within the notion of the electoral system, and so it also was properly delegable under article 38. The second argument was much more concerned with the scope of article 38 delegations. It was suggested by the opposition that the Government was resorting to an enabling law in order to avoid the Conseil constitutionnel assuming immediate control of its reforms. The Conseil accepted that its inability to review the *ordonnances* was a necessary consequence of the constitutional procedure of article 38. All the same, the Conseil noted

[98] CC decision no. 86-207 DC of 25, 26 June 1986, *Privatizations*, DECISION 12.

that resort to this procedure did not dispense the Government from respecting the rules and principles of constitutional value.[99] This point had already been made in the 1982 decision, and was a clear invitation to the exercise of control by the President and by the Conseil d'État. The first article of the decision in both 1986 cases begins: '[S]ubject to the strict reservations of interpretation contained above . . . [the *loi*] is not contrary to the Constitution.' This way of reducing the apparent scope of the *loi* to constitutionally acceptable levels is the most effective means by which the Conseil has sought to control the delegation of power.[100]

Controls other than the Conseil constitutionnel

Until an *ordonnance* has been ratified, it remains an administrative decision and is not subject to control by the Conseil constitutionnel, but, as a *règlement*, it is subject to other controls. The *Conseil d'État* reviews *ordonnances* passed under the authority of an enabling law in two ways. Article 38 §2 requires that an *ordonnance* be made by the Council of Ministers after seeking the opinion of the Conseil d'État. This prior control is the most important method of review, and prevents most unconstitutionality in the exercise of power. It is supplemented by subsequent judicial review (which may also be exercised by the criminal courts). This jurisdiction, established by the Conseil d'État in 1907, was reaffirmed under the Fifth Republic in 1961, when it annulled an *ordonnance* that sought to give retrospective validity to elections to an administrative body that the Conseil d'État itself had previously quashed.[101]

Judicial review is, however, limited by the fact that *ordonnances* are ultimately subjected to a procedure of ratification by Parliament. Article 38 simply requires the laying of a bill for ratification, not its actual approval by Parliament. It is when the bill is approved that judicial review by the Conseil d'État is limited. Once the measures in the *ordonnance* are ratified, they cease to be simply administrative decisions and become parliamentary decisions, which are, as such, excluded from review by the ordinary courts. As under previous Constitutions, it has been held that ratification may be implicit—for example, when a

[99] CC decision no. 86-208 DC of 1, 2 July 1986, *Electoral Enabling Law*, *Rec.* 78; *GD*, no. 42.

[100] On this technique , see O. Beaud and O. Cayla, 'Les Nouvelles Méthodes du Conseil constitutionnel', *RDP* 1987, 677, esp. p. 684.

[101] CE Ass., 24 Nov. 1961, *Fédération nationale des syndicats de police*, *Leb.* 658, applying the principles laid down in CE, 6 Dec. 1907, *Chemins de fer de l'Est*, S. 1908.3.1, *concl.* Tardieu, note Hauriou. This jurisdiction is well illustrated in the privatization case, see CE Ass., 2 Feb. 1987, *Joxe et Bollon*, *RFDA* 1987, 176.

subsequent loi merely modifies some of the provisions of the authorized *ordonnance* or decree.[102]

The second control is exercised by *the President*. Article 13 states that the President 'signe' the *ordonnances*. Such a present tense normally connotes 'shall sign' rather than 'does sign'. However, during the term of office of de Gaulle and, more importantly, during the cohabitation period under Mitterrand, the President took the view that he had an option on whether to sign *ordonnances* or not. In making his decisions, Mitterrand relied on the grounds provided by interpretations given by the Conseil constitutionnel. So, in the case of privatization, he refused to agree to the *ordonnance* permitting wide-ranging privatization, and the measure had to be passed by a *loi* of 6 August 1986 (not referred to the Conseil constitutionnel). Likewise, in the case of electoral reform, Mitterrand did not consider that the Government proposals contained in the draft *ordonnances* conformed to constitutional requirements, and he refused his signature. The draft *ordonnance* was then passed by Parliament as a *loi*, and this received no objection from the Conseil constitutionnel.[103]

In making such 'interpretations' of the Constitution, the President is not subject to any reviewing court.[104] All the same, as in these cases, his interpretation can be circumvented if Parliament passes the necessary legislation to enact the proposed *ordonnances*, and the Conseil constitutionnel does not demur. The President is safest when he is following a lead that the Conseil constitutionnel has given as to what the Constitution requires. The Conseil constitutionnel thus functions both to orient the approach of other bodies and to take its own position with regard to a specific text.

Favoreu has argued that the current trend is towards making *ordonnances* little more than decrees of application. The degree of precision required in the text has been steadily increased to ensure that control can be exercised by the Conseil constitutionnel.[105] All the same, the recent decisions leave significant scope for administrative action. The situation is not the same as it was in the Third Republic, with Parliament delegating to the Government the power to act in any way it wished. On the other hand, it could hardly be said that the privatiza-

[102] See CC decision no. 72-73 L of 29 Feb. 1972, *Employee Participation in Business Profits*, *AJDA* 1972, 638, note Toutlemonde, which is consistent with the approach of the Conseil d'État in CE, 25 Jan. 1957, *Société Charlionais et Cie*, *RDP* 1957, note Waline.

[103] CC decision no. 86-218 DC of 18 Nov. 1986, *Electoral Reform II*, *Rec.* 167; *GD*, no. 42. On this period, see generally A. Stone, 'In the Shadow of the Constitutional Council: The "Juridicisation" of the Legislative Process in France' (1989) 12 *West European Politics* 12.

[104] See R. Romi, 'Le Président de la République, interprète de la Constitution', *RDP* 1987, 1265.

[105] L. Favoreu, 'Ordonnances ou règlements d'administration publique?', *RFDA* 1987, 686.

tion of sixty five companies, or the creation of a new competition law are merely the application of some clearly determined legislation. The difference between the powers contained in article 37 and article 38 may be a matter of degree, but that degree is none the less significant.

3.5. Rule-Making by Bodies other than the Government

In the late 1980s, in particular, a significant problem for the separation of powers in relation to lawmaking has been the growth of independent regulatory agencies (*autorités administratives indépendantes*), especially in the financial and media sectors. The creation of such agencies has been justified by the desire to insulate regulation from traditional political pressures, and to afford greater freedom to actors in those sectors.[106] But, of course, such regulation can only be effective if the agency has the power to establish standards of conduct. The 1958 Constitution does not envisage regulatory power being exercised other than by the Government, so the issue arises as to whether ordinary legislation (as opposed to a constitutional amendment) can give legislative power to such agencies.

Early decisions of the Conseil constitutionnel on this matter related to legislation that gave the agency very specific powers to implement a *loi*, for example, by issuing licences. Provided that the framework for implementation was sufficiently defined, the Conseil was happy to accept such a delegation of functions to an administrative body, even if it were not part of the Government.[107] In relation to the regulatory body covering broadcasting, the CNCL (La Commission nationale de la communication et des libertés), the Conseil ruled that article 21 of the Constitution 'does not prevent the legislature conferring on a body other than the Prime Minister the task of determining . . . the norms for implementing a *loi*', provided that this occurs 'within a defined area and in a framework established by *lois* and *règlements*'.[108] This was taken further in 1989 in relation to its successor, the Conseil supérieur de l'audiovisuel (CSA). Here the Conseil stated that:

> [the provisions of article 21] confer on the Prime Minister . . . the exercise of regulatory power at a national level; that they do not prevent the legislature from conferring on a State body other than the Prime Minister the task of fixing norms to implement a *loi*, provided that this authorization concerns only

[106] See generally C.-A. Colliard and G. Timisit (eds.), *Les Autorités administratives indépendantes* (Paris, 1988); Vedel and Delvolvé, ii. 446–56.

[107] See CC decision nos. 83-167 DC of 19 Jan. 1984, *Credit Establishments*, *Rec.* 23, 84-176 DC of 25 July 1984, *Audio-Visual Law*, *Rec.* 55, 86-217 DC, *Freedom of Communication*, DECISION 27 §§55–61; and cf. CC decision no. 84-173 DC of 26 July 1984, *Cable Television*, *Rec.* 63.

[108] DECISION 27 §58.

measures limited in scope both by their area of application and by their content . . .[109]

In this case the *loi* gave the CSA the power to issues codes of practice on publicity relating to institutional communication, sponsorship of programmes, and analagous practices. The Conseil struck the power down because the scope of the delegation was too wide. By contrast, the Commission des opérations de Bourse (COB) merely issued rules in relation to the stock exchange that were limited in scope, had to respect established legislative provisions, and were subject to approval by the Minister of Economy and Finance. In such a situation, the power to make rules was sufficiently circumscribed.[110] The contrast here is not simply about levels of detail in the framework within which the regulatory agency has to work. The fact that the CSA had to circumscribe the freedom of communication was also significant. As with the issue of rules and principles seen earlier, the more important the civil liberty or constitutional principle that the inferior lawmaking body affects, the greater must be the detail laid down in the *loi*.

The position thus arrived at is that regulatory agencies may be given some rule-making power that is independent of ministerial approval. All the same, such an independent power must be very carefully circumscribed in the enabling legislation.

The picture that emerges from analysis of the actual functioning of the Constitution and from the way in which it has been interpreted belies the suggestion that the division of power between *loi* and *règlement* has been a radical revolution in the operation of French government. Parliament, far from being marginalized, has remained a central lawmaking body. At the same time, the executive has grown more powerful, not so much in terms of its power to bypass parliamentary processes with reliance on its inherent legislative attributes, as by its domination of Parliament, and its effectiveness in being able to impose its will on it. The fact that the domination comes from an enhanced party system has replaced the necessity for an independent source of legislative power. Indeed, recourse to delegated legislative power under article 38 is increasingly a source of authority to make legislation.

Governmental practice has reinforced the position of Parliament. The formal confrontation envisaged by article 41, whereby the Government would object to a bill or clause during the parliamentary deliberations, has fallen into desuetude. There has been no recourse to it since 1980,

[109] CC decision no. 88-248 DC of 17 Jan. 1989, *Conseil supérieur de l'audiovisuel*, DECISION 28 §15.

[110] CC decision no. 89-260 DC, *Commission des opérations de Bourse*, *RFDA* 1989, 671, §31.

and no case before the Conseil constitutionnel since 1979.[111] The Government normally tries to settle such matters in the business committees of the chambers of Parliament.[112] The Government also smooths the path by submitting a preliminary draft of a decree to one of the standing committees of the National Assembly, though Parliament has no formal status in any procedures of declassification under article 37, and its members would appear to have no right to challenge decrees before the Conseil d'État.[113] The process is more co-operative and parliamentarian than the Constitution might appear to suggest, and, indeed, it is often parliamentarians themselves who raise the issue of the division between the powers under article 34 and article 37. Since there is now no objection in most cases to allowing a *loi* to cover matters falling within the province of *règlement*, the scope for challenges under article 61 after the bill has been passed has become very small.

As Favoreu has noted, the use of article 37 is much less than would be expected if attention were paid to the controversy it has aroused. In 1985 some 125 *lois* and 1,435 ordinary decrees were passed, yet only 18 decrees that relied on article 37. In recent years barely fifteen decrees a year are based on this article.[114] For the most part, these have been decrees consolidating legislation and instituting minor reforms in areas such as social security or agriculture.

Even where article 37 powers are used, there exist controls. The administrative sections of the Conseil d'État will have to give prior advice, and may end up ruling on legality at a later stage. The preliminary advice is rarely neglected, since, among other reasons, the Government risks successful challenges to its decrees in the judicial sections of the Conseil d'État. The Conseil constitutionnel will usually see a draft of the Government's proposed decree when considering the declassification process, and this has the function of ensuring that the decision is designed to deal with the matters that are relevant to the Government's intentions. This practice has the effect of removing many problems. The controls exercised over article 38 *ordonnances* are very similar. The Conseil constitutionnel can set out guide-lines for courts like the Conseil d'État to apply when they are faced with specific *ordonnances*. The control operated by the latter is broadly the same, whether it is dealing with a decree or an *ordonnance*.[115]

If the institutional structures have reverted much more towards the

[111] See Pezant, 'Loi/règlement', 937–8.

[112] See J. Foyer, in *Le Domaine de la loi*, 83, 106–8.

[113] Ibid.; on this, see the *concl.* of Massot on CE Ass., 2 Feb. 1987, *Joxe et Bollon*, *RFDA* 1987, 176.

[114] Favoreu, '"Les Règlements autonomes n'existent pas"', 871, 884.

[115] See CE Ass., 2 Feb. 1987, *Joxe et Bollon*, *RFDA* 1987, 176.

norm of previous Constitutions, the same could also be said of the case-law. The decisions of the Conseil constitutionnel on article 37 are very much case-by-case determinations of what is fundamental and what is not, what is novel and what is not, what is a principle and what is a rule.[116] The executive is no more hampered than under previous Constitutions, but its freedom of action does not seem to have increased greatly as a direct consequence of article 37 itself. What seems lacking in the case-law—as, indeed, in the constitutional provisions themselves—is a sense of general principle dividing the spheres of executive and parliamentary lawmaking. The best that authors can offer is to suggest that it all boils down to the importance of the issue on which a rule is being made,[117] or to whether the new provision is there to implement (*la mise en œuvre*) or to challenge (*la mise en cause*) the existing *loi*.[118] But such criteria really take one little further than those established before the 1958 Constitution. Indeed, leading administrative lawyers from the Conseil d'État have suggested that its approach has been very much in line with the constitutional tradition to be found from Romieu onwards.[119] That tradition has been developed to meet new problems, but it has not taken any radical new directions as a result of the provisions of articles 34 and 37.

All the same, there has been a significant rationalization in the operation of government. There has been less detail expected in *lois*, and the Government feels more secure in getting on with enacting decrees by the simple procedure of interministerial co-operation.[120] In addition, there has been a major change in the constitutional theory of the separation of powers. It is now clear that the legitimacy of executive legislation is not derived from a delegation from Parliament (and thereby from the will of the people), but from the nature of governmental responsibilities. Parliament and Government each derive their legitimacy from the Constitution. *Loi* and the Constitution are, in many ways, predominantly controls on freedom of executive action.

The recognition that Parliament cannot control all the rule-making that the 1958 Constitution enshrines is being taken further than the provisions of articles 34 and 37 envisaged. It is the delegation of legislative power to the Government under article 38, and to administrative agencies outside the Government that is the central issue for the modern division of these powers.

[116] See de Villiers, 'La Jurisprudence', 395; B. Nicholas, '*Loi, règlement* and Judicial Review in the Fifth Republic' [1970] *Public Law* 251 at 269–70.

[117] Favoreu, in *Le Domaine de la loi*, 33.

[118] Rivero, ibid. 264.

[119] See Fougère, ibid. 145, and Chardeau, ibid. 152.

[120] See J. Fournier, *Le Travail gouvernemental* (Paris, 1987), 238–43.

4

PARLIAMENTARY PROCEDURE

NEUBORNE[1] argues that constitutional law is, in the end, all about procedure. There is always a way to pass a reform, but it may require a constitutional amendment to do so. Constitutional law explains which procedure is necessary to effect such a reform. In relation to parliamentary procedure, constitutional law also has the function of structuring the process of political decision-making, and thus the influences and legitimate means of pressure that may be exercised in that process. In the French Fifth Republic such rules are all the more important, since the whole object of the 1958 constitutional revisions was to institute fundamental changes in the way in which Parliament operated, and to change the balance of power between the executive and the legislature. The role of the Conseil constitutionnel is to act as arbiter between the opposing interests, and to shape the political process in ways that conform to its ideas of the principles of government under the Fifth Republic.

4.1. SOURCES OF RULES ON PARLIAMENTARY PROCEDURE

The United Kingdom draws a distinction between parliamentary procedure, the law of Parliament, and conventions.[2] Rules of procedure direct the working, the machinery, and the forms of proceeding in each House. Such rules may be laid down by statute, parliamentary standing orders, or by convention. The first two establish the formal rules of procedure, but they are supplemented both by conventional rules and by customary practices. Unlike the situation in the United Kingdom, the rules of French parliamentary procedure are set out to an important extent in legislative form, such that they constrain what Parliament may do. It is not simply a matter of internal self-regulation by Parliament. The scope and importance of these constraints affects not simply the powers of Parliament, but also the way in which the

[1] B. Neuborne, 'Judicial Review and the Separation of Powers in France and the United States' (1982) 57 *New York University Law Review* 363 at 368–9.

[2] See Erskine May's *Treatise on the Law, Privileges, Proceedings and Usages of Parliament* (20th edn. by Sir C. Gordon, London, 1983), 207–8.

executive can dominate the political process. But, in many ways, what may at first sight seem important constraints on Parliament turn out to be significant practical limitations on what the Government can achieve.

The Constitution

The principal interest of the drafters of the Constitution of 1958 was to create a workable set of institutions. The ideal of a rationalized Parliament was not just to give the executive freedom of action. It was also to provide Parliament with effective, if limited, ways of influencing the process of legislation. Its function was not so much to block proposals, as to make them workable, or to fit proper political objectives. Apart from the division of legislative competence between the Government and Parliament discussed in Chapter 3, the Constitution establishes the rights to introduce legislation (articles 39 and 40), rights of amendment (article 44), legislative procedure (articles 42, 43, 45, and 46), finance procedure (article 47), the parliamentary timetable (articles 48 and 51), votes of confidence (articles 49 and 50), and confirms the rights of the Government to address the chambers (article 31). The detail of these provisions sets out a very precise framework for relations between the Government and Parliament, which seems to be premissed on a basic antagonism rather than on the spirit of co-operation that is so essential to a system based, as in the United Kingdom, on arrangements made 'through the usual channels', in other words, behind the Speaker's chair.

Legislation

Legislation also supplements these basic rules. While ordinary *lois* can be repealed easily with a bare majority, article 46 of the Constitution sets out a special procedure whereby organic laws (*lois organiques*) require an absolute majority of the National Assembly if the Senate does not concur, and must be approved by the Conseil constitutionnel. Organic laws cover areas such as finance bill procedure, electoral law, and the status of the judiciary, as discussed in Chapter 2. As will be seen in relation to finance bills, organic laws can effectively regulate the legislative process, even though it is constitutionally possible for Parliament to alter them. Although not in theory a higher norm than a *loi*, an organic law may operate as such simply because of the difficulty in altering it.[3]

[3] The transitional legislation passed under art. 92 of the Constitution also had some constitutional status in relation to various interim measures: CC decision no. 66-28 DC of 16 June 1966, *Standing Orders of the Senate*, *Rec.* 15.

Parliamentary Standing Orders (le règlement)

Under the French parliamentary tradition, much like the British, Parliament had the right to regulate its own procedure. Indeed, the eventual acquisition of this right in the Third Republic marked the beginnings of a sustained parliamentary democracy. Parliamentary standing orders were not just made by Parliament, they were enforced by it. Self-regulation was the key to freedom. The 1958 Constitution could not allow the delicate balance between the powers of the Government and Parliament to be thwarted by the uncontrolled activities of the latter. It has already been noted in Chapter 1 that the competence of Parliament to judge electoral matters was removed by the 1958 Constitution. In addition, under article 61 §1, parliamentary standing orders have to be submitted to the Conseil constitutionnel before they come into force.

The changed situation was brought to the attention of both chambers in the early months of the Fifth Republic. The first standing orders of the National Assembly and of the Senate were submitted to the Conseil constitutionnel in June 1959, and both were declared unconstitutional.[4] The Senate was so infuriated at this interference with its erswhile prerogatives that it refused to discuss a revised version of the standing orders until 27 October 1960.[5] This control is abstract rather than concrete, and authors have noted that the Conseil constitutionnel tends to adopt a rather literal form of interpretation towards standing orders when they are presented for approval, though it is more flexible when deciding whether there have been procedural irregularities in the passing of particular legislation.[6]

Parliamentary standing orders only have an effect within Parliament; they do not provide a basis on which legislation can be struck down for procedural irregularity. Thus, if deputies do not ask for the President of the National Assembly to enforce the standing orders during the parliamentary process, the breach cannot form the basis of a reference to the Conseil constitutionnel to have the resulting *loi* declared unconstitutional.[7] This contrasts with the views of Debré when present-

[4] See CC decision no. 59-2 DC of 17, 18, and 24 June 1959, *Standing Orders of the National Assembly*, DECISION 13, and CC decision no. 59-3 DC of 24 and 25 June 1959, *Standing Orders of the Senate*, *Rec.* 61.

[5] The *rapporteur* in the Senate justified this by arguing that 'Your committee thought that such a reasonable attitude was necessary, that is to say, one marking its deference to the Conseil constitutionnel and its disagreement in substance.' (Cited in C.-L. Vier, 'Le Contrôle du Conseil constitutionnel sur les règlements des Assemblées', *RDP* 1972, 165 at 201.)

[6] Avril & Gicquel, 9; Genevois, §214: see the decision on personal voting: CC decision no. 86-225 DC of 23 Jan. 1987, *Séguin Amendment*, DECISION 16, §4.

[7] See CC decision no. 78-97 DC of 27 July 1978, *Criminal Procedure Reforms*, *Rec.* 31: an amendment concerning the *juge d'application des peines* (the judge supervising the criminal

ing the draft Constitution. He was of the opinion that: 'Anything that concerns legislative procedure, anything that concerns the relations between the chambers, anything that concerns the relations between the chambers and the Government constitutes provisions that have more than a regulatory character in the strict sense: they are constitutional in inspiration, they affect the mechanism of the institutions.'[8]

Customs and Conventions

It has already been noted in Chapter 2 that the Constitution of 1958 operates in practice by way of a large number of conventions. In part, these are necessary simply to make the framework of rules effective. After all, it is not to be expected that politicians should choose rules as a guide—they merely provide limits and opportunities for political action. The way in which the parliamentary agenda is drawn up would reflect the concern with conventions and customs, rather than legally binding rules.

The resort to unwritten rules has been attributed to the very narrow framework within which the chambers must work. Unwritten rules escape the control of the Conseil constitutionnel, and thus allow a degree of autonomous flexibility to the chambers themselves.[9] But it would also be true to say that any institution needs some traditions to provide continuity within a changing membership. Predictability enables planning and saves time. Some of the members, particularly of the Senate, participated in equivalent bodies under earlier Republics: the President of the Senate in 1991, for example, M. Alain Poher, has been President since 1968, and a member of the Senate and its predecessor since 1946.

Among the matters that are predominantly governed by convention and custom, one might highlight the composition and operation of the Conference of Presidents within each chamber, bringing together the party group leaders who then thrash out the details of the timetable for debates, the allocation of bills to committees, and so on. But, as discussed in Chapter 2, the issue of proxy voting shows that parliamentary customs may well be contrary to the express text of the Constitution and to standing orders. On the whole, it is especially

penalty) in a bill covering the judicial police and the jury at the Cour d'assises fell outside the amendment procedure within the standing orders, but was not challenged by any deputy at the time. More explicitly, in CC decision no. 80-117 DC of 22 July 1980, *Nuclear Installations*, *AJDA* 1980, 479, the Conseil stated that 'the provisions of the standing orders of parliamentary assemblies do not have constitutional value'. See also CC decision no. 75-57 DC of 23 July 1975, *Business Tax*, *Rec.* 24.

[8] Cited in Avril & Gicquel, 9.

[9] R. Chazelle, 'Continuité et tradition juridique au sein de la seconde chambre: Le Sénat et le droit parlementaire coutumier', *RDP* 1987, 711, 712.

outside the area of legislative procedure that conventions have an important place. Thus, the way in which parliamentary questions to ministers are answered—and, indeed, the very institution of questions in the Senate and their frequency in the National Assembly—are governed almost entirely by convention. In short, these conventions are a very important part of the Constitution. As Avril and Gicquel put it, they are 'a non-legal means of creating laws inherent in political law'.[10]

4.2. Restricting the Government's Power to 'Streamroller'

The problem of handling coalitions caused enormous governmental instability under the Third and Fourth Republics. The 1958 Constitution tried to meet this in two ways. Under article 44 §3, the Government can insist that its text of the bill is voted upon as a block, rather than article by article at the end of debate. It can also go further, and make its measure an issue of confidence under article 49 §3. This curtails debate and requires the opposition to table and vote upon a censure motion in order to prevent the bill's passage. Since a censure motion requires an absolute majority of the membership of the National Assembly (not simply those voting), all abstentions count in the Government's favour.

Such powers are Draconian, though not too far removed from what can happen in practice in the United Kingdom. The Conseil has consistently refused to put any restrictions on how these powers are exercised, nor will it allow rules of parliamentary procedure to do so.

The reason for granting the power to insist on a 'block vote' on a text was to enable a series of obstructive amendments to be rejected, and so help the Government to get legislation passed. In view of this objective of an effective legislative process, the Conseil constitutionnel refused to allow parliamentary standing orders to limit its use in an early decision of 1960.[11] Standing orders could not prevent the Government from choosing which articles or parts of articles it wished to treat as a text to be subjected to a block vote, nor could it prevent the Government from asking first for such a vote on part of the text, and then on the text as a whole.[12] The purpose here was to provide the Government with an alternative to risking its political life by way of a motion of confidence, as in the Fourth Republic. The provision had therefore to be appliable

[10] Avril & Gicquel, 19.

[11] CC decision no. 59-5 DC of 15 Jan. 1960, *Amendments to Articles 95 and 96 of the Standing Orders of the National Assembly*, *Rec.* 15.

[12] The reason for the proposed change was the use made of the power in early 1959 to block together a specific unpopular proposal on the pensions of former combatants with other popular elements of the budget: Genevois §224.

at any moment in the legislative process. But the Conseil was careful to circumscribe this power. The provision did not prevent discussion before a vote was taken, and standing orders could ensure that this was the case. Similarly, a later decision confirmed that the Senate could not restrict the blocking together of amendments and sub-amendments.[13]

The power to make a matter an issue of confidence has been the subject of much controversy. The President of the Comité consultatif constitutionnel, Paul Raynaud, argued that it would be used to steamroller legislation through: 'every time that an important text is involved, the Government will make it an issue of confidence and the National Assembly, prevented from discussing the text, will have no more than a right of veto. Uniquely in the world, the National Assembly will no longer pass the *loi*, it is the Government that will do it by its own authority.'[14] Triboulet wanted to limit the use of the power to once a year, but these reservations were overruled. While Debré saw it as the ultimate safeguard for when no compromise could be reached in the case of deadlock between the chambers, Waline gave perhaps the clearest defence when he asked: 'can one at the same time keep a Government in power and then refuse it the means of governing?'[15]

The power has been the subject of much criticism ever since. It has been used most often by minority Governments. Thus, in the Barre administration after the 1978 elections the Rassemblement pour la République (RPR) could not be counted on to support the Government's measures, and it had recourse to article 49 §3 some eight times in the autumn session of 1979 (the power had only been invoked four times in the previous twenty years). In DECISION 15, *Finance Law for 1980*, opposition deputies challenged its use to pass a budget that the RPR were unwilling to support without expenditure cuts of two milliard francs. While there were problems (as will be seen) with other aspects of the passage of the bill, the Conseil raised no objection to the use of article 49 §3, although it did not give specific arguments for its view that recourse to this procedure was proper. A further challenge to the use of the procedure also failed when it was argued that the Prime Minister should attend in person to request a vote of confidence. It sufficed under the Constitution that the procedure had been approved by the Council of Ministers, and a personal appearance on the podium in the National Assembly, though customary, was not a constitutional requirement.[16] This reference came at the end of a session in which

[13] CC decision no. 73-49 DC of 17 May 1973, *Senate Standing Orders*, *RDP* 1973, 1031.

[14] *Avis et débats*, 180.

[15] Ibid. 184.

[16] See CC decision no. 89-268 DC of 29 Dec. 1989, *Finance Law for 1990*, *RFDA* 1990, 143.

another minority Government, that of Michel Rocard, beat all records by invoking article 49 §3 on thirteen occasions in the autumn session of 1989 alone, and the frequent application of this procedure continued until the end of his administration.[17] Invoking article 49 §3 stops the debate and, unless the opposition lays down a censure motion, the bill is passed. Indeed, there have been occasions when this procedure has been applied the moment a bill is introduced into Parliament.[18] Effectively, therefore, the fears of Raynaud are fully realized in the practice of the Fifth Republic, and the Conseil constitutionnel does nothing to prevent it, adopting very much the line of Waline.

4.3. Protecting the Functioning of Parliament

In the area of legislation, Parliament has three main tasks: to initiate proposals (*propositions de loi*), to amend texts submitted to it, and to debate and vote on their merits. As in the United Kingdom, most legislation comes from Government initiatives. Bills initiated by individual members of Parliament are unlikely to succeed, if only because they will not appear on the main parliamentary agenda, but only on the supplementary agenda (*l'ordre du jour complémentaire*), unless they receive Government support. In France Government bills (*projets de loi*) must be vetted by the Conseil d'État before they are submitted to Parliament. This procedure is designed to avoid infelicities in drafting, inconsistencies with other legislation, and incompatibility with the Constitution.[19] The Government is not bound to adopt the advice of the Conseil d'État, and it does not have to make this advice public. As a result, the function of Parliament is essentially to examine the political merits of the proposal. By contrast, and unlike the situation in the Second Empire, amendments and *propositions de loi* are not submitted to prior vetting by the Conseil d'État, and so the task of Parliament may be more legalistic in such cases. This is not to say that challenges to the constitutionality of a bill or a provision by means of an *exception d'irrecevabilité* are confined to such cases.[20] They are

[17] In the last full parliamentary session of his Government (autumn 1990) art. 49 §3 was used 8 times: *RFDC* 1991, 108.

[18] Avril & Gicquel, 209, cite the examples of Pierre Mauroy, in relation to the price-freeze bill on 24 June 1982, and Jacques Chirac, in relation to the enabling law on electoral reform on 20 May 1986.

[19] On this process, see B. Ducamin, 'The Role of the Conseil d'État in Drafting Legislation' (1981) 30 *International and Comparative Law Quarterly* 882.

[20] This parliamentary procedure of the *exception d'irrecevabilité* is independent of a challenge before the Conseil constitutionnel. It is not a prerequisite for a reference to be made, though some decisions have given a different impression: see P. Terneyre, 'La Procédure législative ordinaire dans la jurisprudence du Conseil constitutionnel', *RDP* 1985, 691 at 712–13.

frequently made to all kinds of legislation. Detailed amendments are proposed and discussed first in committee (usually a standing committee, but sometimes an *ad hoc* committee), and then the committee's written report is debated in the full chamber.[21]

Although Parliament has significant freedom in the areas of initiative, amendment, and debate, the Constitution introduced certain prerogatives of the Government in these contexts. Thus, article 39 gives the Government the right to initiate legislation, and article 40 makes this right exclusive in finance matters. As an innovation of the Fifth Republic, article 44 gives the Government the some power to amend legislation as members of Parliament have.[22] But, once the chambers are deadlocked, the Prime Minister can require that a joint committee (the Commission mixte paritaire) meet to draw up proposals on the disputed provisions, and only those amendments approved by the Government can then be made to this compromise text. Where the chambers continue to be deadlocked, it is the Government that requests that the National Assembly make a final decision. In this way, the Government dominates procedures for the initiation and amendment of texts, and the Conseil constitutionnel has been concerned to ensure that these powers are not abused.

The Right of Initiative

The right to initiate legislation comprises the right not only to choose the subject-matter, but also to choose the opportune moment to introduce it. Can this be restricted? The issue has arisen in those instances where parliamentary amendments have tried to coerce the Government into laying legislative proposals before Parliament within a prescribed time-limit. Though it is not infrequent for parliamentary amendments to require subsequent reports on the operation of legislation, this form of amendment has been consistently rejected by the Conseil constitutionnel.

Resolutions requiring the Government to take certain actions were common in previous Republics, but the Conseil constitutionnel struck down standing orders of both the National Assembly and the Senate providing for such a procedure in its decisions of June 1959.[23] Two arguments were advanced. First, to the extent that such resolutions aimed to direct or control Government action, this was contrary to article 20 of the Constitution, which entrusted the Government with

[21] See Avril & Gicquel, ch. 6.

[22] Until then, the President could present legislation, but members of the Government had only the same right to amend as any other deputy: see Luchaire & Conac, 865–6.

[23] See CC decisions nos. 59-2 and 59-3 DC *Standing Orders of the National Assembly*, and *Standing Orders of the Senate*.

the determination and the conduct of national policy. Second, to the extent that they formed part of the right of parliamentary initiative, such resolutions were doing the same job as *propositions de loi*. Such a duplication was not provided for in the Constitution, and it could not be allowed. The only function that parliamentary resolutions could have was to affect the internal operations of the chamber in question.

This decision did not, however, prevent attempts to achieve the same result by way of amendments to bills. Thus, the 'Vallon amendment' to the finance law of 12 July 1965 required that the Government lay a bill on the protection of employees' rights when companies were buying their own shares.[24] The current Minister of Finance, Valéry Giscard d'Estaing, did not object. But when he became President of the Republic, and a similar amendment was made to the Seventh National Plan, his Prime Minister referred the issue to the Conseil constitutionnel, which struck down the amendment as unconstitutional.[25] When RPR deputies passed an amendment requiring the Government to produce a coherent package of measures on families, and especially mothers, in the next session of Parliament in order to remedy the crisis in the French birth-rate, the Conseil argued that it was unconstitutional, since such an injunction to the Government did not fall within the categories of Parliament's legislative competence set out in article 34, and was contrary to the freedom of the Prime Minister to initiate legislation given by article 39 of the Constitution.

A less stern approach was taken in 1982, when the injunction was not the result of a parliamentary amendment, but was part of the Government's own bill, The Socialist Government wanted to amend the whole planning process in conjunction with its Eighth National Plan, and its bill included a timetable of other legislation to be introduced over the next two years. Opposition deputies argued that this was unconstitutional in the light of the Conseil's 1979 decision. On the argument that the bill restricted the freedom of legislation, the Conseil remarked that 'the legislature cannot bind itself; and a *loi* can always and without restriction repeal or amend an earlier *loi* or derogate from it.'[26] Any provision that sought to bind the legislature in the future was ineffective. However, even if the freedom to propose legislation to meet changing conditions was therefore respected, the legislature could seek to organize its legislative work so as to ensure that it was spread out over time and that continuity was achieved, and in this way the provision did not interfere with the right of initiative of the Government. This latter approach of ineffectiveness, rather than un-

[24] See Luchaire & Conac, 882.

[25] CC decision no. 78-102 DC of 17 Jan. 1979, *Seventh National Plan*, D. 1980, 233, note Hamon.

[26] CC decision no. 82-142 DC of 27 July 1982, *Reform of Planning*, *Rec*. 52.

constitutionality, is perhaps more satisfactory as a way of handling the problem.

The Right of Amendment

Given its power to steamroller legislation, particularly finance legislation, many references to the Conseil turn on abuse of the power to amend a text. The issue of 'budgetary riders' will be dealt with in the next section. It is sufficient to look here at the process of amendment in ordinary *lois*.

Although an essential part of the parliamentary process, the right of amendment is not always seen as an unqualified boon. Vedel, a member of the Conseil constitutionnel from 1980 to 1989, and a key participant in decisions on this issue, wrote in 1949 that:

> The right of amendment is indispensable to the exercise of the legislative function. It should be recognized, however, that its use leads to deplorable results. Generally, amendments destroy the harmony and the coherence of a proposed text, and often render its interpretation impossible. Sometimes they are the opportunity for demagogic outbursts, as authors of amendments often seek to ensure in the eyes of the law a favoured situation for certain groups who particularly concern them.[27]

The distinctive feature of the Fifth Republic is the attempt to regulate the use of amendments by legal means rather than just by convention and custom, when this latter control was previously seen as the hallmark of the French parliamentary tradition.[28] The practice of submitting Government bills to the Conseil d'État means that there is less need for Government amendments to be made to texts once they have been submitted to Parliament than there is in the United Kingdom. As a result, well over half the amendments proposed to texts in the National Assembly come from deputies, and over a quarter come from the parliamentary committee; much the same is true in the Senate. The Government contributes around 10 per cent of all amendments (though some 70 per cent of these are actually successful)[29] Amendments may, of course, be proposed for a number of reasons—to obtain Government assurances, to provide an opportunity to put forward different political views, as well as simply to obstruct the Government's programme. They can be tabled in large numbers. Thus, in the case of the 1990 law privatizing Renault, some 2,399 amendments were tabled, and the Government was able to make 2,225 of these fall by the simple expedient of dropping the programmatic, but substantively unnec-

[27] *Droit constitutionnel* (Paris, 1949; repr. 1984), 484.
[28] See Luchaire & Conac, 863.
[29] See Luchaire & Conac, 876–9.

essary article 1 from its bill.[30] Since the right of initiative is, in practice, so limited, the right to put amendments is one of the few ways in which deputies and senators can control debate and exert some form of pressure on the Government to take note of their views.

The role of the Conseil constitutionnel has been to preserve this freedom whilst safeguarding an effective legislative process. The Conseil has first been concerned to ensure that the right to amend a text is available at any moment in the legislative process, though (as has been pointed out) in cases of deadlock between the chambers the Government's approval for an amendment will be necessary before it is allowed to go forward for debate.

In its decisions in the 1970s and early 1980s the Conseil was primarily concerned to establish what constituted an amendment and when it could be made. Thus, in 1973 it decided that the right of amendment included the right to make sub-amendments (amendments to amendments), and that these could not be restricted by standing orders to those which did not undermine or contradict the amendment to which they were attached.[31]

The timing of amendments has been of particular concern when there has been deadlock between the chambers, which was especially true when the Socialists were in power, since the majority in the Senate belonged to the opposition. In such a situation, the Government usually approves any amendment to be proposed, which effectively stifles any further opposition intervention. Very soon after the Socialists came to power in 1981, the Conseil ruled that, where the Commission mixte paritaire has failed to come up with a compromise text, the right of amendment still exists, and there was no objection to the Government proposing amendments to its bill at this stage, before it was finally voted upon by the National Assembly.[32] Even where a compromise text on disputed provisions has been agreed by the joint committee, the Conseil has allowed amendments not merely to this text, but also to other provisions that were settled by both chambers.[33] Such a freedom to propose amendments enables the results of any deadlock to be assessed and taken into account in relation to the whole text, and not just parts of it. All the same, it lets the Government have the last say on the scope of the bill by controlling which amendments will be discussed in the final stages of its passage.

By contrast, the Conseil has taken a stricter view of what counts as

[30] *RTDC* 1990, 474. The principal proposers of the amendments were the Communists in this case.

[31] CC decision no. 73-49 DC of 17 May 1973, *Senate Standing Orders*, *Rec.* 15.

[32] CC decision no. 81-136 DC of 31 Dec. 1981, *Third Rectifying Finance Law for 1981*, *Rec.* 48.

[33] CC decision no. 86-221 DC of 29 Dec. 1986, *Finance Law for 1987*, *Rec.* 179 §5.

an amendment and what is really a new initiative. Such a distinction is difficult to draw because of the amorphous character of much legislation. Titles such as 'miscellaneous social measures' give so much scope for relevant amendments to be made that it is very difficult to say what is extraneous. Indeed, bills have a habit of growing considerably. In the 1978 case on *Criminal Procedure*[34] deputies tried to argue that an amendment on the powers of the *juge de l'application des peines* was contrary to the standing orders of the National Assembly, in that it did not relate to the subject-matter of the bill (the judicial police and the jury at the Cour d'assises). The challenge was rejected on the ground that the standing orders had no constitutional status, and no other ground was invoked by the Conseil itself. But a series of decisions in 1985–6 began to suggest, albeit *obiter*, that there were limits to the power of amendment.[35]

The matter came to a head at the end of the parliamentary session in December 1986. The Government, having obtained a power to make *ordonnances* on various social matters, proposed to alter the working day and the restrictions on working hours that currently existed for women and others. On 10 December the *ordonnance* proposed by the Minister of Labour, Séguin, was agreed by the Council of Ministers, but on 17 December President Mitterrand said that he was going to refuse to sign it because it went back on 'social acquisitions' of the previous half-century. The parliamentary session was to end just before Christmas, and the next would not start until April. The Government simply tabled the seventeen articles of the *ordonnance* as an amendment to the bill on miscellaneous social provisions, then in its final stages in Parliament, and used the issue of confidence procedure of article 49 §3 to ensure that it was passed. The Conseil struck the provisions down as exceeding the power of amendment both in their scope and content.[36] Numerous articles of the Labour Code were altered by the Séguin amendment, which instituted a major change by breaking the compulsory link between flexibility and reduction in working hours. A new bill should have been presented, although everyone realized that this would inevitably delay the whole process.

It was widely rumoured that the Conseil was evenly split on this issue (4:4), and, in the absence through illness of Louis Joxe, father of the proposer of the reference, but, unlike his son, a Government

[34] CC decision no. 78-97 DC of 27 July 1978, *RDP* 1979, 504.

[35] See CC decisions nos. 85-191 DC, 85-198 DC, 85-199 DC of 10 July, 13 Dec. and 28 Dec. 1985, culminating in CC decision no. 86-221 DC, *Finance Law for 1987*, where it is stated that 'additions or modifications . . . cannot exceed in their purpose and their importance the limits inherent in the right of amendment without breaching articles 39 §1, and 44 §1 of the Constitution' (§5); see *GD* 704–6.

[36] CC decision no. 86-225 DC of 23 Jan. 1987, *The Séguin Amendment*, DECISION 16.

supporter, the Socialist President of the Conseil, Badinter, had the casting vote. The Speakers of both chambers objected not simply because they were Government supporters, but also because the decision represented an interference by the Conseil constitutionnel in what they perceived to be the internal affairs of Parliament.[37]

The Conseil has persisted in its approach in its decision on *Local Government*.[38] According to its title and the explanatory memorandum accompanying it, the bill proposed by the Government concerned the reorganization of the external services of the State and of the local government service. Articles 16 and 17 of the *loi* concerned respectively the finances of the city of Paris, and the electoral system in communes with populations of between 2,500 and 3,500 inhabitants. The change in title did not affect the validity of the procedure, and the clearly extraneous provisions were struck down as unconstitutional. To be constitutional, an amendment must bear an ancillary relationship to the rest of the text. This is clear from the decision on the *Finance Law for 1990*.[39] Amendments here sought to add a contribution by the State to the railways (the SNCF), to amend various provisions of existing tax law, to amend the search powers of the tax and customs authorities, and to introduce two local taxes to replace the residence tax currently in force. Being mere corrections to existing tax law within a very heterogenous measure on tax law, there would have been great difficulty in declaring these provisions as outside the scope of the right of amendment.

A further example of the limits of the right of amendment is the decision in *Local Direct Taxation*.[40] A bill on the general revaluation of property serving as a basis for local taxation was presented to Parliament. An amendment to article 56 of the text, altering taxes levied by the State for the benefit of local authorities, was considered not to be unconnected with the subject-matter of the bill, since it was still about the funding of local authorities. On the other hand, an amendment in the Senate, article 16, sought to alter the Planning Code to permit the construction of new tourist developments close to mountain lakes, by way of a derogation from the ban on such building in article 145-5 of that Code. This provision was introduced to reverse a recent decision of an administrative court adjudging developments around Lake Fabrèges to be illegal. Of its own motion, the Conseil held that this had no connection with the subject-matter of the bill, and thus it struck the provision down as an abuse of the right of amendment.

[37] See *Le Monde*, 25–6 Jan. 1987.

[38] CC decision no. 88-251 DC of 12 Jan. 1989, *AJDA* 1989, 322, note Wachsmann.

[39] CC decision no. 89-268 DC of 29 Dec. 1989, *RFDA* 1990, 143, note Genevois.

[40] CC decision no. 90-277 of 25 July 1990, *RFDA* 1991, 354, and notes by J.-C. Douence, *RFDA* 1991, 344; L. Favoreu and T. Renoux, *RFDC* 1990, 729.

The attitude of the Conseil has not been altogether consistent in relation to what is an amendment. What counts as an amendment, and what is an entirely new measure, is judged by size, importance, and relevance, but the application is often difficult. Thus, as we shall see below with regard to finance measures, the Conseil has held that for the Government to introduce a completely new tax by way of amendment was improper, since it was really a new initiative.[41] On the other hand, this has recently turned to the Government's advantage when its 'rectificatory letter' introducing a form of wealth tax (the *contribution sociale généralisée*) was treated not as an amendment, but as an exercise of legislative initiative.[42]

Such an approach marks a significant change from the past. As has been remarked, the republican character of the Third Republic owes its existence to a parliamentary amendment in 1875!

The Right to Debate

An essential feature of any parliamentary system is the right to debate issues and to vote on them. The decision not to debate an issue is one that should be left to the chamber itself, though even here the need to respect the rights of minority groups to have their say also needs to be taken into account.[43]

Priority in setting the parliamentary agenda goes to the Government in ordinary sessions, and in extraordinary sessions Parliament sits entirely on an agenda determined by the Government. The timetable for debates is set by the Conference of Presidents within each chamber. The Government's agenda takes priority over the timetable of the Conference, so that the former can insist that an ordinary session continues with items left over from a preceding session.[44]

The standing orders of the National Assembly and the Senate do contain provisions whereby the chamber can vote on a preliminary issue (*question préalable*) of whether to debate a bill at all. This concerns the political merits of a bill, and was used by the Senate in 1981 to express its outright opposition to Government bills on nationalization and decentralization. By contrast, in 1986 it was used by the Senate positively to accelerate the passage of a bill proposed by the Chirac Government on electoral reform. In a guarded phrase, the Conseil

[41] CC decision no. 76-73 DC of 28 Dec. 1976, *Charges on Meat Markets*, *Rec.* 41.

[42] CC decision no. 90-285 DC of 28 Dec. 1990, *Finance Law for 1991*, *RTDC* 1991, 106.

[43] E.g. the Conseil constitutionnel struck down provisions of the standing orders of the National Assembly that would have allowed the Assembly to refuse to recognize a parliamentary grouping on the ground that its policies were not consistent with the democratic character of the Constitution: DECISION 13.

[44] CC decision no. 81-130 DC of 30 Oct. 1981, *Higher Education*, *Rec.* 31.

constitutionnel decided that it did not find this practice unconstitutional 'in this case', leaving scope for a challenge in future.[45] Indeed, a recent proposal that the President of the Senate should be able to submit a bill for a vote as a whole—as a shortened form of proceeding—was struck down as inhibiting the right of amendment.[46]

At the very least, there must be some form of debate in the full chamber. Recent proposals to curtail parliamentary procedures by allowing some bills to be remitted to decision, and not merely preparatory discussion, in a parliamentary committee were struck down by the Conseil.[47] There can be no introduction of the Italian procedures for *leggine*, which are ordinary laws passed by parliamentary committees. On the other hand, the Conseil has never struck down provisions in the standing orders of the National Assembly, and recently adopted by the Senate, whereby the speeches of members of Parliament in debate are restricted in their length.[48]

4.4. Finance Legislation

The procedure for Parliament's scrutiny over finance legislation has been subjected to strict controls in the Fifth Republic. As a result, the mechanisms for passing finance laws contribute to the largest body of case-law developed by the Conseil. Why should that be? First, there are several finance laws passed every year—the main budget, rectifying laws, amending laws, and laws settling accounts (*lois de règlement*). Given the number and frequency of these, it is not surprising that the Conseil has had the opportunity to pronounce upon a wide variety of aspects of the procedure. The frequency of litigation on finance laws is almost encouraged by the technicality of the *ordonnance* of 2 January 1959 that regulates the procedure for the discussion and adoption of finance laws by Parliament. It has similar status to a *loi*, being a measure passed by the Government under transitional powers granted by the Constitution of 1958. As such, it is binding on both Government and Parliament.

Secondly, the Government has very broad powers to secure the passage of its financial legislation. It has the right to initiate proposals

[45] CC decision no. 86-208 DC of 1–2 July 1986, *Electoral Enabling Law*, *Rec.* 78; *GD*, no. 42; Genevois, 147.

[46] CC decision no. 90-278 of 7 Nov. 1990, *Reforms to Senate Standing Orders*, *RFDC* 1991, 113. But a change in standing orders imposing time-limits for the tabling of amendments was upheld by the Conseil constitutionnel: CC decision no. 91-292 DC of 23 May 1991, *Reforms to the Standing Orders of the National Assembly*, *RFDC* 1991, 501.

[47] Ibid.

[48] Ibid. But in DECISION 13 the Conseil did strike down limits on Government interventions as contrary to the freedom of the Government to address Parliament whenever it wishes, set out in art. 31 of the Constitution.

for expenditure and the raising of revenue—Parliament is limited to amendments that do not involve new charges on public expenditure or a diminution in resources. Article 47 of the Constitution subjects the process of parliamentary debate to a rigid timetable, such that, if the finance law is not voted upon within seventy days, the Government may bring it into force by decree.

Given the great political significance of taxation and public expenditure, procedural questions are often crucial in enabling parliamentarians of all parties to exercise influence on the financial decisions taken by the Government. But, carrying over reforms introduced in June 1956, the Fifth Republic has established a large number of rules to ensure financial discipline and restraint by Parliament in its scrutiny of finance bills. This deals with the problem identified earlier, namely, the willingness of deputies to vote expenditure without adequately addressing the question of how revenue is to be raised. This has incurred a number of criticisms, such as those of Jèze in 1929:

> deputies form an irresponsible body . . . They have an irreversible tendency to consider the Minister of Finance as exclusively concerned with finding revenue for the expenditure that they propose . . . An effective way to make only useful expenditure and to have a balanced budget would be to deny to deputies altogether the power to propose expenditure and to demand the opening of credits . . .[49]

The existence of such restrictions on parliamentary debate is common in many countries. In Britain proposals involving a charge upon the public revenue have had to be made by the Crown alone since 1706. In addition, there are other special rules of procedure for money bills, not least that laid down in the Parliament Act of 1911, restricting the powers of the House of Lords.[50]

What is a Finance Bill?

As is the case in Britain, rules on finance bill procedure have carried over from one regime to another. The distinctive feature of finance bills is that they involve *charges publiques* (charges on public revenue) and the raising of taxes to pay for them, although such bills are not exclusively confined to these items. This central notion was defined in an early decision of the Conseil to fit in with earlier ideas.

> Considering that the expression 'charges on public revenue' should be understood as covering, beyond the charges of the State, all those previously mentioned in article 10 of the decree of 19 June 1956 on the form of presentation of

[49] Cited in Luchaire & Conac, 864.
[50] See Erskine May, *Treatise on the Law of Parliament*, ch. 28.

the State's budget and, in particular, the various schemes of public assistance and social security; that this interpretation is confirmed both by the debates of the Comité consultatif constitutionnel and by a comparison between the terms of article 40 cited above and those of the bill presented on 16 January 1958, whose purpose was to amend article 17 of the Constitution of 27 October 1946.[51]

This wide conception of 'charges on the public revenue' enables a broad spectrum of matters to be incorporated in a finance bill, and limits the scope for irrelevance. But it also has important consequences for what Parliament can do by way of proposals or amendments. In particular, control over the financing of public services poses problems for parliamentarians. 'Parafiscal taxation' has to be voted every year for it to continue to be validly collected. Unlike general taxation, which is designed to fill the coffers of the Fisc, parafiscal taxation provides revenue for particular services, rather like water-rates in England. The distinctive feature of such parafiscal taxation is that the payment is not remuneration for services rendered, but is in the form of a licence or facility fee—for example, the licence fee for television and radio reception paid to RTF.[52] Attempts by Parliament to delay the implementation of such taxation, and to impose conditions for it were deemed invalid as against financial bill procedure.

Principal Consequences of the Classification of a Bill or a Proposal as 'Financial'

The first consequence of a measure being classified as 'financial' is that the right to make proposals belongs to the Government. Parliamentarians are confined to making amendments. As in other areas, the line between an 'amendment' and a 'new initiative' may be fine. Replacing one or more taxes with a new one may be counted as a new initiative, even if the broad financial effect is neutral.[53] On the other hand, the Conseil is not too rigid in its approach. In the *Business Tax* decision[54] it stated that its function was 'to decide whether, in the course of the legislative procedure, [article 40 of the Constitution] has been applied in conformity with its letter and spirit'. The 'spirit' was to ensure that provisions with financial consequences were not voted without a proper consideration of all their effects on public finance. The Government has ultimate responsibility for this, so amendments diminishing resources or increasing financial burdens on taxpayers are the

[51] CC decision no. 60-11 DC of 20 Jan. 1961, *Law on Sickness Insurance . . . for Farmers*, *Rec*. 29.
[52] CC decision no. 60-8 DC of 11 Aug. 1960, *Rectifying Finance Law for 1960*, *Rec*. 25.
[53] CC decision no. 76-73 DC of 28 Dec. 1976, *Charges on Meat Markets*, *Rec*. 41.
[54] CC decision no. 75-57 DC of 23 July 1975, *Rec*. 24.

obvious measures implicated. Here the abolition of one tax and its replacement by another was seen as acceptable. The real point is that fiscal neutrality, which mere substitution requires, looks only at the revenues of the State, and not at either public sector revenues as a whole or the position of taxpayers.[55]

The second consequence of classifying a measure as 'financial' concerns the priority of the National Assembly. Under article 39 of the *ordonnance* of 2 January 1959, strict time-limits are imposed for presenting budgetary measures to the National Assembly, and it has the largest share of the allotted time. As a result of this priority, new taxes cannot be introduced in the Senate, even by Government amendment.[56]

A third consequence is a very rigid timetable. The Government must present a bill before the third week in October. The National Assembly has forty-five days to consider it, the Senate fifteen, and if no bill is voted within seventy days, the Government may enact its proposals by way of decree. In 1986 the Government allowed the National Assembly debate to exceed the stipulated period. The Conseil noted this, but, since the Senate was not prejudiced by having its own time for debate curtailed, this did not affect the validity of the *loi*.

A fourth consequence is that various procedural restrictions exist. This is well illustrated in DECISION 15, *Finance Law for 1980*. The Christmas gift of the Conseil to the nation in 1979 was to strike down the entire budget, due to come into force on 1 January, for procedural irregularity. The decision is, at first sight, merely setting a technical obstacle to the passing of the budget, but underlying it is the extent to which the Government needs to build political support for its financial measures.

Since the reforms of 1956, the finance law contains two parts. The first part authorizes the receipt of resources and sets out measures to ensure a balanced budget (for example, from borrowing). The second then authorizes credits for particular services. Prior to this reform, the procedure involved (as in the United Kingdom) two separate laws—an appropriation (spending) law, and a taxation law. Article 40 of the *ordonnance* on finance laws of 2 January 1959 requires that 'the second part of the finance law for the year cannot be put to discussion in an Assembly before the vote on the first part.' The purpose of this article—and, indeed, of the reform of 1956 carried over into the *ordonnance* of 1959—is to ensure a balanced budget. This idea is central to the definition of a finance law that the *ordonnance* gives in the first

[55] CC decision no. 63-21 DC of 12 Mar. 1963, *Stamp Duty*, *Rec*. 23: no duty could be imposed to pay money to local authorities to make up for replacing a local tax by a national tax.

[56] CC decision no. 76-73 DC of 28 Dec. 1976, *Charges on Meat Markets*.

paragraph of article 1: 'Finance laws determine the character, the amount, and the destination of the resources and charges of the State, taking account of the economic and financial balance that they establish.'

The finance law for 1980 was debated at a time of conflict within the governing majority between the RPR, headed by the former Prime Minister and presidential hopeful, Jacques Chirac, and the President's party, the UDF, and its Prime Minister, Raymond Barre. The RPR demanded expenditure cuts of two milliard francs to deal with the worsening economic situation, but refused to specify where these cuts should come from. Since the cuts were not forthcoming, it withheld its support from the last article (article 25) of part I of the finance bill, which was the general provision on the budgetary balance. The President of the National Assembly decided that, despite this, he should allow debate on part II. His principal reasons were that the bill had not been withdrawn by the Government; it still remained on the parliamentary agenda, and so had priority. In any case, if debate did not continue, the time-limits set in article 47 continued to run, after which the Government was entitled to give effect to the budget by decree. In such a situation Parliament would be deprived of any opportunity to contribute further to discussion of the budget. Finally, the standing orders of the National Assembly did not permit a second vote on an article of a bill until the rest of the text had been discussed. Debate therefore continued on part II. The Government then insisted that the National Assembly vote on part I (including the original article 25) and part II (with such amendments as it agreed) as a block, invoking article 49 §3 for the two parts taken together. In the face of Senate opposition, the bill returned to the National Assembly for a final vote. To ensure that it was passed despite continued resistance from the RPR, the Government insisted on using the block vote procedure, and again invoked article 49 §3 to make the passing of these two parts an issue of confidence. The vote of no confidence was rejected on 17 December. A reference to the Conseil was made by the President of the National Assembly without specifying any particular articles to be considered, but simply to secure a ruling on his handling of the procedure. The opposition was more concerned with the repeated use of the confidence procedure of article 49 §3 to get the law passed, but also invoked the failure to vote part I before discussion began on part II. It was only on this latter point that the opposition gained satisfaction, and the President of the National Assembly obtained a ruling that he had acted properly.

The Conseil ruled that observance of the rule in article 40 of the *ordonnance* was mandatory, and that failure to comply led to the nullity of the procedure. The interpretation given was not a simply literal approach. That would effectively have meant that all discussion of revenue would have to be complete before expenditure was considered, thereby considerably reducing the contribution of Parliament to the whole process. Instead, the Conseil insisted merely on the adoption of those provisions of part I 'which constitute its reason for existence and are indispensable for it to fulfil its purpose' before there was discussion of part II. But among such provisions was the general outline of the budgetary balance. Thus, although invoking the 'spirit of the law' and not merely its letter, the Conseil considered the failure to adopt article 25 at all as irregular. The use of article 49 §3 did not cure this, since the Government's issue of confidence related both to part I and to part II at the same time.

The flexibility of the Conseil's approach to procedure was then shown in handling the budget crisis that followed. Reconvened immediately after Christmas, Parliament voted a single article bill to authorize the continued collection of existing taxes after 31 December. The Communists referred this second *loi* to the Conseil, on the ground that it was not a finance law within the procedures set out in the Constitution and the 1959 *ordonnance*, and that it authorized the continued collection of parafiscal taxation, which should not be in a finance law in any case. The Conseil rejected these arguments. Applying a traditional principle of administrative law accepted as constitutional earlier that year by the Conseil itself,[57] it ruled that Parliament and the Government were entitled to take all necessary financial measures to ensure the continuity of national life. Thus, although this was a procedure not provided for in the *ordonnance*, it was still a proper finance law that could validly authorize taxation and expenditure. Since all the provisions were such as normally appeared in a finance law, the procedure was constitutional.[58]

Parliament was then reconvened after the New Year to vote on the full budget. The Government simply insisted that the National Assembly vote on part I as a block, and invoked the confidence procedure of article 49 §3 in relation to it on 7 January. A no confidence motion was duly tabled and defeated on 9 January. The same was done to part II, and the no confidence motion was defeated on 11 January. No challenge was made to this *loi*.

The saga might appear one of pettiness and pedantry. After all, the Government passed the legislation it wanted, more or less in the form it wanted. At the end of the day, the refusal of the Conseil to interfere with the use of the confidence procedure of article 49 §3 effectively enabled it to steamroller its legislation through. The Conseil did

[57] CC decision no. 79-105 DC of 25 July 1979, *Strikes in Radio and Television*, DECISION 23.
[58] CC decision no. 79-111 DC of 30 Dec. 1979, *GD*, no. 30.

nothing to prevent that happening. By insisting on a fairly strict observance of the rules on budgetary procedure, the Conseil merely set out the points at which Parliament could expect to apply pressure on the Government, and where the latter would have to take a political gamble to push its measures through. After all, faced with RPR opposition, the Government did try initially to make political concessions. Only when those failed, did it move to exercise its more Draconian powers. It is in this light that insistance on the rules can give Parliament some power in relation to the Government.

Despite the undoubted rigidity of the *ordonnance*, the Conseil has shown some flexibility in its interpretations, much along the lines of the decision on the *Finance Law for 1980*. Thus, even though article 40 insists that part I be voted before part II, this is not taken to mean that no deliberation on part II can have an effect on part I, as long as the basic outline of the budgetary balance settled in part I remains the same.[59] Equally, though very specific time-limits are laid down, these are not all treated as mandatory as long as the spirit of the text is observed. Thus, although the Government is supposed to provide deputies with the various documents forming annexes to the finance bill by the first Tuesday in October, failure to do so will not invalidate the procedure as long as deputies are not deprived of this necessary information throughout their consideration of the finance bill.[60]

Cavaliers budgétaires

Given the special character of budgetary procedure, and the restriction on parliamentary debate and input that it imposes, the Conseil has been concerned to prevent abuses by way of 'budgetary rider' provisions that are not connected directly with the raising of revenue or the authorization of expenditure.

The problem is not a new one. Indeed, very important provisions have been hidden away in finance bills (for example, the rule that a civil servant can have access to his file in disciplinary proceedings from article 6 of the *loi* of 22 April 1905, and freedom of education from article 91 of the *loi* of 31 March 1931). While the approach to finance laws has expanded to include further aspects of economic policy, there are significant limits.

This is well illustrated by the difficulties of the Ministry of Education in preparing for major changes in the relationship between public and

[59] CC decision no. 82-154 DC of 29 Dec. 1982, *Finance Law for 1983*, *Rec.* 80: distributing the revenue of a tax to local authorities rather than to the State, or changing the basis for a tax could be approved even after part I had been passed, as this did not affect the budgetary equilibrium significantly.

[60] Ibid.

private sectors. The proposal was that up to 15,000 teachers presently in the private sector should be made public employees as part of a package of Government changes not yet approved by Parliament. In anticipation, two new heads were inserted in the education budget, because new posts could only be created under the 1959 *ordonnance* if the budget so provided. The Conseil declared article 44 of the finance law as unconstitutional (and not merely ineffective), on the ground that the new heads did not themselves authorize expenditure. New legislation would be needed both to move money from other heads into these accounts and to authorize payment from them. A budget could not be used simply to announce new policies of expenditure; it must actually authorize expenditure.

On the whole, the line between a budgetary and a non-budgetary measure is clear, and this will usually be sorted out during the parliamentary process. For example, in 1973 the Senate successfully objected to a proposed amendment to a finance bill that altered the Code of Civil and Military Pensions.[61] Indeed, this area has become important as a method of limiting the way in which the Government drives legislation through Parliament.

A recent example was the *Finance Law for 1991*,[62] which in article 47 amended the division of the grant to communes (the *dotation globale de fonctionement*) without altering the amount. Since this did not affect the expenditure or income of the State, the provision was inappropriate in a finance law. Equally, sanctions imposed on public finance officers for accounts in deficit are nothing to do with expenditure as such.

In brief, measures with financial implications may not always be permissible in a budget; the Conseil has reduced its function to the raising of revenue and the authorization of immediate expenditure.

4.5. Referendum Laws

The use of a referendum to pass legislation constitutes both the low point of the review operated by the Conseil constitutionnel and the greatest abuse of procedure in the Fifth Republic.

After the assassination attempt at Petit-Clamart on 22 August 1962, General de Gaulle announced at the Council of Ministers on 12 September that he was going to seek to change the Constitution's method of appointing a President of the Republic and to introduce popular, rather than indirect, election. The procedure he adopted was not that for amending the Constitution provided for in article 89,

[61] See M.-C. Bergères, 'Les Cavaliers budgétaires', *RDP* 1978, 1373 at 1384–5 and 1386–7.

[62] CC decision no. 90-285 DC of 28 Dec. 1990, *Finance Law for 1991*, *RTDC* 1991, 106.

which involves a bill being passed by both chambers. The Pompidou Government was already facing problems in Parliament, so that such a bill was unlikely to pass. Drawing on an idea floated on 10 July in relation to nuclear weapons policy, and given his success in the referendum of April 1962 that put an end to the Algerian crisis, the President relied on article 11, which enables the President to put a bill 'concerning the organization of public authorities' to a referendum, 'on the recommendation of the Government'. The President of the Senate and the opposition criticized the Government for this 'outrageous breach of the Constitution', and it was defeated in a confidence motion on 5 October 1962. Parliamentary elections were called on 9 October, and fixed for 18 and 25 November. In the meantime, the President decided by a decree of 2 October to submit the proposed law in relation to the election of the President to a referendum on 28 October. The decree calling the referendum did not mention the proposal from the Government, merely that the Conseil constitutionnel had been consulted. True enough, it had been consulted at a five-hour meeting on 1 October, but it had advised against the use of this procedure.

De Gaulle was successful in the referendum, but the President of the Senate, the Socialist Gaston Monnerville (later to be a member of the Conseil constitutionnel), referred the *loi* to the Conseil constitutionnel as unconstitional. The Comité consultatif constitutionnel had stated: 'The Committee takes note, with satisfaction, of the spirit in which the referendum is conceived, as in no case can it be a means of setting the Government against the chambers.'[63] Clearly, this was what had happened in this case.

Exceptionally, a former President, Auriol, came to sit for the decision. By a vote of 6 to 4, the Conseil decided that it was not competent to entertain such a reference.[64] In its view, it only had competence to give judgments on *lois* voted by Parliament, and this was not the case here. The Conseil then went on to announce the results of the referendum. Auriol was disgusted and never sat again. The decision was justified by Noël, the President of the Conseil, by suggesting that it was not possible to contradict its author on the meaning of the Constitution.[65] But the legal justification and political wisdom of the judgment have been much debated, as the list of articles referred to in the *Grandes Décisions* demonstrates. It was an understandable decision at the time, but one which contrasts markedly with the interventionist stance taken in later decisions such as the *Séguin Amendment*.

[63] *Avis et débats*, 203.

[64] CC decision no. 62-20 DC of 6 Nov. 1962, *Referendum Law*, DECISION 14. On the voting details, see *GD* 173, and J. Boudéant, 'Le Président du Conseil constitutionnel', *RDP* 1987, 589 at 638.

[65] Boudéant, 'Le Président', 638.

The discussion of procedure may appear very technical, and it is. The number of times that such issues are raised in relation to bills is large, and this has led to a wealth of case-law with sometimes fine distinctions.

But it is neither boring nor unimportant. It is the very stuff of politics and the democratic process. As Llewellyn suggested, arranging for 'the say and the manner of the saying' is a key task of law.[66] How much more so in politics. Ensuring the opportunity for political debate and pressure to be put on the Government, rather than permitting a stage-managed theatrical performance, is surely vital to any democracy.

Procedural law enables pressure to be exercised and a review of the Government's options to occur. As is the case in the United Kingdom, most legislation in France is at the Government's initiative; the Government holds the procedural trump cards in timetabling what will be debated. There is a real danger that the French Parliament has become merely a 'registration chamber' rather than a debating chamber.

What has the Conseil constitutionnel done to help? It starts with texts that are firmly committed to a strong executive. The Conseil has not undermined any of the prerogatives and powers that the 1958 Constitution gave to the executive. As a result, the Rocard Government survived for three years without a parliamentary majority and with only lukewarm support from the President and the Socialist party. The Conseil's job has been to limit abuses of power. It also plays the roles of guardian of the new powers of the executive to govern effectively, and, at the same time, guardian of the traditional functions of Parliament as a debating and influencing chamber.

As guardian of the executive, the Conseil has been loath to see restrictions placed on either the executive's powers to force a decision from Parliament by the use of articles 49 §3 and 44 §3, or on its freedom to choose what it will propose to Parliament as a *projet de loi* and when it will do so, even if it is only by way of amendment. As guardian of Parliament, it is the right to have the time for a say that has been uppermost in the Conseil's mind. Constraints introduced on Government amendments and on *cavaliers budgétaires* reflect a concern to ensure that there is adequate opportunity to discuss major initiatives, and that this is not curtailed by last-minute and substantial amendments to bills, or by recourse to the rapid finance bill procedure in preference to that of ordinary legislation. On the other hand, the failure to restrict the use of article 49 §3 in any way, and the endorsement of referendum laws, which escape parliamentary scrutiny altogether, show that the Conseil has only gone a limited way towards ensuring effective debate in Parliament.

[66] 'The Normative, the Legal and the Law Jobs' (1940) 49 *Yale Law Journal* 1355.

The Conseil's decisions reflect a balance between these objectives, a balance between institutions that are, fundamentally, rivals. But, having arranged for the say and the manner of the saying, it keeps well clear of the merits of issues. It will not question the reasons for the exercise of powers. These are essentially acts of State, which are left to politicians to determine. The Conseil will only decide whether there is power to act, and here the careful control over budgetary and amendment procedure shows a willingness to ensure that no abuse takes place. The character of control has to be understood in relation to the person against whom it is addressed. As Loïc Philip suggests:

> This control should not be analysed, as it is too often, as a control directed against Parliament, and so, in the eyes of some, as an attack on the principle of the sovereignty of the nation implemented by its representatives. This classical criticism is now outdated, and hardly corresponds to reality any more. Nine times out of ten it is directed against the real author of the contested text, that is to say, the Government or even the administration. That appears clearly from the present decision.[67]

The legal technique is typically to look at the spirit of the rules. Procedural rules are designed with care, to ensure that the variety of often conflicting interests have a fair share of the time for debate, consistent with the overall objective of effective government. At the same time, procedure is there to allow debate and reflection to happen. It is not an obstacle course for legislation to pass through. Too rigid an interpretation of the rules would cause legislation to be felled frequently by technicalities in its progress through the legislature. The Conseil has tried, therefore, to adopt interpretations that enable the key participants to have their say, while not rendering the legislative process over-cumbersome. The careful balance of interests in the case of the *Finance Law for 1980* (DECISION 15) is a good example. Another is the obligation under article 74 of the Constitution for the assembly of an overseas territory or department to be consulted before any law is passed that affects the organization of its territory. It is obviously important that such distinct communities within France should have their own input into the legislative process. If this opportunity to be heard is denied or neglected, then the legislative procedure is rightly invalid. For example, when a *loi* applying the Criminal Code to overseas territories was passed by Parliament without having consulted the assemblies of the territories concerned, it was declared unconstitutional.[68] On the other hand, Parliament is not expected to halt its deliberations every time a proposal is made that will affect such areas.

[67] Note under 83-164 DC of 29 Dec. 1983, *AJDA* 1984, 97 at 100.

[68] CC decision no. 80-122 DC of 22 July 1980, *Law Applying the Criminal Code to Overseas Territories*, *Rec.* 49.

Thus, there is no need to consult the overseas assemblies on amendments that are passed in the course of the legislative process.[69]

The exercise of control by the Conseil constitutionnel in this area is criticized in France for arrogating power that belongs to the Presidents of both legislative Assemblies. But the 1958 Constitution gave the Conseil the job of vetting the standing orders of Parliament, so vetting the observance of constitutional rules of parliamentary procedure is not too much of an extension of the Conseil's authority.

In the United Kingdom we are used, as the French were, to political control of parliamentary procedure being exercised through the Speaker. But the French experience is that the most frequent task of the Conseil is to protect Parliament against dictatorship by the executive.

[69] CC decision no. 79-104 DC of 23 May 1979, *New Caledonia, Rec.* 27.

5

FUNDAMENTAL FREEDOMS

LIBERTY is the most persistent of the trilogy of values that stem from the Revolution of 1789. Freedom from restraint and freedom to act are at the heart of liberal values, more so than equality, and definitely more so than fraternity. That said, the Revolution was not the source of all liberties. Although in many ways the touchstone of French constitutional orthodoxy, the Declaration of 1789 was limited in scope, and has been supplemented by later provisions, especially in the Third Republic.

As we noted in Chapter 1, it was long thought that fundamental freedoms offered political rather than legal standards against which the performance of the legislature could be judged. The legislature was itself the guardian of freedom. Even the division of legislative power between the executive and the legislature in the 1958 Constitution reinforced the importance of the legislature in the context of fundamental freedoms. Among the areas in which *loi* must lay down the rules, article 34 includes 'fundamental safeguards granted to citizens for the exercise of civil liberties'.

The classical attitude was well expressed by Barthélemy in his thesis of 1899, when he said that: 'most often, declarations of rights are no more than solemn proclamations of principles, rules for the conduct of the State, pure maxims of political morality, promises whose force lies solely in public opinion and whose solemn inscription alone is made by the Constitution, without the possibility for individuals to enforce their observance or their practical realization.'[1] It was only after the 1958 Constitution came into force that the Conseil constitutionnel perceived the necessity to give legal force to the fundamental freedoms proclaimed in the 1789 Declaration, the 1946 Preamble, and the fundamental principles recognized by the laws of the Republic. DECISION 1, *Associations Law*, of 1971 was the starting-point for a whole new development of fundamental liberties law in France, even if the actual

[1] *Essai d'une théorie des droits subjectifs des administrés* (Paris, 1899), 141. See also J. Robert, *Libertés publiques* (Paris, 1982), 109.

decisions of the Conseil rely heavily for their inspiration on the case-law of public and private law courts.

The function of the Conseil has been threefold. First, it has to decide whether certain freedoms are 'fundamental' at all in the constitutional sense. The legislature declares numerous freedoms that are transient or, at least, not fundamental. The Conseil only concerns itself with fundamental values. Secondly, the Conseil has to decide the extent of the freedoms in question. Are they absolute, or do they admit of qualifications on specific grounds? Thirdly, it has to decide how far the existence of a freedom or right limits the scope of legislation. In the case of the more vague and general rights proclaimed by the Preamble to the 1946 Constitution, the freedom of the legislature is almost total. In other cases, it is greatly restricted. For the most part, the texts to be interpreted are not very clear. They set out principles, and, in the case of the fundamental principles recognized by the laws of the Republic, they have to be contructed by the Conseil itself before they are interpreted. Although certain freedoms are long established, their constitutional status has often remained unclear. The Conseil has sought to establish a framework around which the legislature can act. Given the way in which references are made to the Conseil, it has to act in circumstances of controversy, where any new departure is seen as a victory for one political side or another. This is not the easiest context for serious debate on fudamental freedoms. All the same, through successive changes in Government, the Conseil has endeavoured to build a coherent and sufficient set of principles. To some extent, it has seized upon cases as opportunities to establish principles, though it has not always done so.

Following classical expositions of French civil liberties, I will group the decisions of the Conseil under the headings of (1) freedom of the person, (2) freedom of association, (3) freedom of education, (4) freedom in employment, (5) freedom of communication and information, (6) freedom of property, and (7) freedom of enterprise and commerce. This is by no means an exhaustive list, but it does cover the main issues that have been considered by the Conseil constitutionnel.

5.1. Freedom of the Person

The preservation of liberty is declared by article 2 of the 1789 Declaration to be one of the ultimate purposes of every political institution. The obvious concern of that period was the power of the administration to impose arbitrary physical constraints on freedom of action, especially through detention without trial. English ideas of habeas corpus as a challenge to such powers were taken by many as a model of what the new France should achieve. This is reflected in article 7 of the Declaration: 'No individual may be accused, arrested, or detained except where the law [*loi*] so prescribes, and in accordance with the procedures it has laid down . . .' At the same time, such liberty is not absolute. Articles 4 and 5 of the Declaration limit all freedom of action by the harm principle—that no one may act in such a way as to harm the rights of others or society as a whole. The limits imposed on freedom in order to prevent such harm are laid down by *loi*, to which citizens must conform. In one of the few duties defined by the Declaration, article 7 states that 'any citizen charged or detained by virtue of a *loi* must obey immediately; resistance renders him culpable.' Even the most basic freedom of the person was, from the outset of the modern constitutional tradition, subject to limitations in the interests of social peace and order, as well as to protect the rights of others.

To a great extent, the tension between freedom and public order has remained a recurrent issue in civil liberties. Legislation passed in the 1970s and 1980s to meet an increasing crime-rate and especially a rise in terrorist offences provided the Conseil with a chance to settle the constitutional limits on freedom of individual action. In the area of immigration, controls were tightened in the 1970s in the wake of a rising tide of unemployment, and the pursuit of clandestine immigrants became an important public order issue. A third area pushed in the opposite direction. The freedom of the individual to do his or her own thing gave rise to a concern to protect private lives against increased State intrusion or an intrusion made possible for others by modern technology.

Established principles needed to be refined considerably to meet the new challenges. In effect, the Conseil has created two broad principles that bring together the values involved in the rights of the person. On the one hand, there is the principle of individual liberty that first surfaced in DECISION 17, *Vehicle Searches*, in 1977, and is based, fundamentally, on article 2 of the 1789 Declaration. This has been invoked as a justification for restricting the actions of the State in relation to individuals in the main issues on civil liberties discussed above. More recently, the Conseil has invoked the value of 'personal freedom', which consists of the independence of the individual in relation to other private persons. This arose in the *Redundancy* case, where the freedom involved the right of the individual to decide how to defend his own rights and interests, rather than having this determined by his trade union alone.[2] The individual instances mentioned below are illustrations of the ways in which these general freedoms have been developed in crucial areas.

[2] See below, ch. 5.5.

Freedom of the Person and Police Powers

As the *commissaire du gouvernement*, Corneille, stated in 1917:

police powers are always restrictions on the liberties of individuals, and the starting-point of our public law is the body of freedoms of the citizen. The Declaration of the Rights of Man is explicitly or implicity the frontispiece of republican constitutions, and any controversy of public law must, in order to base itself on general principles, start from the point of view that freedom is the rule, and restriction by police is the exception.[3]

The Conseil constitutionnel has taken this further by requiring that police powers should be adequately defined in advance by the legislature. In DECISION 17, on *Vehicle Searches*, the Conseil stated that police powers should be defined with certainty both with regard to their extent and to the conditions for their use, as had not been the case where a bill permitted the police to search vehicles on the highway without restriction as to reasons or procedure.

Freedom of Movement The clause in question in the *Vehicle Searches* case had been introduced by parliamentary amendment to give more powers to the police at a time of kidnappings by terrorist gangs. Though the authority to search vehicles was not questioned in principle, its unrestricted scope posed too great a threat to personal liberty. Freedom of movement, more specifically recognized in a subsequent decision,[4] was given constitutional status by this decision.

The limited nature of the constitutional principle involved here is seen in the leading case of *Security and Liberty*, DECISION 18, of 19 and 20 January 1981, where the Conseil took a more public order line. Paragraph 56 of its decision stated that 'the pursuit of criminals, and the prevention of threats to public order, especially of threats to the security of persons and property, are necessary for the implementation of principles and rights of constitutional value . . .' This justified the police having the power to detain people in order to check their identity. The balance between the freedom of movement and the prevention of crime was to be struck by the legislature, and the Conseil would only intervene where the interference was 'excessive'. Unlike the *Vehicle Searches* case, the power was confined to identity checks in the conduct of criminal investigations or in handling threats to public order. It did, however, authorize the detention for up to six hours of those who refused to, or could not, provide proof of identity. In paragraph 54 the Conseil held such powers to be sufficiently limited and justified as not to be an unconstitutional infringement of freedom

[3] CE, 10 Aug. 1917, *Baldy*, *Leb*. 637.
[4] CC decision no. 79-107 DC of 12 July 1979, *Toll Bridges*, *Rec*. 31.

of movement, but only when it had 'interpreted' a number of its provisions. The text under scrutiny provided that a person could prove their identity by 'any means'. The Conseil elaborated this into a rule that a person could provide proof of identity by any means they chose, when asked, and that they could only be taken to a police station in case of necessity. Equally, it interpreted the provision that a person who had been taken to a police station 'could request at any moment that the *procureur de la République* [State prosecutor] be informed' as establishing a right to that effect. For the most part, the Conseil as content to enjoin the judiciary and the administration to respect the numerous safeguards contained in the text and in the Conseil's interpretation of it.

The changing parliamentary majorities were thus left with significant freedom to impose their own policies on identity checks. In 1983 the Socialists limited such checks to situations where the security of persons or property was immediately under threat. When the right returned to power in 1986, it legislated for increased police powers in this area. In 1986, it legislated for increased police powers in this area. In *Control and Verification of Identity*[5] the Conseil considered that greater police powers to establish identity were not contrary to 'the reconciliation that should be effected between the exercise of constitutionally recognized freedoms and the need to pursue the authors of crimes and to prevent threats to public order, each of which is necessary to safeguard rights of constitutional value'. The Conseil's control and influence in this area was limited. It relied on the administrative and private law courts to enforce the legislative limits on the powers granted, as it interpreted them. But, as noted earlier,[6] the Conseil's influence over the criminal courts was not guaranteed.

The Conseil's conception of freedom of movement is predominantly concerned with the rights of citizens. The legislature is free to define the freedom of entry of foreigners. All the same, paragraph 4 of the Preamble to the 1946 Constitution gives those who are persecuted for the sake of freedom the right of asylum in France. With a rising tide of claims for refugee status following restrictions imposed in 1974 on ordinary immigration, legislation in 1980 and 1986 sought to limit the right of abode and to give immediate effect to the refusal of entry at a port. In 1980 the Conseil found that there was no intention to infringe the right of asylum. In 1986 the Conseil went further, and relied on article 55 of the Constitution, which gives effect in domestic law to reciprocal treaties ratified by France. It noted that this applied in

[5] CC decision no. 86-211 DC of 26 Aug. 1986, *Rec.* 120.

[6] See Cass. crim., 25 Apr. 1985, *Bogdan* and *Vukovic*, D. 1985, 329, discussed above, ch. 2.

the absence of specific legislative provisions, and this ensured that the terms of the Geneva Convention of 1951 on refugees had to be respected.[7]

Powers of Detention Detention is one of the severest restrictions on personal freedom, and needs the most justification. The Conseil has offered limited resistance to attempts by the public order lobby to increase powers of detention. It has done this in two ways: by ensuring adequate scrutiny either before detention or the authorization of its continuation, and by examining the necessity of a detention power in the first place.

Adequate scrutiny usually means that detention beyond twenty-four hours must be authorized by a judge, since article 66 of the Constitution makes her the guardian of civil liberties. For example, in the *Security and Liberty* case the Conseil ruled that it was not unconstitutional for a judge to be empowered to authorize the extension of custody while investigations continued and before a charge was laid where the offence in question involved attacks on persons or robbery. Thus, a person could be detained for seventy-two hours, instead of the normal maximum of forty-eight. All the Conseil required was that the judge authorizing the extension of detention read the file on the suspect before making an order.

In other cases, the procedural safeguards have to be greater. For instance, in the *Immigration Law* decision[8] the Conseil upheld the power of the authorities to detain illegal immigrants who had been refused entry. The power was not arbitrary, since it was of limited duration, and its extension could only be authorized by the President of a Tribunal de grande instance (the equivalent of a Circuit judge), whose reasons were reviewable by the Cour de cassation. 'Interpreting' the text, the Conseil held that there were sufficient safeguards for the detainee, who was entitled to seek the assistance of a legal adviser, an interpreter, or a doctor, as required. But the Conseil struck down the provision permitting the detention of an immigrant for seven days before judicial authorization was required. It maintained that 'individual liberty could not be considered as safeguarded unless the judge intervenes within the shortest possible period', which was clearly not

[7] CC decision nos. 80-109 DC of 9 Jan. 1980, *Immigration Law*, *AJDA* 1980, 356, 88-216 DC of 3 Sept. 1986, *Entry and Residence of Foreigners*, *Rec.* 135.

[8] CC decision no. 80-109 DC of 9 Jan. 1980. Even where a would-be immigrant or refugee is refused entry and is held in a transit area prior to deportation, this constraint on individual freedom must be authorized by a judge of the ordinary courts. Consequently, a parliamentary amendment authorizing such detention without such a safeguard was struck down as unconstitutional: CC decision of 25 Feb. 1992, *Entry of Foreigners*, *Le Monde*. 27 Feb. 1992.

the case here. Parliament duly provided that, in cases of necessity, the President of the Tribunal de grande instance could authorize the detention of an expelled person in a non-penal institution for the period necessary for his departure, up to a maximum of six days. In 1986 the Chirac Government sought to reinforce this by permitting the judge to extend the detention by up to three days where special difficulties prevented the detainee's departure. The Conseil considered that this was too wide. Only 'absolute urgency and a threat of special seriousness to public order' could justify such an interference with the liberty of the individual. Such a condition was not required by the *loi*.[9] This point is reinforced by DECISION 19, *Terrorism Law*, where an extension of detention before charge by up to ninety-six hours could be ordered by a judge in terrorist cases if this was necessary for police inquiries or for the judicial investigation of the offence (*l'instruction*). The Conseil read the new provision in the light of existing safeguards in the Code of Criminal Procedure, especially the requirement of a medical examination every twenty-four hours, and the supervisory role of the *procureur de la République*. Article 66 was thus respected. But the imposition of these exceptional procedures could not be extended beyond the strict confines of terrorist offences to include other crimes against the State.

The Conseil is willing to go beyond merely requiring a judicial review of detention in accordance with article 66 of the Constitution. Only certain grounds may justify detention beyond limited periods, even if the person in question has no right to remain in France. Personal liberty is a basic human right that is extended even to illegal immigrants and terrorists. All the same, outside observers have noted the apparent laxity of the Conseil's control over Parliament in this area. An American commentator, Vroom, compares the Conseil's approach unfavourably with the US Supreme Court.[10] It would be fair to say that the Conseil has not been much of an activist in this area, has deferred to the legislature's assessment of public order, and has only sanctioned provisions (often introduced by parliamentary amendment) that stray outside the purpose of the legislation. As the Conseil stated in 1986: 'it is for the legislature to determine the terms for the exercise of [freedom of the individual], taking account of the public interest.'[11]

[9] CC decision no. 86-216 of 3 Sept. 1986, *Entry and Residence of Foreigners*, *Rec*. 135, §§21, 22.

[10] C. Vroom, 'Constitutional Protection of Individual Liberties in France: The *Conseil constitutionnel* since 1971' (1988) 63 *Tulane Law Review* 266.

[11] CC decision no. 86-216 of 3 Sept. 1986, *Entry and Residence of Foreigners*. See also CC decision no. 89-261 DC of 28 July 1989, *Entry and Residence of Foreigners*, *RFDA* 1989, 699 §15.

Personal Freedom and the Criminal Law

Article 8 of the Declaration of 1789 provides that '*loi* may only create penalties that are strictly and evidently necessary. No one may be punished except in accordance with a *loi* passed and promulgated prior to the offence, and lawfully applied.' Although this might appear to create an opening for strong scrutiny of criminal statutes, the Conseil has followed an approach similar to that on the requirements of public order, and has adopted a limited control over both the necessity of a penalty and the creation of offences. As we shall see later, its concern is much more with equal treatment.[12]

The Necessity of Penalties The Conseil basically defers to the legislature's assessment of the necessity of criminal penalties. In the *Security and Liberty* decision it went out of its way to deny that it had an appropriate role in challenging the estimation of the severity of criminal penalties:

> Considering that, within the framework of [its] mission, it is not for the Conseil constitutionnel to substitute its own judgment for that of the legislature concerning the necessity of the penalties attached to the offences defined by the latter, provided that no provision of title I of the *loi* is manifestly contrary to the principle laid down by article 8 of the Declaration of 1789 [that penalties must be necessary] . . .

The notion of 'manifest error' in evaluation occurs here for the first time, and requires that there be some proportionality between the penalty and the offence. The proviso included to that effect envisaged manifestly disproportionate penalties, such as those imposed under the Vichy regime—for example, the death penalty for killing a pregnant cow. Most later cases on this point have taken a similar course of referring to the principle of proportionality, while finding that, in fact, there was no manifest error in evaluation.[13] In later cases the proviso has been construed more strictly. It has been used actually to strike down an excessive penalty imposed by the legislature, admittedly in relation to tax penalties rather than to those arising from the criminal law. In 1987[14] the Conseil decided that a *loi* that imposed on any person (for example, a journalist or an official) a tax penalty equivalent

[12] See below, ch. 6.3.

[13] See e.g. CC decision no. 86-215 of 3 Sept. 1986, *Criminality and Delinquency*, *Rec.* 130 §7.

[14] CC decision no. 87-237 DC of 30 Dec. 1987, *Finance Law for 1988*, *RFDA* 1988, 350. See also CC decision no. 89-260, *Commission des opérations de Bourse*, *RFDA* 1989, 671, where the Conseil approved a provision only on a strict reservation of interpretation that no double jeopardy would arise between criminal and administrative penalties.

to the income of the taxpayer, which he had disclosed in breach of obligations of confidentiality, was excessive. The penalty might well be disproportionate to the harm caused.

In addition, the Conseil has given a wide interpretation to the notion of a 'penalty', applying it not only to sentences imposed, but also to their implementation.[15] Accordingly, any requirement that an offender serve a period in prison before he is eligible for parole is judged by the requirement that it be 'strictly and evidently necessary'. All the same, the Conseil operates a limited control on the basis of manifest error. Here the normal rule that the period before parole should be half the sentence (or fifteen years for life sentences) was considered not to be contrary to article 8, especially as it was reviewed by judges and limited to serious offences against the person.

The Creation of Offences If the penalties imposed are subject to minimum control, the creation of offences is more strictly regulated Article 8 requires that the penalties not only be necessary, but that the offences be defined by *loi* in advance. Apart from the implications of *loi* and *règlement*, discussed in Chapter 3, the Conseil also inferred that the constituent elements of an offence must be stated in a manner that is 'sufficiently clear and precise'. Thus, in DECISION 18, *Security and Liberty*, paragraph 5, it examined whether the terms used to defined offences involving the destruction or deterioration of property were adequate, and found that they were. By contrast, to create an offence of 'malversation' (misappropriation of funds, or embezzlement) for those engaged in the receivership of a company, without providing a legal definition of this term, was held to be too vague to be constitutionally acceptable.[16] But reliance on ordinary language definitions may well be acceptable in other areas. Thus, in the *Press Law* decision of 1984[17] it was sufficient for the *loi* to identify the person who commits the offence as the one who 'controls' an organ of the press, or a person who is 'a director in law or in fact'. What is essential is that the individual should be able to act knowing the potential criminal liability to which he may be subjecting himself. For that reason, a strict form of control by the Conseil is appropriate.

Article 8 outlaws retrospective criminal legislation. This has been interpreted strictly by the Conseil constitutionnel to prevent increased severity in penalties being applied to crimes committed before the legislation came into force, and this includes greater restrictions on the

[15] See CC decision no. 86-215, *Criminality and Delinquency*, *Rec.* 130.

[16] CC decision no. 84-183 DC of 18 Jan. 1985, *Receivership of Companies*, D. 1986, 425, note Renoux.

[17] See below, ch. 5.5.

granting of parole.[18] By contrast, in paragraph 71 of its *Security and Liberty* decision it insisted that *less severe* criminal penalties must be imposed retrospectively. This principle of mercy had been regularly applied by the criminal courts, but was elevate to a constitutional principle by the Conseil.

Personal Freedom in One's Private and Family Life

Privacy is not, as such, a constitutionally recognized value. It owes its legal status to the *loi* of 17 July 1970, although it might have been considered implicit in the earlier law. As a consequence, its constitutional recognition has been rather oblique, though the end-product is something approaching a constitutional principle of privacy.

Paragraph 8 of DECISION 20, *Abortion Law*, in 1975 talked of 'the freedom of the persons who resort to, or participate in, a termination of pregnancy', but this was based on the general principle of individual liberty found in article 2 of the Declaration of the Rights of Man. The same value was invoked in the *Vehicle Searches* decision to restrict police powers.

More specific mention has come in later decision, mainly in response to arguments in references to the Conseil based specifically on the principle of privacy. Privacy in the sense of a right to keep one's private affairs confidential has not fared very well, though the arguments in which it has been invoked have not been very strong. For example, in the *Press Law* decision of 1984[19] it was argued that the requirement to publish the names of shareholders and the transactions involving shares in companies owning newspapers infringed the privacy of the shareholders. This argument was rejected by the Conseil, following the established distinction in French law between interference with a person's private life and the confidentiality of business affairs or trade secrets, which are less strongly protected.

The Conseil has, however, been willing to protect the home, an important part of an individual's private life.[20] Article 89 of the finance law for 1984 permitted searches to be made on private property to combat tax evasion, but made no mention of a requirement of prior judicial authorization. The Conseil struck down the article because it considered that entry to domestic premises must be authorized specifically by a judge, not simply by a general power given to officials. A similar approach is found in the *Eiffel Tower Amendment* decision,[21]

[18] See CC decision no. 86-215, *Criminality and Delinquency*, *Rec.* 130 §23.

[19] See below, ch. 5.5. See also CC decision no. 84-172 DC of 26 July 1984, *Structure of Agriculture Law*, *Rec.* 58 (provisions requiring farmers to provide information on the structure of their farms and their rents).

[20] CC decision no. 83-164 DC of 19 Dec. 1983, *Finance Law for 1984*, *Rec.* 67.

[21] CC decision no. 85-198 DC of 13 Dec. 1985, D. 1986, 345, note Luchaire.

where the Conseil held that rights of entry to inspect premises and to instal equipment for the transmission of radio and television broadcasting could not be left to basic powers of the administration, in so far as they affected not only the building, but also the people living there.

Despite such general statements of constitutional principle, France has been condemned by the European Court of Human Rights (ECHR) for the inadequacy of its legal criteria for the authorization of telephone tapping.[22] Indeed, the criminal courts have weakened the protection involved in judicial authorization by sanctioning delegations of authority to Telecom officials with the status of officers of the judicial police, a view that the Conseil constitutionnel has upheld.[23] The Conseil confined itself to ensuring that there were satisfactory limits on the search powers granted. Despite Government assurances, the power to search for the use of unauthorized telecommunications equipment was not adequately confined to business premises, and the provisions did not insist on the same safeguards—the presence of the owner, and the taking of a record—as existed in the Code of Criminal Procedure, so the provisions were struck down.

The development of a general principle of individual freedom may provide a more solid foundation for the idea of privacy as a constitutional value. More broadly, the right to family life proclaimed in article 10 of the Preamble to the 1946 Constitution has yet to give rise to the same limits on the freedom of the legislature as it has in relation to the administration, where provisions on immigration and naturalization have been declared illegal, as contrary to a general principle of law in favour of the right to a normal family life.[24] But the response of the Conseil constitutionnel has been more limited. In its view,

> it is for the legislature to determine the terms under which the rights of the family can be reconciled with requirements of the public interest; that even where it does permit the authority designated to decide on the expulsion of a foreigner to take account of all relevant factors, including, if necessary, his family situation, it does not transgress any constitutional provision by making the necessities of public order prevail.[25]

Freedom of the Individual and the Right to Health

Paragraph 11 of the Preamble to the 1946 Constitution guarantees to all 'the protection of health, material security, rest, and leisure'. This

[22] See ECHR, 24 Apr. 1990, *Kruslin and Hervig*, D. 1990, 354.

[23] See Cass. Ass. plén, 24 Nov. 1989, *RFDC* 1990, 139; Paris, 1er Chambre d'accusation, 18 Oct. 1990, *RFDC* 1991, 331; CC decision no. 90-281 DC of 27 Dec. 1990, *Telecommunications*, *RFDC* 1991, 118.

[24] See CE Ass., 8 Dec. 1978, *GISTI*, *GA* no. 110, D. 1979, 661, note Hamon.

[25] CC decision no. 86-216 DC of 3 Sept. 1986, *Entry and Residence of Foreigners*, *Rec.* 130 §10.

guarantee of the material conditions for freedom has only been treated as being of constitutional value in relation to health. This right was first recognized, albeit in an oblique way, in DECISION 20, *Abortion Law*, where it was held not to be infringed by the cases in which the *loi* permitted abortions. On the other hand, it has not been used by the Conseil to impose positive prescriptions.[26]

The leading case here is the 1991 decision on the *Tobacco and Alcohol Advertisements*.[27] The *loi* banned all direct or indirect advertising of tobacco products from 1993. It was argued that this infringed the property rights of the tobacco manufacturers, who were unable to make use of their brand names to attract customers, and also that it was against their freedom of enterprise. The Conseil considered that the right to health provided a public interest justification for restricting these rights, and so the legislation was not unconstitutional. The same arguments were used to justify restrictions on alcohol advertising.

As was stated in the *Miscellaneous Social Measures* case of 1990,[28] paragraph 11 of the 1946 Preamble lays down principles that require further legislation through *loi* and *règlement* to create concrete norms. Parliament has considerable discretion in determining the 'appropriate rules designed to achieve the objective defined by the Preamble'. Indeed, the legislature could, as in that case, simply determine the mechanisms and principles for ensuring that the medical fees of doctors were paid, but leave the details to be settled by agreements between doctors and the social insurance funds.

As a result, the principles of paragraph 11 contain only limited constraints on what the legislature can enact. Moreover, it has been argued that the Conseil is insufficiently strict in assessing whether the restrictions imposed on other fundamental rights in the name of the right to health are necessary.[29]

5.2. Freedom of Association

While individual freedoms, such as those of the person, were recognized by the authors of the Declaration of the Rights of Man and of the Citizen of 1789, collective rights to form associations, unions, and so

[26] CC decision no. 90-287 DC of 16 Jan. 1991, *Public Health and Social Insurance, RFDC* 1991, 293: the right did not prevent a lowering of the entry qualifications into certain medical professions. See also CC decision no. 77-92 DC of 18 Jan. 1978, *Medical Investigations, Rec.* 21: the Conseil accepted that an individual had the right to choose his or her own doctor, and the doctor had the right to choose his or her patients. But this did not prevent employers from insisting that employees be examined by doctors that they had appointed as a condition for obtaining sickness benefit.

[27] CC decision no. 90-283 DC of 8 Jan. 1991, *AJDA* 1991, 382, note Wachsmann. See also CC decision no. 80-117 DC of 22 July 1980, *Nuclear Installations, AJDA* 1980, 479, where the right to health restricted the right to strike (see below, ch. 5. 4).

[28] CC decision no. 89-269 DC of 22 Jan. 1990, *RFDA* 1990, 406, note Genevois.

[29] See Wachsmann, *AJDA* 1991, 387–8.

on came much later. The State's concern with subversion, no doubt reinforced by the history of republican clubs in the revolutionary period, caused strict checks to be imposed on groupings of all kinds. Article 291 of the Penal Code of 1808 forbade the formation of associations of more than twenty persons without Government authorization. Despite some greater tolerance in the first year of the Second Republic, freedom of association was only established in the liberal Third Republic. Even then it took until 1901 to lay down a general principle of freedom of association, though the freedom of union and trade associations was recognized in 1884, and that of co-operatives in 1898. All the same, the *loi* of 1901 that established freedom of association went on, in title III, to state that religious congregations could only be formed if they were specifically authorized by a *loi* (and, in practice, this authorization was refused in the ensuing decade). The legal sanctioning of such associations had to wait until 1942, when a *loi* of the Vichy regime imposed on them substantially the same requirements for formation as other associations.

The Conseil's decision of 16 July 1971 in the *Associations Law* case (DECISION 1) was therefore quite a bold step in giving constitutional value to a principle of such uncertain pedigree, even if, by then, the principle was not really contested. As was noted in Chapter 2, the decision was based not on a specific constitutional text, but on a fundamental principle recognized by the laws of the Republic:

> Considering that, among the fundamental principles recognized by the laws of the Republic and solemnly reaffirmed by the Constitution, is to be found the freedom of association; that this principle underlies the general provisions of the *loi* of 1 July 1901; that, by virtue of this principle, associations may be formed freely and can be registered simply on condition of the deposition of a prior declaration; that, thus, with the exception of measures that may be taken against certain types of association, the validity of the creation of an association cannot be subordinated to the prior intervention of an administrative or judicial authority, even where the association appears to be invalid or to have an illegal purpose . . .

From the practical application of the principle contained in the 1901 *loi*, the Conseil drew the conclusion that freedom of association involves the absence of prior restraint, a view laid down by the 1789 Declaration in relation to freedom of the press, but one which is not adopted in relation to other freedoms. This reinforces freedom of association. On the other hand, the Conseil's decision, rather obliquely, notes that the principles are not general, since it states: 'with the exception of measures that may be taken against certain types of association'. As Genevois[30] notes, this would still permit special provisions to be made

[30] Genevois, 225–6.

for religious congregations, associations of public utility, and associations of foreigners (a measure repealed in 1981), each of which would have some public interest justification.

The Conseil has had little further to say on associations. What is clear from the cases on broadcasting in the early 1980s, is that the freedom to form an association does not involve the right to seek financial assistance from any source.[31]

One special group of associations has a firmer constitutional status. Article 4 of the 1958 Constitution recognizes the right of formation of political parties. In more recent years this has been seen as justifying, though not requiring, the legislature's financing of political campaigns. At the same time, the Conseil constitutionnel has insisted that State aid should not compromise the pluralism of ideas, and thus that a provision imposing too high a threshold of votes on political parties to qualify for such State aid was contrary to this value, and therefore unconstitutional.[32]

5.3. Freedom of Education

Freedom of education has been a notorious battleground between political parties for over a century. The principal reason for this has been the secular character of French Republics. State schools do not teach religion, so it is either an extra-curricular activity or the preserve of independent schools. Whereas conservatives have generally favoured religion and private education, Socialists and Communists have been against them, a sentiment that reached its height in the law of 7 July 1904, which banned religious orders from teaching at all in France.

Given this background, it is not surprising that freedom of education is not enshrined in any constitutional text. Indeed, the Constituent Assembly, which drafted the 1946 Constitution, actually defeated a proposal to include such a right in the Preamble by 274 votes to 272.[33] Instead, paragraph 13 of the Preamble merely states: 'The nation guarantees the equal access of children and adults to instruction, to professional training, and to culture. The organization of free and secular public education at all levels is a duty of the State.' However, with the Fifth Republic and, in particular, with the *loi Debré* of 31 December 1959, the principle of freedom of education is clearly stated.

[31] See CC decision no. 81-129 DC of 30, 31 Oct. 1981, *Free Radio*, *Rec.* 35: a broadcasting association can be prevented from receiving revenue from advertising: see below, ch. 5. 5.

[32] CC decision no. 89-271 DC of 11 Jan. 1990, *Electoral Expenses*, *RFDC* 1990, 332 (DECISION 37 §§11–14). But to impose scrutiny by the National Commission on Campaign Accounts on such recipients of State aid was valid, as it merely prevented fraud. See also CC decision no. 88-242 DC, *Organic Law on Transparency in Political Life*, *Rec.* 36.

[33] See *GD* 359.

Article 1 of that *loi* provided that 'The State proclaims and respects the freedom of education, and guarantees its exercise to lawfully opened private establishments.' Of course, this text, being a *loi* of the Fifth Republic, has no constitutional status as such.

But, to appreciate what was already common ground by 1946, it is necessary to look more closely at what is involved in the freedom of education. If Napoleon created a State monopoly of education, the freedom to open private educational establishments was recognized for primary schools in 1833, for secondary schools in 1850, for higher education in 1875, and for technical education in 1919. It was this freedom that was repeated by various *lois* during the Third Republic. It has never been clearly separated from the freedom of parents to choose the schooling that they wish for their children. What was more contentious was the right of private education to receive subsidies from the State, which would provide an effective freedom to operate private schools. By permitting private educational establishments to obtain contracts of association with the State, giving them a kind of direct-grant status, the *loi Debré* ensured that many private schools could enjoy considerable State subsidies, at the cost of some limits to their independence. It is freedom of education in this strong sense that was so hotly contested in 1946. By contrast, the freedom of conscience of teachers has a less contested constitutional pedigree. The issue of freedom of pupils is an explosive one in the era of Islamic fundamentalism, but this has yet to be raised as an issue before the Conseil constitutionnel.[34]

Freedom of Private Education

If the *loi Debré* set out the policy at the beginning of the Fifth Republic, this by no means meant that the Socialists and Communists were happy with it. The questions raised before the Conseil both before and after May 1981 concerned the extent to which the *loi Debré* represented a constitutionally untouchable status quo.

This issue came before the Conseil in DECISION 21, when considering the *loi Guermeur* in November 1977. During the run-up to the 1978 National Assembly elections, in which private education was an issue, a deputy, Guermeur, sought to reinforce the values of the *loi Debré* by making it easier for private establishments to obtain contracts of association with the State, which would given them a kind of direct-grant status, with all the financial advantages that this entailed. In addition,

[34] See, however, the advice of the administrative sections of the Conseil d'État: J. Bell, 'Religious Observance in Secular Schools: A French Solution' (1990) 2 *Education and the Law* 121.

teachers paid and appointed by the State to such schools had to respect the character of the private establishment. This 'updating' of the *loi Debré* was challenged by the Socialist deputies, who argued that it infringed the 1946 Preamble's emphasis on State education. The Conseil replied that freedom of education was a constitutional value, and appealed to article 91 of the finance law of 31 March 1931 as embodying this 'fundamental principle of the laws of the Republic'.

The argument was not uncontroversial. The members of the Constituent Assembly nearly rejected the inclusion of a reference to such principles in the Preamble precisely on the ground that it was feared to be a back-door method of giving constitutional force to the freedom of education.[35] However, in settling this controversy, the Conseil was careful to pick a text that made no mention of the right to a State subsidy. It contented itself with noting that the 1946 Preamble, in creating a duty of the State of organize education, 'does not exclude the existence of private education, nor the granting of State aid to this education in circumstances defined by the law'. Contrary to the contentions of some on the left, State aid was permitted by the Constitution, but the Conseil did not go so far as to say that it was required by the Constitution, as some on the right contended.[36] All in all, the Conseil only affirmed the most uncontroversial element of the freedom of education, so that it played only a limited role in constitutionalizing this principle.

This decision set the constitutional framework of the Fifth Republic, but it did not end the social controversy. The events of 1984, and the fate of the proposals of the Minister for Education, Savary, to include the teachers in private schools within the category of civil servants, to which State teachers already belonged, need not be fully recounted here.[37] Suffice it to say that the attempt was perceived as the first step towards the abolition of private education, and the massive demonstrations led by the Catholic hierarchy and the opposition eventually led to the abandonment of the proposals and the resignation of Monsieur Savary. The subsequent *loi Joxe–Chevènement* merely sought to introduce a few changes in the status of private education. First, it went back on the *loi Guermeur* in requiring private schools to adhere to the criteria of operation set for State schools, and not merely to 'general criteria' established for both. Secondly, the Government simply had to obtain the agreement of private schools for the appointment of

[35] *GD* 359. Freedom of education had already been recognized as a 'general principle' by the Conseil d'État: CE Ass., 7 Jan. 1942, *UNAPEL*, *Leb.* 2; CE Sect., 27 Apr. 1945, *Lauliac*, D. 1945, 282, *concl.* Lagrange.

[36] See also L. Favoreu, 'La Reconnaissance par les lois de la République de la liberté de l'enseignement comme principe fondamental', *RFDA* 1985, 597.

[37] See D. S. Bell and B. Criddle, *The French Socialist Party* (2nd edn., Oxford, 1988), 173–4.

teachers, rather than wait for their proposals and either agree to them or reject them. Both changes were a return to 1959, and reaffirmed the dominance of the State education sector. Despite opposition arguments that, in the light of the Conseil's decision of 23 November 1977, the *loi* infringed constitutional values, it was upheld by the Conseil in a decision of 18 January 1985 (DECISION 22). The judgment expressly cites the *loi Debré,* and, though the Conseil fell short of according it any constitutional status, it did seem to consider it as declaratory of constitutional values. The reforms were upheld, but with the reservation that they were not to be interpreted as infringing the specific character of private establishments, which was a constitutionally protected value, nor could there be said to be any constitutional objection to State funding of such establishments. With these reservations, the Conseil stated that there was no constitutional principle that prevented the legislature from repealing earlier legislation. Thus, although Parliament had a free hand to repeal and amend earlier legislation, this could not affect the existence and specific character of private educational establishments, since these values, given constitutional status, were beyond the reach of ordinary *lois*.[38]

Freedom of Teachers

The quarrel over the right of existence of private schools has been intimately bound up with the rights of teachers to freedom of conscience. Under the contracts of association created by the *loi Debré,* the teachers in a private school thus linked to the State system are public employees, either as civil servants appointed to the school, or attached to the State by a public law contract. In addition, all the top rank teachers (*professeurs agrégés*) are trained by the State. The question is, what freedom do the State teachers retain when teaching in private schools? The freedom of the teacher has to be balanced against that of the pupil and of the institution. The *loi Guermeur* of 1977 imposed on teachers an obligation to respect the specific character of the establishment. Despite objections that this infringed the freedom of conscience established by article 10 of the Declaration of the Rights of Man and by the 1946 Preamble, in paragraph 6 of DECISION 21 the Conseil held that these constitutional values did not displace the teacher's duty to be reserved in expressing opinions, given the character of the establishment in which he or she worked. Attempts to revoke this duty by the legislation of 1985 were effectively frustrated by a remarkable piece of 'interpretation' by the Conseil. Applying the traditional Conseil d'État

[38] On the issue of State funding for private education, see CE, 6 Apr. 1990, *Ville de Paris and École Alsacienne, AJDA* 1990, 563.

method of 'taking the venom out of the statute', it practically emptied this provision of all content. In paragraph 10 of DECISION 22 the Conseil stated that 'the recognition of the specific character of private educational establishments [by article 1 of *the loi Debré* of 1959] merely implements the principle of the freedom of education, which has constitutional value.' Given this view, it was difficult to see what effect the repeal of the provision in the *loi Guermeur* imposing the duty to be reserved in expressing opinions could have:

> Considering that thus the repeal of the provision of the *loi* of 25 November 1977 imposing on teachers taking classes under a contract of association the obligation to respect the specific character of the establishment does not have the effect of exempting teachers from the duty deriving from the last paragraph of article 1 of the *loi* of 31 December 1959; that such an obligation, even if it cannot be interpreted as permitting an infringement of the freedom of conscience of teachers, which has constitutional value, requires them to observe a duty of reserve in their teaching . . .

In other words, although the duty is repealed, it survives as a constitutionally required obligation on teachers in private schools!

A more curious episode in the history of the freedom of teachers is the decision in *University Professors*[39] on the representation on administrative bodies in higher education. The freedom of higher education was proclaimed by the *loi* of 13 July 1875, one of the foundation texts of the Third Republic, and article 11 of the Declaration of the Rights of Man had proclaimed the freedom of thought and opinion to be one of the most precious rights of mankind.

In the aftermath of May 1968 higher education was reformed, and the universities were broken up into smaller units. Instead of large faculties, the new universities are made up of a series of 'units of study and research', run by the tenured professors. Teaching is done by both tenured professors and non-tenured *maîtres-assistants* and *assistants*. The universities are governed by elected councils and committees, composed initially of academic staff and students. As part of their plans to 'democratize' the public sector, the provisions on representation in the universities were extended by the Socialists to include all categories of university employees and students. The role of the academic staff was reduced, and there were many who considered that such changes would endanger academic freedom, since decisions would be taken to comply with pressures from politically motivated staff or students, rather than on academic grounds.

[39] CC decision no. 83-165 DC of 20 Jan. 1984, *University Professors*, D. 1984, 593, note Luchaire; L. Favoreu, 'Libertés locales et libertés universitaires', *RDP* 1984, 687. See also CC decision no. 89-249 DC of 12 Jan. 1989, *Miscellaneous Social Measures*: the presence of non-teaching doctors on a merely advisory committee of a university teaching hospital did not affect the promotion or freedom of university professors teaching in it.

Under the new *loi* on universities, professors, non-tenured teachers and researchers, and personnel from libraries and museums were to form one electoral body to choose representatives for the governing council of the university. Given the structure of the universities, the teachers and researchers would outnumber the professors, and, it was claimed, this would leave the appointment of the representatives of the latter group in the hands of the former, contrary to the principle of equality in elections. The Conseil struck down the provision not only on this ground, but also on a ground that it raised of its own initiative, the freedom of teachers. One of its members, Professor Vedel, had written that the independence of university teachers and professors rested more on tradition and custom than on texts. Obviously, the Conseil thought that the traditions needed some reinforcing. Its basic argument was that, because professors were outnumbered by the other groups, 'the independence of the professors was threatened' in a number of respects. Professors required freedom in their tasks of preparing programmes, attending to students, co-ordinating teaching teams, and in their necessary involvement in decisions on promotions of teachers and researchers. This freedom would be impaired by belonging to the same electoral group as the teachers and researchers, presumably because they would feel beholden to their potential electors. In any case, the professors would not be able to choose their members on the basis of their own assessment of their colleagues, and they could not be sure of an unbiased representation, since the delegates would be chosen by the majority group, the teachers and researchers. Equally, the independence of the researchers could be threatened by the presence of the professors in the electoral body. Overall, the genuineness of the elections was not guaranteed.

Although this decision has implications for equality, which will be examined in the next chapter,[40] it also reflects a desire to ensure freedom in the exercise of academic thought. The same concern was present in the Conseil's examination of other provisions.[41] The high value placed on academic freedom may be well established, but, as Luchaire suggests, this is normally thought to apply to what is said in class rather than to the way in which administrative tasks are carried out.[42] If it can be justified, the decision perhaps reflects the highly political character of French universities, and the extra care required to protect academic freedom within them, especially where the Government retains at least a formal role in appointments.

The Quality of Education

If the treatment of private education and teachers reflects a concern to guarantee their particular identity, the Conseil's attitude to freedom of

[40] See below, p. 208.

education equally contains a concern to ensure that the 'equal access' of the 1946 Preamble entails an equal standard of quality. Thus, in paragraphs 16 and 17 of DECISION 22 in 1985 the Conseil dealt with a provision whereby the opinion of local councils was to be sought before the State signed contracts of association with private schools. Although the Conseil upheld the provision, it 'interpreted' it in a very centralist way: 'even if the principle of the free administration of local authorities has constitutional value, it should not lead to the result that the essential conditions for the application of a *loi* organizing the exercise of a civil liberty should depend on the decisions of local authorities, and thus might not be the same throughout the country . . .' This tension between equality of opportunity and diversity in presentation cuts across all aspects of freedom of education, particularly the debate on private education. The Conseil seems more willing to permit diversity arising from the internal character of the education system—namely, the different teachers and institutions who provide the service—than that imposed from the outside by others, such as local authorities.

5.4. Freedom in Employment

Unlike the rights of association and education, employment rights[43] might be thought to have at least a foundation in constitutional texts. The 1946 Preamble contains four paragraphs affirming successively the right to work, the right to join a union, the right to strike, and the right to participate in the running of a business. However, given the difficulties that were noted in Chapter 2 with the 1946 Preamble, it has been by no means obvious that these are to be treated as constitutional rights. Many authors certainly regarded such provisions more in the way of pious hopes and statements of intention than of constitutionally respected rights.[44]

The overall approach of the Conseil has been to accept that the vague and non-self-executing character of the provisions of the Preamble creates constitutional values, but to leave the legislature a

[41] e.g. the composition of the academic board was approved, because it did not contain any representatives of administrative, technical, or manual staff, and the only student representatives were at the doctorate level, and were thus engaged in research, so that it could be said that all members of the board were engaged in academic work.

[42] See D. 1984, 598.

[43] See L. Hamon, 'Le Droit du travail dans la jurisprudence du Conseil constitutional', *Droit social*, 1983, 155; J.-F. Flauss, 'Les Droits sociaux dans la jurisprudence du Conseil constitutionnel', ibid. 1982, 645; P. Terneyre, 'Droit social et la Constitution', *RFDC* 1990, 339; A. Jeammaud, 'Le Droit constitutionnel dans les relations du travail', *AJDA* 1991, 612.

[44] See L. Hamon, 'Le Droit du travail dans la jurisprudence du Conseil constitutionnel', *Droit social*, 1980, 431.

large measure of discretion as to their implementation. Only where a clearly inadequate consideration has been given to the principles in the Preamble will a *loi* be declared unconstitutional. This was unequivocally stated in the first of the major employment cases,[45] where the Conseil said that 'whilst respecting the principles declared . . . in the Preamble, it is for the legislature to determine the conditions for their implementation'. As has been seen in relation to criminal peralties, such an approach means that the Conseil will only intervene on grounds of manifest error in evaluation of the situation, or of unequal treatment. However, it is also clear that not all the values in the 1946 Preamble enjoy the same status, and decisions as to the importance of a particular value are another method of tailoring the constitutional protection to each specific employment interest.

The Right to Work

Paragraph 5 of the 1946 Preamble proclaims that 'Everyone has the duty to work and the right to obtain a job.' Among jurisprudents, this raises a whole host of problems: in Hohfeldian terms, is this a liberty-right, a claim-right, or merely an immunity-right? Equally, one may ask to whom is the duty owed to work, and against whom does the right exist to be given a job? Is unemployment unconstitutional? They very vagueness of the right thus proclaimed is enough to deter the jurist from according it the status of a legal rule. At best, it is a legal principle or goal that should not be neglected by the legislature.

Given this difficulty, it is not surprising that the right has given rise to few decisions and to no declaration that a legal provision was unconstitutional. The three references to the Conseil on this matter have all concerned essentially the same legislation, that preventing the cumulation of a retirement pension and a salary. The measure was part of the Socialist strategy to combat youth unemployment by reducing the retirement age and by penalizing those who receive both a pension and a salary and those who employ them. An enabling law of January 1982 had given the Government the right to pass *ordonnances* to limit such accumulations of income. The opposition challenged this as contrary to the right to work, but the Conseil rejected the argument in an indirect fashion. It recalled that article 34 of the 1958 Constitution made it the prerogative of Parliament, who would have to ratify the *ordonnances*, to lay down the fundamental principles of labour law, and it stated that the provisions of the *loi* did not dispense the Government from respecting constitutional values.[46] Since the reply was so oblique,

[45] CC decision no. 77-79 DC of 5 July 1977, *Youth Employment*, *Rec.* 35.

[46] CC decision no. 81-134 DC of 5 Jan. 1982, *Enabling Law on Miscellaneous Social Measures*, *AJDA* 1982, 85.

it is not surprising that the opposition tried the same argument when Parliament ratified the *ordonnances* of 30 March 1982, which had established the complete termination of any professional links with an employer or with any public body as a pre-condition for obtaining a State pension, and which had imposed a 'solidarity payment', to go to an unemployment fund, both on employees who were in receipt of a pension and a salary and on their employers. This latter payment clearly reduced the pensioners' incentive to work and the employers' incentive to employ them. Again the Conseil rejected the argument that this was contrary to the right to work. Recalling article 34, the Conseil stated that 'it is up to it [the legislature] to lay down the appropriate rules for ensuring to best advantage the right of everyone to obtain a job, with a view to permitting the exercise of this right by the largest number of relevant persons.'[47]

The small degree of importance attached to this principle outside the realm of political rhetoric can be seen from the total absence of any reference to it in connection with the *loi* on redundancies.[48] Given its indeterminate nature, the low constitutional status accorded to it by the Conseil is not surprising.

The Right to Join a Union

The existence of the closed shop and restrictions on union activities have not had the same impact in France as in the United Kingdom. Union membership is much lower, and the unions are divided on political lines, which reduces their effectiveness.[49] The legislation on union activities has been the product not of conservative attacks on union powers, but of efforts by left-wing Governments to make union activity more effective.

Paragraph 6 of the Preamble to the 1946 Constitution provides that 'every individual may defend his rights and his interests by union action, and belong to the union of his choice'. As with other such provisions in the Preamble, the legislature is free to settle the means for giving effect to this, within the general power given by article 34 of the Constitution to pass *lois* on the fundamental principles of labour and union law. The Conseil will only challenge the exercise of this discretion where constitutional values are infringed. In the rare cases where union rights have been discussed, the concern of those making the reference has been to protect the individual against union organizations and the special rights given to them.

[47] CC decision no. 83-156 DC of 28 May 1983, *Pensions Law*, *Rec.* 41, a phrase repeated in CC decision no. 85-200 DC of 16 Jan. 1986, *Cumulation of Pensions and Salaries*, *Rec.* 18.

[48] CC decision no. 89-257 DC of 25 July 1989, *AJDA* 1989, 789, discussed below.

[49] On unions, see generally J. Ardagh, *France Today* (London, 1990), 98–118.

The leading case is that on the *Redundancy Law* in 1989.[50] One provision of this *loi* authorized unions to litigate for the benefit of employees without having a specific mandate from them, thus enabling speedy action in the face of redundancies. It was sufficient that the employee was notified in writing of the intended action, and had not objected in a fortnight. The opposition argued that such union actions would infringe the freedom of the individual. In order to safeguard personal freedom, the Conseil observed that such collective litigation was only constitutional if the employee was in a position to give a fully informed consent to the union action, and if he preserved the freedom to conduct the defence of his interests in person. Accordingly, the Conseil insisted that the letter sent by the union should provide the necessary information about the scope of the litigation and the employee's rights to put an end to the union's mandate at any time. Furthermore, before it could proceed with the litigation, the union had to prove that the employee was personally aware of this required information.

Merely giving unions special rights is not in itself objectionable, even though this may, in reality, be disadvantageous to those employees who do not belong to one. Thus the Conseil did not take exception to provisions that obliged employers to negotiate with unions on additional ways of exercising union rights (especially on the collection of subscriptions), and to communicate with unions before proposing collective agreements.[51] In neither case was the position of the individuals made worse.

The Right to Strike

Paragraph 7 of the 1946 of Preamble proclaims that 'the right to strike shall be exercised within the framework of laws that regulate it', and article 34 of the 1958 Constitution allows Parliament to determine the fundamental principles of labour law. But how far can Parliament go in circumscribing the right to strike? The issue has arisen in three areas: Does public necessity justify limits on the right to strike? Can unions be civilly liable for going on strike? And can employers impose financial penalties on strikers?

Public Necessity and the Right to Strike Although the right to strike contained in the Preamble appears unlimited, French public law has never accepted this view. In the leading case of *Dehaene*[52] the Conseil

[50] CC decision no. 89-257 DC of 25 July 1989, *AJDA* 1989, 736.
[51] CC decision no. 83-162 DC of 19, 20 July 1983, *Democratization of the Public Sector*, *Rec*. 49.
[52] CE, 7 July 1950, *Rec*. 426; *GA*, no. 78.

d'État interpreted the Preamble's words, 'within the framework of laws that regulate it', in a wide sense: 'the Constituent Assembly intended to invite the legislature to make the necessary reconciliation between the defence of professional interests, of which the strike is a means, and the safeguarding of the public interest, which a strike might infringe . . .' Even in the absence of legislation on the question, the Conseil d'État held that the Government could limit the right to strike of the staff of a *préfecture* where this was liable to interfere with the good functioning of the public service.

In DECISION 23, *Strikes in Radio and Television*, of 1979 the Conseil constitutionnel adopted the same approach. It held that the right to strike could be limited in order to preserve the *continuity* of the public service, but not to ensure a *normal* service. The French 'winter of discontent' of 1978–9 included strikes in the television services. Because of the unwillingness of the employees to co-operate with management in advance of the strike, a minimal service could not be maintained, and the service was disrupted. As a result, the RPR had a private member's bill passed that sought to modify the right to strike in the television services, and to reinforce the requirement that a minimum service be maintained, as provided in earlier legislation, by issuing a decree to define both the essential features of the service to be kept going and the categories of personnel required for them. In addition, if the number of relevant employees available was insufficient to ensure a 'normal service', the Presidents of each radio and television corporation were entitled to requisition those persons that they needed to remain in their posts. In reply to a reference based on the 1946 Preamble, paragraph 1 of the Conseil's decision substantially repeated the arguments of the Conseil d'État in the *Dehaene* decision:

> in enacting this provision [of the Preamble], the drafters of the Constitution intended to recognize that the right to strike is a principle of constitutional value, but within limits, and empowered the legislature to define these by making the necessary reconciliation between the defence of professional interests, of which the strike is a means, and the safeguarding of the public interest, which a strike might infringe . . .

In the public sector this enabled the legislature to limit the right to strike 'in order to ensure the continuity of the public service, which, just like the right to strike, has the character of a principle of constitutional value'. All the same, the principle only required the continuity of those aspects of the public service which ensured the essential needs of the country, and so the granting of a power to ensure a 'normal service' was excessive and unconstitutional.

Although the notion of the continuity of the public service finds some support in article 5 of the 1958 Constitution, which deals with the powers of the President, it owes its origins more to the case-law of the Conseil d'État. The Conseil was therefore expanding the effective Constitution by its 1979 decision.

So what does justify restriction of the right to strike? Continuity of the public service is limited to certain sectors of activity. The 1979 decision, reinforced by DECISION 27 in 1986, suggests that a minimum service in broadcasting is of sufficient importance. At the same time, these decisions make it clear that the requirements of the public service are limited both in terms of the personnel concerned and the range of their activities. As the Conseil noted in the 1986 case: 'these limits [on the right to strike] can go so far as to ban the right to strike for employees whose presence is essential for ensuring the functioning of parts of the service whose interruption would harm vital needs of the country.' Inevitably, such persons are few in number. Which public service requires such constraints on the freedom of its employees is for the legislator to decide, but as yet few general principles have emerged on this question.

Other values may also justify restrictions on the right to strike. The health and security of persons and property have been held to warrant such a restriction among employees in nuclear installations.[53] But, on the whole, few values justify the curtailment of the right to strike. As in Britain, however, public order constitutes a constraint on how a strike is conducted. Thus, paragraph 16 of DECISION 18, *Security and Liberty*, rejected arguments that the right to strike was in any way infringed by making it an offence to block railway lines (however popular this might be as a tactic in industrial and other disputes).

Civil Liability and the Right to Strike If the right to strike is accepted as a constitutional value, it is not unconditional. Even within the limits approved by Parliament and the Conseil constitutionnel, it is still capable of abuse. This raises the important question of civil liability.

The 1980s saw a marked increase in the use of civil actions for damages brought by employers and non-striking employees for what they considered to be abusive strikes and factory occupations. In 1982 there were over a hundred such actions pending before the courts, and the unions wanted some protection lest the strike weapon be deprived of much of its effectiveness. The Government originally thought of defining the right to strike, but gave up at the complexity of the task. Instead, it adopted the traditional English approach of creating a statutory immunity, whereby no one was liable for damages in respect of a trade dispute unless there was a breach of the criminal law, or acts

[53] CC decision no. 80-117 DC of 22 July 1980, *Nuclear Installations*, *AJDA* 1980, 479. For an analysis of the Conseil constitutionnel's interpretation of the right to strike, see S. Dion-Loye and B. Mathieu, 'Le Droit de grève', *RFDC* 1991, 509.

'manifestly incapable of relating to the exercise of the right to strike or the right to belong to a union'. In DECISION 24, *Trade-Union Immunity*, the Conseil held that this went too far. Recalling that, 'in principle, every human act that causes loss to another obliges him by whose fault it occurs to pay make reparation' (which is the basic principle in article 1382 of the Civil Code), it noted that the legislature had introduced more limited schemes of liability in certain areas. However, no such scheme had granted an immunity without regard to the seriousness of the fault involved. The immunity here gave the unions an excessive advantage over other subjects of the law and their civil liability. Thus, taking account of considerations of equality, the Conseil considered that the *loi* went too far.

By coincidence, the Chambre sociale of the Cour de cassation solved the question of how to reconcile article 1382 with the right to strike in a decision of 9 November 1982, some three weeks after the Conseil.[54] The Conseil constitutionnel's judgment was referred to at the beginning of the arguments of the Advocate-General as setting out the general principle in this area, but it was not mentioned in the decision itself. The Chambre sociale held that a union would not be liable unless there were a breach of the criminal law, or the actions were not 'attributable to the normal exercise of the right to strike'.

The net result is that all liability for fault cannot be excluded. This was confirmed in the *Amnesty Law* decision of 1988 (DECISION 3b), where dismissal for serious fault in the course of an industrial dispute could not be amnestied without regard to the consequences for the employer and fellow workers. As a minimum, the law must punish serious abuses of the right to strike.

Financial Penalties for Employees It has long been a principle of French public and private law that workers are, in theory, remunerated only for work completed. But there has been considerable divergence over the years on the issue of whether a full day's pay should be deducted in the case of a strike during part of the day. Administrative practice, subsequently confirmed by legislation, permitted one-thirtieth of a month's salary to be deducted for every whole or part of a day on which work was not performed. In 1977 the Conseil constitutionnel upheld Parliament's approach as consistent with the Constitution.[55] The Conseil considered that, provided the failure to carry out the

[54] *Syndicat CGT de l'usine Trailor de Lunville* c. *Dame Abadine*, D. 1983, 531, note Sinay; *JCP* 1983, II. 19995, *concl.* Gauthier. See further M. Forde, 'Bill of Rights and Trade Union Immunities: Some French Lessons' (1984) 13 *Industrial Law Journal* 40, and id., 'Liability in Damages for strikes: A French Counter-Revolution' (1985) 33 *American Journal of Comparative Law* 447.

[55] CC decision no. 77-83 DC of 20 July 1977, *Services Rendered*, *Rec.* 39.

duties was sufficiently clear, the deduction of salary was merely an accounting procedure, not a disciplinary measure, so that the right to a hearing was not required.

The Socialist Government amended this in 1982, introducing a less severe scale of deductions in the case of collective strike action. During strikes by air traffic controllers in 1987, several deputies took the opportunity to introduce amendments to a bill on miscellaneous social measures to return to the previous state of the law and to harsher penalties.

In DECISION 25, *Public Service Strikes*, the Conseil constitutionnel reconsidered the matter (since the *loi* was, technically, a different legislative item from the 1977 text), but came to the same solution. As in other aspects of the right to strike, it announced that it was for the legislature to determine the scope of that right, and what actions amounted to its abuse. In addition, the legislature could determine the consequences of a strike, and could review not only the work done, but accounting rules on paying salaries, 'practical constraints', in determining the extent of a stoppage of work and in discounting the period not worked when fixing pay. In other words, there need not be any strict correlation between the hours not worked and the proportion of pay deducted. Given that strikes of short duration 'abnormally affect the regular functioning of the public service', the legislature could seek to impose rules that discouraged the repeated use of short strikes. All the same, it did consider that an automatic deduction applying to all public employees was too wide. It failed to take account of the differences between public services, and the way in which short-term strikes would affect them. As a result, extending the deduction rules to all employees of public services, rather than just to the main activities subject to public accounting procedures (civil servants in the strict sense), was declared unconstitutional. Here one does see an idea of proportionality creeping in, albeit in a limited way, and preventing excessive penalties.

The Right to Participation

Paragraph 8 of the 1946 Preamble provides that 'Every worker shall participate, through his representatives, in the collective determination of conditions of work and also in the running of businesses.' Rather like the right to work, this appears more of an exhortation to the legislator than a specific and immediately enforceable right like the right to strike. Although it was the first of the employment rights to be discussed by the Conseil, this did not result from the obvious importance of its content. As a consequence, no provision of a *loi* has been held unconstitutional specifically for failing to observe this principle.[56] As with the right to work, the approach of the Conseil has been to treat the question of participation as a matter to be left to the competence of Parliament under article 34 of the 1958 Constitution. Indeed, the definition of key concepts in legislation, such as 'aged workers' or 'employees with special social characteristics', and other concrete matters of implementation may be left to negotiation by way of collective agreements between workers and employers.[57] Furthermore, the legislature can act without consulting those affected.[58]

Unless there is a manifest error or breach of the principles of equality, as in the *University Professors* case, the Conseil is unlikely to censure legislation. By contrast with the Conseil constitutionnel, the Conseil d'État has resisted the idea that paragraph 8 of the 1946 Preamble contains a principle of constitutional value, and considers it merely to be an ideal to be realized.[59] In practice, there is little real difference between the two positions. If the Conseil d'État is stricter in using the term 'principle of constitutional value' rather than 'constitutionally declared objective', this does not mean that it substantially rejects the conclusions of the Conseil constitutionnel.

Even if the freedom to participate, like many of the employment freedoms discussed by the Conseil, is best seen as having more of the character of an exhortation than a self-executing right, this does not mean that it is without influence in judging the constitutionality of legislative provisions. Apart from acting as a principle of interpretation, as just described, it may also affect the division of legislative roles between Parliament and the executive. By declaring a value to be a fundamental principle of labour law, or even a fundamental right, its limits and application will have to be determined by Parliament, and not by an executive decree. Thus, whereas the content of the contract of employment can be modified by decree, questions on participation, even going down to the detail of fixing the number of employee representatives, belong to Parliament.[60] The recognition that a provision has constitutional value can take various forms, therefore, depending on its status.

[56] See CC decision no. 77-79 DC of 5 July 1977, *Youth Employment*, *Rec.* 35 (to exclude young people on short-term work experience from employee participation was held to be valid).

[57] CC decision no. 89-257 DC of 25 July 1989, *Redundancy Law*, *AJDA* 1989, 769, §§11, 12.

[58] CC decision no. 77-83 DC of 20 July 1977, *Services Rendered*, *Rec.* 39 (failure to consult civil service unions did not invalidate the legislation).

[59] See CE Sect., 13 Dec. 1978, *Manufacture française de pneumatiques Michelin*, D. 1979, 327, *concl.* Latournerie, but note P. Terneyre, *RFDC* 1991, 317 at 321.

[60] Cf. CC decision no. 779-11 FNR of 23 May 1979, *Rec.* 57 (whether 8 May ought to be a public holiday), and CC decision no. 83-162 DC, *Democratization of the Public Sector*, *AJDA* 1983, 614 §3.

5.5. Freedom of Communication and Information

The area of media law demonstrates some of the most creative work performed by the Conseil constitutionnel. The problems handled by recent legislation were barely envisaged when the constitutional texts were written, and the established principles are thus of a sketchy and unsatisfactory form. Even in 1958 broadcasting was only a State monopoly in relation to both television and radio. As a result, no thought had been given to the principles governing competition and the concentration of media ownership. Such problems in relation to the press had been recognized at the Liberation by the *ordonnance* of 26 August 1944, but no constitutional principles were ever enunciated.

The central constitutional texts were thus from an earlier period, when the main problem was that of State censorship of the press. Article 11 of the Declaration of the Rights of Man declares that 'The free communication of thoughts and opinions is one of the most precious rights of man; hence every citizen may speak, write, and publish freely, save that he must answer for any abuse of such freedom in cases specified by *loi*.' This is clearly about the right to *publish* opinions. After many restrictions in the intervening period, this was confirmed by the *loi* of 29 July 1881, which abolished prior restraint and declared the freedom of the press.

Modern problems are different. In the area of the press, concentration of ownership has grown since the 1930s, although newspapers remain more regional than they are in Britain.[61] If modern investment and rationalization in the face of competition from other media require a more efficient scale of operation than that of the individual editor-proprietor envisaged in 1789 or 1881, concentration of ownership creates the dangers of private censorship and the restriction of access to a significant medium for freedom of expression. Can the press baron be prevented from influencing the content of the outlets that he can acquire?

As far as broadcasting is concerned, the issue is whether it should be subject to the same regime as the press, or whether new principles are needed. In particular, given the limited range of broadcasting frequencies available, how is regulation in this respect compatible with freedom in broadcasting? Again, is there a right to freedom to broadcast equivalent to that of the freedom to publish, or does the legislature have a free hand to establish any regime (including a monopoly) that it chooses?

The solutions adopted by the Conseil reflect an attempt to draw more specific rules and principles by analogy from the older texts,

[61] See J. Ardagh, *France in the 1980s* (London, 1982), 596–603.

particularly that of 1789. The areas in question were the subject of several *lois* in the 1980s, representing the divergent positions adopted by the political parties. One of the tasks of the Conseil was to establish a body of principles that could provide a consensus framework amid the disparate and changing views of the political parties. Its success in this regard is one of the more important achievements of its adaptation of old texts to modern constitutional realities.

Freedom to Broadcast

The values embodied in the texts of 1789 and 1881 might have been thought to justify the freedom to create broadcasting networks in the same way as newspapers. But the development of broadcasting was very different. The State monopoly of telegraphy was quickly extended to radio and television. Even when it was made a distinct corporation in 1959, the State broadcasting network remained under strong Government influence.[62] Split into a number of radio and television channels in 1974, broadcasting remained a State monopoly, supervised by two bodies with consultative status, the High Authority of Audio-Visual Communication, and a parliamentary Committee for Radio and Television. By contrast with the United Kingdom, advertising on State radio and television was permitted on October 1968.

This State monopoly was contested on two fronts. Liberals argued that the freedom of enterprise and commerce, as well as the freedom of communication, required that private broadcasting be permitted. Indeed, a number of such stations (Radio Andorra, Radio Luxemburg, and Radio Monaco) existed just beyond the French border, broadcasting to France. Socialists argued that the State networks were too much under Government control, and did not permit the opposition a fair share of media exposure. Despite these arguments, successive Governments remained unmoved, and, indeed, the monopoly was strengthened by a *loi* of 28 July 1978, upheld by the Conseil constitutionnel, which made it a criminal offence to broadcast in breach of the monopoly.[63]

Following the Socialist victory of May 1981, prosecutions stopped, and legislation was put forward to break up the State monopoly, starting with radio. The fundamental principles of these reforms was to liberalize radio, to attempt to integrate the peripheral commercial stations, and eventually to liberalize television, but there was an initial hostility to commercial broadcasting. In line with the proposals made

[62] Ibid. 580–5.

[63] The main reason for this decision was that the Conseil could not consider a challenge to the provisions of an already enacted *loi*.

in relation to the press by the Lindon report on editorial collectives in the 1970s, freedom of communication was envisaged as the right to form non-profit-making associations for the communication of ideas, rather than as an aspect of freedom of enterprise.

The early decisions of the Conseil constitutionnel on these reforms made clear the differences between freedom of broadcasting and freedom of the press. In its first decision of 30 and 31 October 1981[64] the Conseil did not consider that the requirement that an operator obtain a licence from the Minister of Culture was either a prior restraint or otherwise unconstitutional. Unlike freedom of association, the freedom to broadcast had no special constitutional status, and thus fell within the general powers of the legislature, subject to the requirements of equal treatment. But the Conseil saw no objection to restricting the freedom to broadcast to non-profit-making associations, and, since commercial companies were objectively different in character, there was no unconstitutional discrimination.

In July 1982 the Conseil rejected the opposition argument that the freedom of communication contained in article 11 of the Declaration of the Rights of Man, like the freedoms of association and movement, could only be limited by requirements of public order.[65] Other considerations could also be relevant:

> it is up to the legislature to reconcile, in the current state of technology and its control, the exercise of the freedom of communication resulting from article 11 of the Declaration of the Rights of Man with, on the one hand, the technical constraints inherent in the means of audio-visual communication and, on the other hand, with objectives of constitutional value—the safeguard of public order, respect for the freedom of others, and the preservation of the pluralistic character of socio-cultural currents of expression—which these forms of communication, by reason of their considerable influence, are likely to infringe . . .'

These considerations were to be balanced by the legislature, and the Conseil would only intervene where there was a manifest failure to do so correctly.

The Conseil here based the newly acknowledged right to broadcast on the freedom of expression principle in article 11 of the Declaration of the Rights of Man. At the same time, it drew analogies from other texts to complete its construction—the safeguard of public order from freedom of association, and respect for the freedom of others from article 4 of the Declaration. But it also extrapolated the value of pluralism from the 1789 text. The principle of pluralism was not expressed in the 1981 decision, and is not easy to trace in any remotely constitutional

[64] CC decision no. 81-129 DC of 30, 31 Oct. 1981, *Free Radio Stations, AJDA* 1981, 595.

[65] CC decision no. 82-141 DC of 27 July 1982, *Audio-Visual Law, Rec.* 48, repeated in DECISION 27 §8.

text. The nearest expression comes in the *ordonnance* of 26 August 1944 on the press, but even there it is an implicit value. An explicit statement of the principle only came in the *loi* of 9 November 1981. Its quick adoption as a constitutional value by the Conseil constitutionnel might, no doubt, be justified by its implicit presence in what had gone before. However, it represents a remarkable extension of the corpus of constitutional values, and one which provides the justification for significant restrictions on the freedom to broadcast. At the same time, the extrapolation remained limited and general—there was no attempt to create a notion of the freedom of commercial speech, which the right-wing opposition would have wished.

Freedom to broadcast is not much use without the money with which to establish the broadcasting operation. The Conseil, however, did not restrict the legislature in its decisions on how the new free radio stations were to be funded: 'Considering that [the freedom of communication and the freedom of enterprise], which are neither general nor absolute, can only exist within the framework of regulation established by legislation, and that the rules limiting the financing of activities by commercial advertising are, in themselves, neither contrary to the freedom of communication nor to the freedom of enterprise . . .' Since the associations seeking to establish broadcasting stations were equally entitled to compete for public funds, this satisfied any constitutional obligations imposed on the legislature. The legislature was equally free to change its mind by allowing advertising at a later date.[66]

Freedom of the Press

If the freedom to broadcast is a relatively new dilemma, that of the ownership of the press is not. The problem of the concentration of press ownership in a few hands was already a problem in the Third Republic, but no legislative solution was found until the Liberation.[67] The *ordonnance* of 26 August 1944 insisted essentially on the transparency of press ownership—that the 'proprietor' of the paper was always the major shareholder or the president of the major shareholding company, and that the names of its owners, directors, and one hundred major shareholders were published. In this way, ownership became public knowledge, but this did nothing to ensure the pluralism of ideas communicated by the press. The mergers of the 1970s, especially the growth of the Hersant empire, caused the Socialist Government to increase the regulation of press ownership.[68] Adopting

[66] CC decision no. 84-176 DC of 25 July 1984, *Advertising on Radio*, *Rec.* 55.
[67] J. Rivero, *Libertés publiques* (4th edn., Paris, 1989), ii. 216–18.
[68] See generally Ardagh, *France Today*, 572–8.

the principles of a report presented to the Economic and Social Council in May 1979, a *loi* in 1984 sought to limit the size of groups owning newspapers, to create a Press Commission on the model of the High Authority of Audio-Visual Communication, and to require further publicity on press ownership. This *loi* and its sequel provoked some of the most creative activity on the part of the Conseil.

The argument against the *loi* was based on the freedom of the press, understood in the classical sense of the freedom to publish the printed word. In order to rebut the argument, the Conseil had to restate the principle of 1789 in a different form from that in which it had been principally understood. The opposition referred the *loi* to the Conseil constitutionnel, which upheld it on the report of Professor Vedel, author of the 1979 report to the Economic and Social Council, and a member of the Conseil since 1980. This *Press Law* decision of 1984[69] rejected the idea that to require publicity of transactions was against the freedom of the press or of commerce:

> Considering that, far from being contrary to the freedom of the press or limiting it, the implementation of the objective of financial transparency tends to reinforce an effective exercise of this freedom; that in fact, by requiring that the public know the real people running press enterprises, the terms under which newspapers are financed, the financial transactions of which they are the subject-matter, and the interests of all kinds that may be involved in them, the legislature is enabling the readers to exercise their choice in a truly free way, and to form their opinion by an informed judgment on the kinds of information offered by the written press . . .

Furthermore, with regard to the measures on pluralism (restricting holdings to 15 per cent of the market) in the legislation, the Conseil stated:

> Considering that the pluralism of daily newspapers dealing with political and general information . . . is in itself an objective of constitutional value; that in fact the free communication of ideas and opinions, guaranteed by article 11 of the Declaration of the Rights of Man and of the Citizen of 1789, would not be effective if the public to which these daily newspapers are addressed were not able to have access to a sufficient number of publications of different leanings and characters; that in reality the objective to be realized is that the readers, who figure among the essential addressees of the freedom proclaimed by article 11 of the Declaration of 1789, should be able to exercise free choice without either private interests or public authorities substituting their own decisions . . .

The plasticity of the constitutional texts is well exhibited here. First, the Declaration's *right of publishers to publish* becomes *the right of the*

[69] CC decision no. 84-181 DC of 10, 11 Oct. 1984, *GD*, no. 38.

audience to receive information. This is in line with article 10 of the European Convention on Human Rights, but has no place within the ideas of 1789 or 1881. Secondly, the Conseil introduced the notion of 'an objective of constitutional value', a means for giving effect to an explicit constitutional principle. As was seen in Chapter 2 new ideas can be slipped in thus, even without any formal text, and can be used as a way of circumscribing the freedom to publish that the Declaration had most explicitly in mind. All the same, the freedom to publish is not thereby ignored. Only careful and 'strict reservations of interpretation' by the Conseil saved the provisions on pluralism from being held unconstitutional. The Conseil acknowledged that, on its own, the ownership limit of 15 per cent, could be unconstitutional, as being against the freedom of enterprise, but a correct interpretation merely limited its scope to expansion due to future take-overs and mergers, not that due to success in the market or to existing holdings.

The force of the newly discovered 'objectives' can be seen from the attempt by the Chirac Government to amend the 1984 *loi* in 1986. The reform wanted to increase the percentage of the market that an individual could own to 30 per cent (thereby protecting the RPR's ally and benefactor, Hersant). In paragraph 2 of its *Press Law* decision of 29 July 1986 (DECISION 26) the Conseil stated that Parliament could always repeal earlier *lois*, and that it was the judge of the merits of such action. But 'the exercise of this power cannot lead to the removal of legal safeguards for requirements having constitutional value'. It repeated the principle on pluralism stated in 1984, and then noted two dangers arising from the law. No provision had been made to combat evasion of the restrictions on ownership by the use of groups of companies or 'front' persons or companies. The new *loi* did not just modify the rules, but removed existing legal protections of pluralism in the press, which was a 'principle of constitutional value'. The new provision on pluralism was held to be unconstitutional. Once an objective has constitutional value, it must be pursued by the legislature, though the means for this task are its own choice, provided that existing protections are not weakened. Clearly, this suggests a rather State-centred view of the protection of constitutional values, and ran contrary to the deregulation philosophy behind the Chirac reform.

A final feature of the 1984 legislation concerned the freedom of journalists. It required a newspaper to have a permanent editorial team large enough to ensure the 'autonomy of conception' of the publication, and this was to be reinforced by a provision that the members of the editorial team should be journalists carrying a professional card. The Conseil upheld these stipulations on the ground that the objectives of transparency and pluralism could legitimately require an autonomy of conception, and, since access to the journalists' profession was free, neither provision infringed the Constitution. Parliament was thus permitted to create a buffer of professional journalists between the press baron and the output of his newspapers.

New Principles on the Media

The development of a right to broadcast, and the reinterpretation of the freedom of communication set out in article 11 of the 1789 Declaration come together to make a coherent picture of the powers of the legislature in relation to the media. The freedom to communicate serves primarily the interest of the consumer in receiving a wide variety of opinions, and then of the communicator who wishes to contribute to that variety. Questions of freedom of enterprise or ownership are very much secondary considerations. Three values emerge in the Conseil's development of the constitutional principles, each of which really echoes the views of Parliament in the early 1980s and of the Economic and Social Council in 1979. The values are competition, pluralism, and transparency of financial control. Each contributes to the existence and effectiveness of the diversity of expressions of currents of socio-cultural thought.

As has been seen in relation to radio, competition in broadcasting was not originally considered to be a constitutional requirement. But once the monopoly was broken, freedom to broadcast had to be allowed on the basis of equal treatment for the chosen category of broadcasters. Full competition need not exist, but at least the existence of the High Authority of Audio-Visual Communication as an independent licensor and regulator did help to avoid arbitrariness in decisions about who was to be allowed to broadcast. The same was seen in relation to cable and satellite television. The Conseil raised no objection to High Authority issuing licences for cable television, and merely insisted that *loi* and not *règlement* fix the maximum limits of the cable network provided by local television companies.[70]

In DECISION 27, *Freedom of Communication*, concerning the privatization of Télévision française 1 (TF1), a major State television channel, the Conseil expressed indifference as to the character of the regime adopted. The legislature could choose to concede a public service, or it could simply turn the channel over to the private sector. In either case, it could impose a system of administrative authorization on the broadcasting channels. Indeed, on the issue of privatization itself, the Conseil refused to acknowledge the 'principle of competition between private activities' to which the opposition appealed in its argument against the transfer of the largest television company to the private

[70] CC decision no. 84-173 DC of 27 July 1984, *Cable Television*, *Rec.* 63.

sector. As a result, competition for its own sake it not a constitutional requirement. As that decision makes clear, the principle of pluralism is the real justification of competition. Since pluralism can be within a public sector (internal pluralism) or between the public and private sectors (external pluralism), competition of this latter kind is only one form of pluralism.

Pluralism is a persistent refrain of the Conseil's reinterpretation of article 11 of the Declaration of the Rights of Man. The content of pluralism is set out in didactic style in paragraph 11 of the *Freedom of Communication* decision:

> Considering that the pluralism of currents of socio-cultural expression is in itself an objective of constitutional value; that the respect for this pluralism is one of the conditions for democracy; that the free communication of ideas and opinions, guaranteed by article 11 of the Declaration of the Rights of Man and of the Citizen of 1789, would not be effective if the public to which the means of audio-visual communication address themselves were not able to have access, both within the framework of the public sector as well as in the private sector, to programmes that guarantee the expression of tendencies of different characters, respecting the necessity of honesty in information; that in reality the objective to be realized is that the listeners and television-viewers, who are among the essential addressees of the freedom proclaimed by article 11 of the Declaration of 1789, should be able to exercise free choice without either private interests or public authorities being able to substitute their own decisions for it, and without it being possible to make them the subject-matter of a market . . .

Essentially, this amalgamates the concerns expressed previously in relation to radio and the press. The form of pluralism, external or internal, is primarily for the legislature to determine, providing the appropriate issues have been addressed. The function of the Conseil is to draw up a check-list of features that any legislation in this area must include.

Pluralism can be internal when a particular media institution is required to include a variety of opinions within its columns or air waves. In the *Freedom of Communication* decision the Conseil was satisfied that the legislature had made adequate provision for pluralism in the public sector by requiring, *inter alia*, the broadcasting of party political statements during elections, of a right of reply to any Government statement, and of religious programmes by the major religions in France on Sunday mornings. Furthermore, the Conseil went on to 'interpret' the provisions governing the private sector so that, where there was only one frequency authorized in a particular area, the National Commission of Communication and Liberties (the CNCL) could require the licensee to undertake obligations to ensure the free and pluralist expression of ideas and currents of opinion.

More frequently, the requirement in broadcasting, as with the press, is that there should be a sufficient diversity of channels to provide the audience with a variety of viewpoints and sources of information. One way of ensuring this is to prevent concentration of ownership, and it is on this, rather than on provisions concerning editorial freedom, that the Conseil has insisted. The Chirac reforms of the press met with objections because they made inadequate provisions in this regard, and the same was true of the broadcasting reforms introduced a few months later. Paragraphs 31 *et seq.* of the *Freedom of Communication* decision (DECISION 27) focused on article 39 of the *loi* that prohibited the acquisition of more than 25 per cent of the shareholding in a private television company, and on article 41 that prohibited the existing licensee of a national radio or television station from becoming the licensee of another. The Conseil considered these provisions as inadequate protection of pluralism. Article 39 did not prevent a shareholding of up to 25 per cent in more than one television company, and it set no limit either on such shareholdings or upon shareholdings by that person in cable television. Article 41 made no attempt to regulate the problem of multi-media concentration, where the same person was licensee of both radio and television stations, or of the ownership of local television licenses in more than one geographical area. 'From the existence of gaps in the *loi*,' stated the Conseil, 'there is the risk that situations of concentration will develop, particularly in the same geographical area, not only in the field of broadcasting, but equally in relation to the totality of the means of communication, of which broadcasting is one of the essential components.' It struck down the two provisions, and, in consequence, curtailed the major powers of the CNCL, effectively forcing new legislation to be passed.

The principle of transparency was established as a means of ensuring pluralism in the *Press Law* decision of 1984, and it was restated in paragraph 16 or DECISION 26, *Press Law*, in 1986. However, the legislature was able to reduce the strictness of the provisions in earlier legislation. As long as the objective of financial transparency continued to be pursued, a weakening in requirements was not in itself a ground for unconstitutionality. There would have to be an manifest error in evaluation, and this was not the case when the legislature did not require the immediate publication of changes in ownership, nor the names of directors of companies that had a major shareholding in a publication. The comparative flexibility in this regard, and in the area of competition, compared with the position adopted in relation to pluralism, merely reflects the lesser importance of these other values.

Regulation of broadcasting is not a new phenomenon. What has developed in the 1980s is the willingness to have an independent administrative agency act as a buffer between the Government and the media in allocating and enforcing licences. While the High Council on Audio-Visual Communication was merely an advisory body, its successors have been given an increasing regulatory authority that has called into question the separation of powers established by the 1958 Constitution. An independent monitoring agency in its own right, the Conseil has been seeking to accommodate to initiatives for the lessening of political interference with broadcasting.

Created in 1982, the High Authority of Audio-Visual Communication was the regulator of the licences that the Minister of Culture issued to the new free radio stations, and it advised on the associations to which they should be granted. Its powers were extended in 1984 to include the issuing of licences for cable television. The Conseil took this opportunity to remark that the High Authority's independence from the Government provided a fundamental guarantee for the exercise of the freedom of broadcasting. Given such a status, its nature and activity became an indispensable part of the broadcasting scene. Two aspects of its operation caused difficulty in terms of the separation of powers—the power to make rules, and the power to impose sanctions—trespassing respectively on the executive and judicial functions. Neither aspect met with objections from the Conseil constitutionnel.[71]

The process here was one of establishing independent regulation for a more pluralistic system of broadcasting. The political will to move in this direction encountered little opposition from the Conseil, whatever constitutional obstacles might seem to lie in its way.

Although establishing independent regulatory bodies, each incoming Government effectively sought to replace those appointed by its predecessor. Thus, in 1982, 1986, and 1989 the composition of the regulatory body on broadcasting underwent complete change. This might appear paradoxical, in that the purpose of such bodies was to place the regulation of the media outside political control. The device used in each case was the abolition of the previous institution, and the creation of a new one with some different powers. The changes in powers of the bodies created in 1986 and 1989 were not radical. All the same, the Conseil took the view that the legislature could abolish previous institutions at any moment, provided that there was no weakening of the guarantees offered for fundamental liberties.[72] Since this was not the case here, the term of office of the members could be terminated without infringing the Constitution. In other words, the deference to Parliament remains very strong in this area.

To summarize: in the context of media law, the Conseil consti-

[71] On rule-making by bodies other than the Government, see above, ch. 3, and on sarctions, see the discussion of due process, below, ch. 5. 8.

[72] DECISION 27 §5. See *RFDA* also J. Morange, 'Le Conseil supérieur de l'audio-visuel', *RFDA* 1989, 235.

tutionnel has essentially been following initiatives from political and other sectors, and selecting items to be recognized as constitutional, thereby creating a consensus. The constitutional objectives of pluralism and transparency, as well as the importance of having an independent regulatory agency as the guarantor of the freedom to broadcast, were all ideas invented by the legislature. Because of their significance, and in an attempt to stabilize the 'social acquisitions' that the legislature was thus creating, the Conseil gave them constitutional value under article 11 of the Declaration of the Rights of Man. At best, its approach is rather statist, making full deregulation rather difficult.

5.6. Freedom of Property

Property is among the four rights that article 2 of the 1789 Declaration describes as 'natural and imprescriptible', rights which political institutions exist to preserve. Article 17 goes on to state that: 'Property being an inviolable and sacred right, none can be deprived of it, except when public necessity, legally determined, evidently requires it, and on condition of a just and prior indemnity.' The rhetoric seems to give property a high social priority. But the right has always been qualified, and the Conseil constitutionnel reflects the many other concerns of modern society in its guarded endorsement of the constitutional importance of property rights:

> Considering that the objectives and conditions for the exercise of the right of property have undergone an evolution characterized by the extension of its range of application to new areas and by the limitations required in the name of the public interest; that it is in relation to this evolution that the reaffirmation by the Preamble to the Constitution of 1958 must be understood.[73]

The notion of property goes beyond 'real estate' as envisaged primarily in 1789, to cover many forms of intellectual and commercial property where the boundaries between 'mine' and 'thine' can be drawn less easily. Furthermore, such developments do require a careful balancing of individual and public interests in the availability of the benefits of modern technology. So far, these aspects have not been the concern of the Conseil constitutionnel, which has confined itself to more classical forms of property. All the same, this is one of the more creative areas of the Conseil constitutionnel's work, as well as one of the most politically delicate.

The Constitutional Value of Property

Although the 1789 Declaration proclaims the importance of property, the constitutional status of the right was questioned by most authors

[73] CC decision no. 89-256 DC of 25 July 1989, *TGV Nord*, decision 31 §18.

until the *Nationalizations* decision of 1982. Thus, in 1979 the foremost constitutionalist, Loïc Philip, wrote that property should not be regarded as a constitutional principle, but merely as an exhortation similar in character to many rights proclaimed by the 1946 Preamble.[74] The main reason for this was that property rights had been so circumscribed by legislation that no 'inviolable and sacred right' seemed now to be arguable. After all, in one of its earliest decisions the Conseil constitutionnel had stated that, although the fundamental principles of property rights had to be defined by *loi*, 'these principles have to be determined within the framework of the limitations of a general character that had been introduced by legislation prior to the Constitution [of 1958]'.[75] Given the scope of previous legislative limitations, this appeared to reduce significantly the importance of property rights.

Limitations on property were of two kinds. On the one hand, various legislative provisions had limited the rights of use, enjoyment, and disposition that article 544 of the Civil Code of 1804 defines as the content of the right of property. That Code itself provides that property may be expropriated for public utility (article 545), that certain uses are prohibited (for example, diverting natural watercourses to the detriment of landowners downstream (article 640)), and that rights of disposition by will or gift are significantly reduced by the rights of specified persons to succession. Twentieth-century legislation on the rights of tenants, restructuring agriculture, and planning all served to undermine further the absolute nature of property rights. On the other hand, constitutional provisions also restricted the values of 1789. Most notable was paragraph 9 of the Preamble to the 1946 Constitution, which states that 'Any property or business whose exploitation has or acquires the character of a national public service or of a *de facto* monopoly should become the property of the community.' This duty to nationalize does not fit easily with the protection of private property. Indeed, article 34 of the 1958 Constitution seems almost indifferent to the question of public or private ownership of property, requiring merely that *loi* define the rules concerning 'the nationalization of undertakings, and the transfer of undertakings from the public to the private sector'. In addition, *loi* is required only to lay down the fundamental principles governing the regime of property rights, which effectively provides significant scope for *loi* and *règlement* to restructure the content of property rights.

At the same time, as Luchaire points out,[76] the right to property is a

[74] 'La Valeur juridique de la Déclaration des droits de l'homme et du citoyen du 26 août 1789 selon la jurisprudence du Conseil constitutionnel', in *Études offertes à P. Kayser* (Aix, 1979), ii. 317 at 332–3; also F. Luchaire, 'Le Conseil constitutionnel et la protection des droits et libertés des citoyens', in *Mélanges Waline* (Paris, 1974), ii. 563.

[75] CC decision no. 61-4 FNR of 18 Oct. 1961, *Rec.* 50.

[76] *Droits et libertés*, 270.

freedom of the individual to own or manage property—it is part of the freedom of individual action. As a result, the very fact that the intervention of a *loi* is necessary at all is seen as a safeguard of a specially protected interest.

DECISION 29, *Nationalizations*, went further than most commentators thought likely. The decision was politically important, as it represented the first significant challenge to the Socialist platform before the Conseil constitutionnel after the change of power in May 1981. The *loi* nationalized major banks, the steel industry, and certain industrial and armaments manufacturers, thus giving the Government command over the heights of industry. Such a policy figured in the 1979 *programme commun* with the Communists, but it was presumed then that a constitutional amendment would be necessary, a move blocked by the opposition majority in the Senate. To be fair, the legislation marked only the continuation of control of banks (some of which were nationalized just after the war), of steel, and of other major industries in which the State already held important shareholdings.

The opposition argued that the policy of the legislation amounted to a measure of collectivization, being too extensive to be just action to meet a public necessity as required by article 17 of the Declaration of the Rights of Man.[77] Equally, they challenged the argument of the *Garde des Sceaux* (the Minister for Justice), defended by the Socialist deputies in their *mémoire* to the Conseil, that the assessment of public necessity was entirely a matter for the legislator. As the Socialist deputies put it: 'In these circumstances, public necessity is that which the holder of sovereign power defines as such: that is to say, the people and, hence, its representatives. The latter are subjected to only one control in the exercise of this competence: that which the people carry out at the moment of elections.'[78] By contrast, the opposition argued that the Conseil had the right and duty to intervene in this matter.

The decision of the Conseil gave partial satisfaction to both sides. While it did not consider that these particular nationalizations were unconstitutional, the Conseil made it clear that it had the right to control the assessment of public necessity, though it would only intervene where there had been a manifest error in evaluation. This seems to reverse the burden of proof set out in the 1789 Declaration, where the onus seems firmly on the legislature to demonstrate the necessity of the interference with property. All the same, the Conseil did affirm the principle of constitutional review on this question.

Equally, on the right to property itself, the Conseil had more

[77] See L. Favoreu, *Nationalisations et Constitution* (Aix, 1982), 212–14, 253 ff.

[78] Ibid. 320.

sympathy with the arguments of the opposition than with the Socialist deputies. Although neither side denied that the 1789 Declaration set out the basic conditions for nationalizations, the Socialists had argued for the priority of article 9 of the 1946 Preamble. The Conseil, like the opposition, preferred to see this as a limited exception. As has been noted in Chapter 2, the arguments were primarily historical—that the 1946 Preamble was adopted to complement, not to contradict, the 1789 Declaration. From this, the Conseil inferred that the 'principles declared by the Declaration of the Rights of Man retain full constitutional value, in so far as they concern the fundamental character of the right of property'. This strong affirmation left a limited place for the 1946 Preamble. As Luchaire pointed out, the Conseil effectively declared that the Fifth Republic, though 'social' as declared by article 2 of the 1958 Constitution, is not a 'Socialist' Republic, and any attempts to introduce wider public ownership would require a constitutional amendment.[79] In practice, this was not a significant restriction on Mitterrand-style Socialism.

Although the arguments on principles might have gone in favour of the opposition, the application to the facts did not. Despite very detailed analyses of how much of the economy was being nationalized and how unnecessary this was, the Conseil refused to find either an abuse of power or a manifest error in evaluation. Since the measures were proposed in order to provide the authorities with the means to face an economic crisis, to promote growth, and to fight unemployment, only in some specifics were the provisions of the *loi* unconstitutional. As will be seen below, these were mainly related to the amount of compensation to be paid for the assets to be nationalized. Some other provisions, notably the exclusion of co-operative and mutualist banks, were held to be contrary to the principles of equality.

The converse case to *Nationalizations* was *Privatizations*,[80] which expanded the scope of the constitutional protection of property to cover public property. The return of a right-wing coalition in the March 1986 parliamentary elections set the scene for a reversal of Socialist policies. Since the President was a Socialist, the cohabitation period could easily be brought to an end. Accordingly, a *loi* enabling the transfer of sixty-five companies to the private sector within five years

[79] Favoreu, *Nationalisations et Constitution*, 71. The opposition deputies (ibid. 215) defined the distinction thus: '"Social" implies here that the Republic is attentive to the situation of individuals. "Socialist" signifies, by contrast, that the Republic condemns the private ownership of the means of production and exchange.'

[80] CC decision no. 86-207 DC of 25, 26 Jun. 1986, DECISION 30. See generally T. Prosser, *The Privatisation of Public Enterprises in France and Great Britain* (EUI Working Paper no. 88/364), and 'Constitution and Economy: Privatisation of Public Enterprises in France and Great Britain' (1990) 53 *Modern Law Review* 304; C. Graham and T. Prosser, *Privatizing Public Enterprises* (Oxford, 1991).

was passed by Parliament. Its scope was very ambitious, reversing nationalizations effected by the Governments immediately after the Liberation of 1944, and returning some 1,454 companies to the private sector. Following advice from the Conseil d'État, the Government had declared its intention of subjecting the assets to be privatized to an independent valuation, ensuring that these would not be sold at an undervaluation. In paragraph 58 the Conseil constitutionnel took the opportunity to insist that such a pricing policy was a constitutional requirement, based on the principles of 1789:

> Considering that the Constitution opposes the transfer of property or enterprises forming part of public property to persons pursuing private-interest objectives for prices lower than their value; this rule follows from the principle of equality invoked by the deputies making the reference to this Conseil; that it is also based on the provisions of the Declaration of the Rights of Man of 1789 concerning the right of property and the protection due to it; this protection concerns not only the private property of individuals, but also, on an equal footing, the property of the State and of other public persons.

'Equality' here refers to equality between citizens, in that, were any undervaluation to occur, the purchasers would obtain a benefit from public funds that other citizens would not gain. This argument is the standard reason given in public law for refusing to grant the administration the power to make gifts or other gratuitous transactions.[81] The novelty lies in using article 17 of the 1789 Declaration as a foundation, since it was drawn up to protect the rights of citizens against the State, not those of the State against predatory citizens.

Unlike the process of nationalization, there is no requirement of public necessity for privatization. It suffices (contrary to recent practices[82]) that Parliament endorses the transfer. In paragraphs 38 to 40 of the *Nationalizations* decision a number of provisions authorizing the administration to dispose of subsidiaries of the nationalized companies to the private sector were struck down for want of adequately precise instructions from Parliament. By all means was it constitutional for Parliament to authorize a public body to conduct such privatization, but it had to lay down the rules for the conduct of such operations, and could not leave it to a wide and uncontrolled executive discretion. The Socialist deputies tried to argue in this case (and later with regard to the privatization of TF1) that article 9 of the 1946 Preamble prevented the transfer of major and almost monopolistic enterprises to the private sector. But in paragraphs 9 and 39 of the decision dealing with TF1, DECISION 27, the Conseil distinguished

[81] R. Odent, *Contentieux administratif* (Paris, 1971), 1749; art. L69 of the *Code du domaine*.

[82] See Prosser, *The Privatisation of Public Enterprises*, 308, for references to decisions holding earlier privatizations by the executive to be unlawful.

between those national public services which constitutional values require to be public, and others which the legislator has created as such. In the latter case, an activity either created as a public service or nationalized can be transferred to the private sector if the legislature considers it appropriate. More specifically, article 9 only applies to monopoly situations, and this is understood restrictively, taking account of the place of the firm within the market in question as well as of its competition. It is not sufficient that the firm has a momentary dominant position. The decision as to whether a monopoly exists is left to the legislature, subject to control of manifest error in evaluation. Thus, neither in the *Privatizations* case nor that of TF1 was such an error discovered. Parliament would effectively have to confer a national monopoly to a private enterprise to act unconstitutionally. Although the Compagnie générale d'électricité and the Compagnie générale de constructions téléphoniques came close to having a monopoly in the supply of telecommunications equipment, as the Socialist deputies argued, the Conseil held that there was no manifest error. Certainly, the minister responsible for the privatizations specifically denied any intention to hand over public services or monopolies to the private sector. Very few of the British privatizations would have been affected by such an interpretation, except possibly that of British Gas.

Although no privatization was prevented by the Conseil, Prosser argues that 'it is evident that constitutional considerations have had considerable importance in shaping the scope of privatisation available to a French Government.'[83] The President of the Conseil, Badinter, certainly left the impression that it might have found the privatization of prisons difficult to countenance.

However, the division between public and private ownership that this discussion represents is too stark. On the one hand, even before 1982 the State had important shareholdings, acquired freely in the market, and controls over the private sector, together with assets in semi-public companies like the oil companies Elf and Total.[84] On the other hand, many parts of the public sector were 'commercial and industrial public enterprises' with considerable managerial autonomy, and subject to many principles of private law in their operations. In both cases, the interpenetration of public and private sectors made either the acquisition or disposal of such public assets of limited significance and legal difficulty. In addition, France has long known the legal regime of the 'public service concession', whereby a private enterpreneur operates the whole or part of a public service—for example, water or transport—but the public authority retains powers

[83] Ibid. 311.
[84] V. Wright, *The Government and Politics of France* (3rd edn., London, 1988), 104–8.

under general principles of public law to redefine the requirements of that service.[85] Real privatization involves the transfer of a public service to the private sector outside such a regime of concession. An example here was TF1, the main public television channel, occupying 40 per cent of the market, which was simply floated as a private company, subject to an administrative licensing arrangement similar to that existing in relation to the other private television stations (recently created by the Socialists). The Conseil constitutionnel did not require the legislature to justify its refusal to use the half-way house of 'public service concession' rather than full privatization.

As a result, it can be seen that Parliament retains a very wide discretion to determine the appropriate regime of property rights in any particular area. Although there is a bias towards private ownership, in that nationalization by way of expropriation needs special justification, constitutional principles place few major restrictions on the way in which Parliament can restructure the property sector. The only substantial restriction lies in the obligation to pay compensation for any forced transfer between sectors. In other words, the public and private patrimonies must remain basically the same at the end of the process of transfer. In this way, the burdens of changes are shared among the community, not borne by some alone.

What Constitutes Deprival of a Property Right?

For article 17 of the Declaration of the Rights of Man to come into play, there must be a property right. The mere expectation that a concession will be renewed has been held not to amount to such a right.[86] Article 34 of the Constitution merely talks about 'property rights', a rather traditional concept of 'property'. Thus property can be split into a number of rights distributed among different people. Even where a real right does exist, not all interference will amount to a 'deprival' regulated by article 17. The Conseil has been loath to consider indirect interference or regulation as sufficient.

Such a policy of the Conseil merely continues the tradition of the Cour de cassation and the Conseil d'État, which refused to give compensation for rights of passage for public services (an 'administrative easement')—for example, telegraph wires, electricity cables, and the like—unless legislation specifically provided for it, which it did not always do.[87] The principles are well illustrated by the *Eiffel Tower*

[85] See L. N. Brown and J. Bell, *French Administrative Law* (4th edn., Oxford, 1993) ch. 8.

[86] CC decision no. 82-150 DC of 30 Dec. 1982, *Transport Law*, *AJDA* 1983, 252.

[87] See F. Colly, 'Le Conseil constitutionnel et le droit de propriété', *RDP* 1988, 135, 174, 180–1.

Amendment decision of 13 December 1985.[88] A Government amendment to a law creating new television channels permitted Télédiffusion de France to instal and use equipment for transmitting radio and television programmes on roofs, terraces, and the superstructures of buildings. The building principally envisaged here was the Eiffel Tower, owned by the city of Paris (run by the opposition leader, Mayor Jacques Chirac). The installation had to be approved by the President of a Tribunal de grande instance. While not challenging the idea that such an administrative easement could be granted, the opposition argued that article 17 of the 1789 Declaration prevented its creation unless public utility so required, and that there should be safeguards to minimize the interference with private property. The Conseil took the view that an administrative easement of this kind did not amount to the deprival of a property right:

> the right granted to the public corporation by article 3 §2 of the *loi* to place equipment on the upper part of buildings, in so far as it only imposes a tolerable inconvenience, is not a deprival of property within the meaning of article 17 of the Declaration cited, but a public interest easement affecting buildings by reason of their height and elevation . . .

The conclusion would have been different, it continued, if the effect of the easement had been to empty the property right of all content, or if it had affected persons occupying the property.

As in this case, much legislation does not deprive the owner of his property rights, nor even the whole enjoyment of them. The law merely introduces a number of restrictions and inconveniences. An obvious example is planning law. Zoning may restrict the use that can be made of property, and may make it necessary to obtain administrative authorization before changes can be made to that use. While permission is being sought, the owner is left with the status quo, and the Conseil has been unwilling to see this as a deprivation of property, even where it may affect a farm for a whole year.[89] Indeed, Parliament may go so far as to restrict the right to dispose of property either temporarily, as in the case of shareholdings in recently privatized companies,[90] or even permanently, to maintain the minimum size of agricultural units. In the latter case it was decided that a transfer could be prevented if it produced an agricultural unit below a minimum size, and that agricultural land could only be exploited by a person with a permit. The opposition argued that this amounted to a deprival of

[88] CC decision no. 85-198 DC of 13 Dec. 1985, *Rec.* 78. For subsequent litigation on the decree allowing an installation on the Eiffel Tower, see CE Sect., 9 July 1986, *Ville de Paris*, *AJDA* 1986, 595.

[89] CC decision no. 89-267 DC of 22 Jan. 1990, *Adapting Agriculture to the Environment*, *RFDC* 1990, 329; CC decision no. 85-189 DC of 17 July 1985, *Town Planning*, *Rec.* 49.

property rights, since the owner might be unable to exploit the holding himself, and could transfer it effectively only to someone with a permit. The Conseil replied that this was not sufficient interference to amount to 'undermining the meaning or importance' of the right, for ownership remained, and the owner could still select his tenant, albeit within the limited category of permit-holders.[91]

To be equivalent to deprival, an interference must be significant. This is well illustrated by the decision on *Democratization in the Public Sector* of 1983. Although the imposition of a requirement that each firm in which the State had a 50 per cent shareholding should have worker-directors was held not to infringe the rights of the private shareholders, a power for the Minister of Commerce and Industry to appoint representatives of private shareholders 'if necessary' was held to do so, since they would potentially lose their say in the running of the company altogether.[92]

Even where restrictions are introduced without breaching article 17 of the 1789 Declaration, the persons affected may be able to obtain compensation by the application of general principles of administrative law—the principle of equality before public burdens. In the *Eiffel Tower* decision, provisions that limited compensation to specific harms caused by the installation of aerials were held to be unconstitutional because they prevented compensation for other harms on the basis of equality before public burdens. In effect, the issue is the basis of compensation. The granting of such compensation as a constitutional requirement seems to be limited to the effect of public works.[93] While deprival of a property right, as will be seen, is compensated at full value, other interferences are recompensed only to the extent that they cause losses that have a direct, certain, serious, and special character, different from those suffered by other members of the community.[94]

Grounds for Interference

The only interference with property rights that requires justification is unilateral action by a public body. An authorization to purchase shares

[90] CC decision no. 89-254 DC of 4 July 1989, *Application of Privatization*, *RFDA* 1989, 786; *AJDA* 1990, 347.

[91] CC decision no. 84-172 DC of 26 July 1984, *Structure of Agriculture Law*, *Rec.* 58.

[92] CC decision no. 83-162 DC of 19, 20 July 1983, *Democratization of the Public Sector*, *AJDA* 1983, 614. See also CC decision no. 90-283 DC of 8 Jan. 1991, *Tobacco and Alcohol Advertising*, discussed above, ch. 5.1.

[93] Colly, 'Le Conseil constitutionnel', 181–2.

[94] On this, see ibid. 180, and CE, 28 Feb. 1986, *Commune de Gap-Romette*, *AJDA* 1986, 318. See further R. Errera, 'The Scope and Meaning of No Fault Liability in French Administrative Law' [1986] *Current Legal Problems* 157, and CE, 2 Oct. 1987, *Spire*, [1989] *PL* 357.

in the open market is not a form of nationalization that falls within the constitutional principles discussed above, since there is no 'decision of a public authority to which the owner or owners are obliged to bow'.[95] The Conseil will ensure that certain formal and substantive conditions are met.

In the first place, article 34 requires Parliament to pass a *loi* in a number of situations where nationalization, privatization, or fundamental principles of property law are involved. One consequence of this is that the Government cannot be left with a wide discretion. The Conseil has been content to approve the granting of administrative powers to interfere with property, such as in restricting the transfer or use of assets, where the grounds for the exercise of the power are sufficiently precise, enabling control over abuse to rest with the administrative courts.[96] In the second place, as we shall see, judicial intervention is a constitutional prerequisite in some cases.

Prior Compensation

Article 17 of the Declaration of the Rights of Man makes it clear that compensation should be provided before expropriation takes place. This represented best practice at the period, and has its origins in medieval law. It also represents traditional French administrative practice, where the determination of the compensation precedes the judicial authorization to the administration to enter the property to be expropriated. Unlike the situation with regard to other aspects of property, the Conseil constitutionnel has been more ready to scrutinize the administration's assessment of the appropriate level of compensation.

In DECISION 29, *Nationalizations*, the Conseil struck down the first basis of compensation proposed by the *loi*, which had been adopted on the advice of the Conseil d'État. The amount to be paid to shareholders in the nationalized companies was to be based on three factors: the average share price between 1978 and 1980, the value of the firm's assets, and its profits. The additional criteria were meant to correct the imperfections of stock-market valuation, but the Conseil considered that this had not been done fairly. In the first place, the assessment of profits did not include those of subsidiaries, which meant that different forms of management structure and accounting created a divergence in the valuation of different firms, and an undervaluation of some of them. Finally, the refusal to give shareholders their 1981 dividend was

[95] CC decision no. 83-167 DC of 19 Jan. 1984, *Credit Corporations, Rec.* 23.

[96] See e.g. CC decision no. 89-254 of 4 July 1989, *Application of Privatisation*, and CC decision no. 89-267 DC of 22 Jan. 1990, *Adapting Agriculture to the Environment*.

a deprival of property. Given the complexity of calculating the effect of the threat on nationalization on the share prices, it is surprising that the Conseil took such strong exception to the way in which the assessment of compensation was made. Its intervention represented more than a correction of a mere error in evaluation, and amounted to a requirement that the compensation be substantially correct. In part, this was in response to the detailed criticisms of the basis of calculation provided by the opposition.

The framework of compensation procedures and criteria laid down by the Conseil in DECISION 30, *Privatizations*, was equally detailed. It required not merely that there be an independent valuation to establish the real value of the property, but that six specific features should be taken into account. It provided ample ammunition for an unsuccessful action in the administrative courts, although the Conseil d'État did not require the Privatization Commission to give reasons for its valuations.[97]

Given this detailed control, DECISION 31, *TGV Nord*, represents somewhat of a relaxation. To link up with the Channel Tunnel, it was proposed to construct a new fast train (TGV) line from Paris to the Tunnel. Local residents at Amiens objected because the route of the new line did not pass through their town, threatening it with economic decline. In protest, some local farmers sold plots of land along the proposed route in one-metre square units to a variety of people from France and elsewhere. The object was to complicate the process of compensation before expropriation. A deputy, Floch, proposed an amendment to the bill on planning currently before Parliament, whereby the procedure of 'expropriation in great urgency' set out in articles L15-9 *et seq.* of the Code of Expropriation could be used. This involved an interim payment of compensation by the administration on the basis of a valuation made by the Government's own land-agents, with the final settlement being awarded by the civil judge in the normal way. The provisional payment alone was to be the prerequisite for expropriation. The Conseil upheld this procedure as constitutional, even though all the details of compensation were not settled before expropriation took place. It did affirm that 'to be just, the compensation should cover the totality of the direct, material, and certain harm caused by the expropriation'. All the same, 'the payment by the expropriating body of an interim award representing the compensation due is not incompatible with respect for these requirements, if such a mechanism responds to imperative reasons of the public interest, and provides safeguards for the rights of the owners concerned', which was the case where the civil judge fixed the final amount of com-

[97] CE Ass., 2 Feb. 1987, *Joxe et Bollon*, *RFDA* 1987, 176.

pensation. A certain realism seems to have characterized this decision. There was no doubt that compensation would be paid, the question was merely whether this had to be done in full before the expropriation took effect. In this case, high-speed trains required high-speed expropriation. The actions of the administration were closely circumscribed by the Code of Expropriation and could always be challenged before the administrative courts, so the citizens affected did have reasonable guarantees.

5.7. Freedom of Enterprise

The freedom of commerce or enterprise was not one of the values set out in the Declaration of the Rights of Man and of the Citizen in 1789. All the same, it was a value that the bourgeoisie cherished, and it was recognized by the *loi Chapelier* of 12 and 17 March 1791, article 7 of which provided that 'any person is free to conduct such business or to exercise such profession, art, or trade as suits him'. Despite this early recognition in a non-republican text, this right has never found its way into any formal constitutional provision. In part, this could be explained on the ground that freedom of enterprise and commerce might appear as either an alternative form of the freedom of work, or a continuation of the right to property. All the same, the constitutional status of the principle was still seriously doubted by jurists when the Conseil constitutionnel came to declare that the freedom of enterprise was a principle of constitutional value in paragraph 16 of DECISION 2, *Nationalizations*, of 1982.[98] Indeed, it was silent about the origin of the principle. At best, it seemed to be an extension of the freedom of the individual enshrined in article 2 of the 1789 Declaration.

Luchaire argues that there are three different values at issue in this area.[99] First, the freedom to exercise a profession was specifically recognized as a civil liberty by the Conseil in 1976.[100] Secondly, freedom of enterprise, recognized in the *Nationalizations* case, is really an extension of this freedom to exercise a profession by establishing a business (though it also includes the creation of non-profit-making enterprises). The freedom of commerce and trade has long been recognized as a general principle of law by the Conseil d'État, and has its origins in the *loi Chapelier*. Thirdly, the Conseil d'État has interpreted this principle as forbidding the public services to compete with the private sector,[101] or the imposition of any requirement of prior author-

[98] See R. Savy, 'La Constitution des juges', D. 1983 Chr. 105 at 106.

[99] *Droits et libertés*, 118–19.

[100] CC decision no. 76-88 L of 3 Mar. 1976, *Hospital Reforms*, *Rec.* 50.

[101] CE Sect., 30 May 1930, *Chambre syndicale du commerce en détail de Nevers*, *GA*, no. 48.

ization for businesses. The Conseil constitutionnel has not given any constitutional force to this latter aspect, and its decisions with regard to the freedom of communication in broadcasting would suggest that it might be unwilling to extend the constitutional value of freedom of enterprise in this direction. It is up to the legislature to decide which activities are performed by the public service, and there can be many good reasons for allowing commercial activities in the public sector or for imposing licensing arrangements on businesses.

Even if the general freedom of enterprise has constitutional value, the Conseil has consistently stated that this freedom is neither general nor absolute, and can only exist within the framework established by the legislature.[102] As a result, a number of restrictions are permitted. The first is the limitation of scope introduced by legislative restrictions before 1958. As stated in paragraph 4 of DECISION 5, such restrictions have their origin in 'interventions judged necessary for public authorities in the contractual relations between individuals'. Secondly, freedom of enterprise is capable of significant constraint in the interests of other constitutional values and the prevention of abuses. Thus, in the *Press Law* case of 1984,[103] a number of restrictions on press ownership were introduced in favour of pluralism in the freedom of communication. All the same, the way in which the Conseil 'interpreted' the ceiling on the market share that could be owned by a single proprietor so that it did not inhibit the expansion of existing titles nor the creation of new ones shows that the principle of freedom of enterprise is not without effect on the constitutionality of legislation. Thirdly, restrictions may be introduced where the public interest requires them—for example, to reduce alcoholism.[104] Finally, as in broadcasting, constraints are justified by the particular subject-matter at issue. Given the shortage of radio frequencies, there must be some restriction on freedom of enterprise in that sector, and this justifies some form of prior authorization by way of licensing.

In the light of its limited scope, freedom of enterprise must be regarded as a less highly protected value than freedom of the press or of personal liberty. It falls short of a self-executing right, and is more of a constitutionally recognized principle that gives rise to specific rights to freedom of enterprise in particular areas. All the same, in some sense, as Savy puts it, France does have an 'economic constitution' that confers constitutional value on some fundamental rules

[102] First stated in CC decision no. 82-141 DC of 27 July 1982, *Audio-Visual Law*, above, p. 169.

[103] CC decision no. 84-181 DC of 10, 11 Oct. 1984. See also CC decision no. 90-283 DC of 8 Jan. 1991, *Tobacco and Alcohol Advertising*.

[104] CC decision no. 67-44 L of 27 Feb. 1967, *Code on Alcohol Sales*, *Rec*. 26, and also CC decision no. 90-283 DC of 8 Jan. 1991, *Tobacco and Alcohol Advertising*.

of the free market, and this is not an uncontrovertible position.[105] Its constitutional status has an effect. It requires that restrictions be imposed by *loi* and not by *règlement*. In addition, it requires Parliament to produce arguments relating to the public interest or to the nature of the subject-matter to justify any restriction. Of course, Parliament enjoys a discretion here, and will only be sanctioned for a manifest error in evaluation. All the same, it does impose some significant protection on freedom of enterprise and on the free-market economy structure in France.

5.8. Defence of Freedoms

The basic issues arise in the defence of constitutional freedoms, particularly freedom of the person. First, article 66 of the Constitution confers responsibility for the protection of the individual on the ordinary judiciary. This limits the right of the administration to act on its own, and also limits the jurisdiction of the administrative courts. Secondly, there is the issue of due process. Is due process to be equated with the application of the same procedure to all, and is the right to a hearing (*les droits de la défense*) to be observed in all instances?

Article 66

Article 66 of the Constitution of 1958 provides that the (ordinary) judiciary is the guardian of the liberties of the individual (*la liberté individuelle*). Since such judges have formal guarantees of independence,[106] this is meant to enhance the protection of the individual from interference by the executive. Legislation that infringes this allocation of power is unconstitutional. This poses problems of interpretation in a number of areas: does it mean that only judges can make decisions about the liberty of individuals, or impose sanctions; and what rights are covered by the idea of 'the liberties of the individual'?

Although the ordinary judge has to take decisions about the detention of individuals, as was seen in Chapter 5.1, this does not prevent the administration from being given powers to order a foreigner to be 'returned to the frontier' if she or he has never been given permission to enter France in the first place, or if that permission has expired.[107] Such a decision by the administration may be challenged in

[105] 'La Constitution des juges', 109.

[106] The independence of the judiciary was the ground on which the Conseil constitutionnel struck down some provisions of the organic law on the judiciary: see CC decision of 21 Feb. 1992, *Organic Law on the Judiciary*, *Le Monde*, 23–4 Feb. 1992.

[107] CC decision no. 86-216 DC, *Entry and Residence of Foreigners*, *Rec.* 135.

the administrative courts, which can suspend the decision and review the direction in which the person is sent.

It is fair, however, to suggest that this legal protection has been improved as a result of the condemnation of French procedures by the European Court of Human Rights in the *Bozano* case.[108] Since a would-be immigrant does not have any right to enter France, she or he is not being deprived of this right by being returned, and so article 66 does not come into play. Indeed, in *Entry and Residence of Foreigners* in 1989 the Conseil constitutionnel went so far as to strike down a *loi* that transferred competence over this administrative expulsion of foreigners to the ordinary courts.[109] The decision limits the 'liberty of the individual' to detention, criminal penalties, and civil status. The only relevance of article 66 to administrative expulsion was where the immigrant could not be expelled immediately. The ordinary judge has to review detention, as the statute provides; but this does not affect the legality of any expulsion order. It had been argued in Parliament that illegal entry might well be a criminal offence, and certainly involved issues of nationality and civil status, both of which fell within the jurisdiction of the ordinary courts. But the Conseil did not consider that such issues arose frequently enough or were of such a character[110] as to justify derogation from the normal rules of the competence of the administrative courts. By contrast, where a *loi* fails to provide that the ordinary judge should review the detention of a would-be immigrant or refugee pending his or her expulsion, this will be struck down, since detention, albeit in a transit area and not in gaol, affects the freedom of the individual, and so falls within the competence given to the ordinary judge by article 66 of the Constitution.[111] In its decision on the *Competition Council*[112] the Conseil had suggested that 'the good administration of justice' might justify a derogation from the normal rules of judicial competence. In *Entry and Residence of Foreigners* it was argued that the proximity of the Tribunal de grande instance, and the immediacy of its oral procedure made it the most suitable body to deal quickly with litigation on expulsions. The Conseil was unconvinced, and took the usual French view that the administrative courts were just as capable of providing effective remedies as the ordinary courts. As a result, article 66 can be seen as having a limited application in those cases where administrative decisions affect civil liberties.

A second concern is the power to impose sanctions for breaches of

[108] ECHR, 18 Dec. 1986, series A, no. 111. See generally 'Le Contentieux de l'expulsion', *RFDA* 1989, 3–45.

[109] CC decision no. 89-261 DC of 28 July 1989, *RFDA* 1989, 691, note Genevois.

[110] The reason is that issues of foreign law are questions of fact, and thus do not involve the administrative courts seeking a preliminary ruling from the civil courts.

[111] CC decision of 25 Feb. 1992, *Entry of Foreigners*, *Le Monde*, 27 Feb. 1992.

[112] CC decision no. 86-224 DC of 23 Jan. 1987, *GD*, no. 43.

regulations. The line between criminal and administrative sanctions is always difficult. In the *Press Law* decision of 1984[113] the Conseil had struck down the power of the newly created Commission for Transparency and Pluralism in the Press to impose sanctions for the breach of provisions on pluralism and for failure to comply with its directions. This was not based on article 66 of the Constitution, but rather on the idea that it constituted prior administrative approval for take-overs, and this prior restraint infringed article 11 of the 1789 Declaration.

The existence of such sanctions is an important part of the efficacy of a regulatory authority. The *loi* of 30 September 1986 gave the CNCL two sanction powers—to suspend a licence for up to a month, or to withdraw it altogether. Both of these sanctions required the Commission to issue default notices. In other cases, the CNCL could simply initiate proceedings before the administrative or criminal courts, in the latter case by referring matters to the *procureur de la République*. In 1989 the CSA (Le Conseil supérieur de l'audio-visuel) was given wider powers, allowing the duration of the licence to be curtailed, and the imposition of a fine, provided that the breach of the terms of the licence did not amount to a criminal offence. The fine was up to 3 per cent of the turnover of the company in question (5 per cent in the case of persistent breaches of the licence terms). The *loi* gave the right to a hearing and the right to challenge a decision before the Conseil d'État, as well as requiring full reasons to be given for any penalty. In paragraph 27 of DECISION 28 the Conseil constitutionnel gave its approval to such a regime. Since, in the case of broadcasting, the legislature was entitled to balance technical requirements with the freedom to communicate and to establish a system of administrative authorization, it could also create an independent body to supervise this system and give it the power to impose sanctions within the limits of its mission. If such sanctions were permissible in principle, the Conseil was careful to point out that the offences must relate to specific provisions of the licence granted within the framework of rules established by *loi* by *règlement*.

Similar powers were also approved in the case of the Commission des opérations de Bourse (COB).[114] Again no objection was raised, provided that no imprisonment was involved and that there were sufficient safeguards for the protection of constitutional rights. As has been noted, the Conseil was careful to interpret these powers to impose sanctions in such a way as to avoid a double jeopardy, where both criminal and administrative penalties might be incurred for the same offence. Although such sanctions are not imposed by ordinary

[113] CC decision no. 84-181 DC of 10, 11 Oct. 1984, discussed above, ch. 5. 5.
[114] CC decision no. 89-260 DC of 28 July 1989, *RFDA* 1989, 671, note Genevois.

courts, they can be, and they are reviewed by the administrative courts.[115]

It is easy to understand why judicial intervention is required where the freedom of the individual from arrest, imprisonment, or criminal sanctions is involved. But the question arises of whether this extends to other civil liberties.[116] This has been specifically considered by the Conseil, and, in relation to property rights, rejected.

It has long been the practice in France that an expropriation order is made by a civil court, and that a prior award of compensation is settled either by this court or by a commission, with appeal to the civil judiciary. The precise status of this practice has been disputed. It is certainly a fundamental principle of property law, but it fits ill within article 66 of the Constitution. In the *Town Planning* decision of 1985[117] the Conseil made it clear that article 66 was limited to personal freedom, and did not apply to property rights. In addition, in the *Eiffel Tower Amendment* decision[118] the Conseil held further that 'no principle of constitutional value requires that, in the absence of dispossession, compensation for harm lies within the jurisdiction of the civil judge.' This apparent limitation of the role of the judiciary came in for criticism, and has been rectified by paragraph 23 of DECISION 31, *TGV Nord*. In that case the *loi* allowed the administration to expedite the expropriation process by taking possession of the property by decree, after a favourable opinion of the Conseil d'État. The normal procedure for an expropriation order had then to be begun before the civil judge within a month of fixing the amount of compensation. The Conseil did not disapprove of this, but set out a new principle: 'that, thus, in any case, the importance of the functions conferred on the judicial authority in relation to immovable property by the fundamental principles recognized by the laws of the Republic is not disregarded'. Commentators have some difficulty in discerning which laws of the Republic are referred to, but it seems clear from this that the traditional functions of the judiciary with regard to the expropriation of immovable property are to be treated as being of constitutional value. Here the Conseil merely attaches a constitutional value to principles established in both administrative and civil case-law.[119]

[115] e.g. see *AJDA* 1991, 357–62, 398–403, 911. See also CE Sect., 15 Dec. 1989, *Société Métropole TV*, *JCP* 1989, II. 21455.

[116] See generally T. Renoux, *Le Conseil constitutionnel et l'autorité judiciaire* (Paris, 1986).

[117] CC decision no. 85-189 DC of 17 July 1985, *Rec*. 49: it is for the administration to decide whether land can be built on.

[118] CC decision no. 85-198 DC of 13 Dec. 1985, *Rec*. 78.

[119] See P. Bon, *RFDA* 1989, 1018, and Colly, 'Le Conseil constitutionel', 188–9: CE Ass., 12 Feb. 1960, *Fédération algérienne des syndicats de défense des irrigants*, *Leb*. 129. The decision merely confirms the published views of Luchaire (then special adviser to the President of the Conseil), *Droits et libertés*, 283, and Genevois (its Secretary-General), §416, as well as Colly, 'Le Conseil constitutionel', 192.

Due Process

The French Government and judiciary have been concerned about the length of delay before the courts. The written character of many proceedings, and the careful gathering of evidence before the judgment hearing can extend the proceedings inordinately. Indeed, France has recently been held to be in breach of the European Convention on Human Rights because of the length of time taken for the hearing of an administrative case.[120] Over a number of years attempts have been made to remedy this by introducing expedited procedures. The Conseil has, on the whole, accepted the legitimacy of this objective, while insisting that equal treatment is secured for both litigant and accused. Generally, concern has centred on the fairness of criminal procedure.

The first allusion to due process came in relation to criminal penalties, with a reference to 'the rights of due process, such as follow from the fundamental principles recognized by the laws of the Republic, applicable in criminal cases'.[121] It has its origins in a general principle recognized by the Conseil d'État in 1944.[122] As such, its requirements could only be altered by *loi* and not by *règlement*.[123] These early reference establish due process at least as a criterion for the division of legislative functions and for the interpretation of legislation.

The content of due process is defined by Luchaire thus: 'none may have a sanction imposed on his person, liberty, rights, or interests without a procedure permitting him to know with what he is charged, and to prepare arguments in his defence.'[124] Defined in this way, it seems of similar scope to the common law principles of *audi alterem partem*. But this would be too narrow an interpretation. It is true that a *procédure contradictoire* (an opportunity for both sides to put their case) is a minimum requirement for administrative procedures, but more is needed when a sanction is imposed, and it is here that the value of due process is invoked.[125] Due process thus extends to include the right to be notified, the right to apply for the suspension of an administrative decision pending litigation, and the right to an independent court.

The *CSA* decision notes that the right of due process applies not only to judgments by courts, but also to sanctions imposed by any administrative body.[126] At the same time, it applies only to decisions that

[120] ECHR, 24 Oct. 1989, *H.* v. *France, RFDA* 1990, 203.

[121] CC decision no. 76-70 DC of 2 Dec. 1976, *Work Accidents, Rec.* 39: penalties for breach of hygiene regulations. See also *Security and Liberty*, DECISION 18 §33.

[122] CE, 5 May 1944, *Trompier-Gravier, Leb.* 133.

[123] CC decision no. 72-75 L of 21 Dec. 1972, *Administrative Court Procedure, Rec.* 36.

[124] *Droits et libertés*, 395.

[125] Ibid. 303, 396.

[126] DECISION 28 §29. *Droits et Libertés*, 395, lists procedures requiring due process as including administrative expulsion (return to the frontier), tax demands and searches, withdrawal of licences, and disciplinary procedures.

finally affect rights or impose a penalty. Thus, the decision about the appropriate procedure through which a case is to be sent for trial is a purely administrative one, in that it does not finally settle rights, and so due process is not required.[127]

The 1976 *Work Accidents* case stipulated that employers should be notifed and be able to make representations before they are required to pay fines imposed on their employees for breach of health and safety legislation.[128] Similarly, where the *procureur de la République* was given the right to appeal against the decisions of the judge in charge of implementing criminal sentences, and the court was required to hear all interested parties, the Conseil interpreted the text as implying, 'consistently with due process', that the prisoner be notified of the appeal.[129] The full content of such notification was given in paragraph 29 of DECISION 28, *CSA*, of 1989, in which the Conseil stated that due process required that no sanction should be imposed on the holder of a licence to run a broadcasting station 'without the holder having been enabled both to present observations on the facts that are alleged against him and to have access to the file concerning him'. This could well involve full access to all the evidence against an individual, similar to that accorded to the accused's lawyers in a serious criminal case before meetings between the accused and the *juge d'instruction*.

The general principle under which the French administration acts is the *privilège du préalable*, the right to act, but, where it acts wrongfully, its decision can be annulled, or compensation may be payable.[130] In the case of financial penalties or decisions affecting a person's livelihood, the effect can be very serious. An appeal or an application for judicial review does not automatically suspend the administrative sanction or decision. But the rules of administrative court procedure provide that the court may order such a suspension, where serious and almost irreparable consequences would follow from failure to do so. In the *Competition Law* case of 1987 the Conseil insisted that such a right to apply for suspension of a sanction must exist against the decisions of an independent regulatory body, such as the Competition Council, since it was 'an essential guarantee of the rights of due process'.[131] Since the ordinary courts cannot suspend administrative decisions, to attribute jurisdiction over the decisions of the Competition Council to them rather than to the administrative courts was unconstitutional.

As in common law, the right to an unbiased court is seen as part of due process, and was acknowledged in paragraph 13 of DECISION 19,

[127] DECISION 18 §33; CC decision no. 86-215, *Criminality and Deliquency*, *Rec*. 130 §13–19.
[128] CC decision no. 76-70 DC of 2 Dec. 1976, *Rec*. 39.
[129] CC decision no. 86-214 DC of 3 Sept. 1986, *Criminal Penalties*, *Rec*. 128.
[130] On this *privilège du prélable*, see Brown and Bell, *French Administrative Law*, ch. 8.
[131] CC decision no. 86-224 DC of 23 Jan. 1987.

Terrorism Law. But in that case the Conseil did not think that the removal of a jury trial for terrorist offences breached the principle, since a panel of seven judges offered sufficient guarantees of independence and impartiality. Since the independence of the ordinary judiciary is guaranteed by article 64 of the Constitution, and the Conseil has treated the independence of administrative judges as a fundamental principle recognized by the laws of the Republic,[132] judges themselves are seen as constituting a guarantee of this aspect of due process.

The right of access to a lawyer is not always regarded as an element of due process. In paragraph 30 of DECISION 18, *Security and Liberty*, the Conseil did not deem unconstitutional a provision that failed to provide the accused with the right to be accompanied by a lawyer when being interviewed by the *procureur* prior to the latter deciding upon the next stage in the process leading to trial. The liberty of the subject was not at issue at this stage, and the absence of a lawyer at that point was current practice in criminal procedure. In part, as was noted in the case, the practice is justified by the fact that the *procureur* and the *juge d'instruction* are members of the judiciary, and their involvement provides the suspect with some guarantees in the pre-trial process. Luchaire argues that the suspect in a criminal or administrative process has the right to a lawyer, and that he should be provided with one if necessary.[133] But these aspects of due process have yet to be endorsed by the Conseil itself.

Equality before justice is often connected with due process. All the same, it is a distinct concept. Equality merely requires that the same procedure be followed for the same kinds of case, while due process can be more exigent in setting a standard for that procedure. On the other hand, equality can be applied to extend the requirements of due process—for example, in stating that the court should not decide a case on the basis of documents to which neither of the parties has had access.[134] In paragraph 27 of DECISION 18, *Security and Liberty*, the Conseil made a declaration of principle:

> Considering that, though by virtue of article 7 of the Declaration of the Rights of Man and of the Citizen of 1789 and of article 34 of the Constitution, the rules of criminal procedure are fixed by *loi*, it is permissible for the legislature to provide different rules of criminal procedure according to the facts, circumstances, and persons to whom they apply, provided that these differences do not proceed from unjustified discriminations, and that equal safeguards for the accused are ensured . . .

[132] CC decision no. 80-119 DC of 22 July 1980, *Validation of Administrative Decisions*, *Rec.* 46.

[133] *Droits et libertés*, 397.

[134] Ibid. 303. On equality before justice, see below, Ch. 6. 3.

Due process thus limits the scope of executive rule-making, but, within its competence, the legislature has a fairly free hand in determining what due process requires, with equality before justice comprising the principal constraint here.

The freedoms considered in this chapter are of widely differing constitutional status, and the protection accorded to them, therefore, is also different. It is important to distinguish between those values which are self-executing and which produce concrete legal rules or implications, and those which have more of the character of constitutional goals and which require further elaboration in order to produce concrete legal obligations. In the former case, the constitutional corpus provides the individual with a right that the legislature can abridge only for a limited number of reasons. Some of the most fundamental rights—the freedom of association, freedom of movement or due process, freedom of the home or property, as well as freedom from detention—can be abridged only on grounds of public necessity. These are subject to what Franck has called 'maximum control' by the Conseil constitutionnel.[135] This is seen, for instance, when the Conseil requires that the grounds for criminal liability be clearly specified. Other, less fundamental freedoms, such as the freedom to broadcast, to publish, to run private schools, to teach, to join a union, or to strike, are more easily abridge in order to reconcile them with a number of principles of constitutional value. Which principles of constitutional value may justify the abridgement of one such right varies according to the right in question, but it would seem that there are certain acceptable grounds in respect of each specific right that justify its abridgement. In addition, the legislature enjoys a significant discretion either to determine when public necessity should override a right, or to reconcile a right with other constitutional values. The Conseil will exercise 'normal' control here, intervening to ensure that the legislative provisions are not actually inconsistent with the constitutional values that they are supposed to be implementing, and cannot be interpreted in that way. All the same, a considerable margin of discretion is left to the legislature.

Finally, in other cases constitutional texts merely proclaim ideals such as the right to health, to work, or to participate in industrial enterprises, and the freedom of enterprise, and individual and

[135] C. Franck, 'L'Évolution des méthodes de protection des droits et libertés par le Conseil constitutionnel sous la septième législature', *JCP* 1986, I. 3256. Franck includes the right to join a union here, as well as freedom of the press. The decisions of the Conseil seem to me to justify a less extensive protection.

On the difference between 'maximum', 'normal', and 'minimum' control, see J. Bell, 'The Expansion of Judicial Review over Discretionary Powers in France' [1986] *PL* 99 at 103–7.

personal freedom have also been extracted from the corpus of constitutional materials. These generate no self-executing rights. It is up to the legislature to determine how to realize these ideals by creating concrete rights, and its freedom is therefore greater than in respect of self-executing rights. The Conseil will rarely strike down a provision because it has breached these ideals. Its 'minimum control' looks to ensure that there has been no manifest error in evaluation, nor any breach of equal treatment. These rights do, however, act as presumptions of what the legislator ought to be trying to achieve, and so will serve as principles for the interpretation of legislation. In addition, because such freedoms form part of the 'fundamental safeguards granted to citizens for the exercise of civil liberties', they fall within the legislative domain of Parliament, and cannot be restricted simply by decree.

Overall, as Bockel has suggested,[136] the Conseil typically exercises a kind of minimum control over legislation affecting fundamental freedoms. What has the Conseil contributed in this area? Its contribution has been threefold: to define the content and character of constitutionally protected freedoms; to specify the grounds on which their restriction may be justified; and to ensure that the legislature has not imposed restrictions on freedoms that are manifestly disproportionate to the objective to be realized. With regard to the first, the Conseil has engaged in major works of construction of a coherent set of constitutional rules and principles in some areas, notably media law and education, where the constitutional texts have been exiguous in content. It has clearly tried to establish some consensus values that stand above the varying demands of political opportunism. As Flauss puts it: '[b]y conferring constitutional status on rights whose legal value until then was still contested, the Conseil constitutionnel undoubtedly engages in a jurisprudential policy not lacking in boldness.'[137] In both the areas mentioned it has laid down principles that were immune to attempted changes by subsequent Governments. The second context is no less significant, in that it forces the legislature to provide reasons for restricting freedoms. Although it will usually be obvious from the preparatory materials when a provision has been proposed by the Government, this will not always be the case where amendments in Parliament are concerned. The requirement of adequate reasons does limit what deputies can achieve more often than it limits the Government, which will have been warned off such provisions by

[136] A. Bockel, 'Le Pouvoir discrétionnaire du législateur', in *Itinéraires: Études en l'honneur de L. Hamon* (Paris, 1982), 43. See also Flauss, 'Les Droits sociaux', 655. This would be comparable to the use by the German Constitutional Court of the principle of proportionality.

[137] Flauss, 'Les Droits sociaux', 647.

the Conseil d'État. The third role of a typically minimum control over the restriction of freedoms is significant, even if the ground for it is rarely invoked. More often than not, manifest error will give cause for review in relation to breaches of principles of equality. In other areas this will be the case more rarely, though there are instances of the penalties on taxation being found to be excessive.

Cynthia Vroom[138] has been more sharply critical of the role of the Conseil constitutionnel in the protection of civil liberties, especially in comparison with the Supreme Court of the United States. Most of her concerns relate to the Conseil's limited jurisdiction—its inability to review the constitutionality of criminal procedure in the ordinary courts, or even to review their application of its decisions, and its inability to review existing laws. These simply reflect the character of the institution, but she is also right to suggest that the French give a high priority to legal certainty. There is, however, another feature of the French scene that resembles that of the United Kingdom. Leo Hamon has suggested that 'the practical necessity of constitutional review appears less in a country where civil liberties are effectively defended by the democratic spirit and the vigilence of public opinion . . . The constitutional judge, however courageous and firm he may be, could not prevent [dictatorship such as that of Hitler]. On the other hand, he can not only save his honour, he can also sound an alarm and call upon public opinion to assess the gravity of the issues at stake.'[139] In the end, French expectations of the role of the Conseil constitutionnel are more limited than American expectations with regard to the Supreme Court.

[138] 'Constitutional Protection of Individual Liberties'.

[139] 'Contrôle de constitutionnalité et protection des droits individuels', D. 1974 Chr. 83 at 84 and 89.

6

THE PRINCIPLE OF EQUALITY

EQUALITY has stood at the core of the French republican tradition ever since the Revolution of 1789.[1] That event also played an important part in giving equality its central place in the western, liberal political tradition as a whole. Reaffirmed in subsequent Constitutions, it is now enshrined in article 2 of the Fifth Republic Constitution of 1958, which proclaims that the French Republic 'ensures equality before the law for all citizens, without distinction as to origin, race, or religion'. Given its long-standing prominence, it is not surprising that equality should have been a constitutional value that has been frequently invoked before the Conseil constitutionnel. The abundant case-law developed since the Conseil first invoked equality in December 1973 reveals both the conception of equality current in the French republican tradition and the extent to which the Conseil engages in the review of *lois* on substantive grounds.

Although 'equality' occurs alongside 'liberty' and 'fraternity' in the motto of the Republic, it does not feature prominently in leading treatises on French constitutional or administrative law.[2] Even in works on fundamental rights it has a limited place, and occupies few pages.[3] Constitutional texts have focused predominantly on the institutional arrangements of the Constitution, not on its political values. Indeed, constitutional writers have been much vexed with the question of whether values such as equality have any constitutional status at all.[4] The vagueness of such a notion has deterred many from treating it as a legally enforceable value rather than just a piece of political

[1] This chapter is a revised version of my article, 'Equality in the Case-Law of the Conseil constitutionnel' [1987] *PL* 426.

[2] It does not occur as a separate entry in the index to many leading treatises on constitutional law and political institutions: e.g. G. Vedel, *Droit constitutionnel* (Paris, 1949), J. Gicquel, *Droit constitutionnel et institutions politiques* (9th ed., Paris, 1987).

[3] e.g. J. Rivero, *Les Libertés publiques* (4th ed., Paris, 1984), i. 72. But see J. Morange, *Droits de l'homme et libertés publiques* (Paris, 1989).

[4] See e.g. L. Philip. 'La Valeur juridique de la Déclaration des droits de l'homme et du citoyen du 26 août 1789 selon la jurisprudence du Conseil constitutionnel', in *Études offertes à P. Kayser* (Aix, 1979), ii. 317; F. Luchaire, 'Un Janus constitutionnel: L'égalité', *RDP* 1986, 1229.

rhetoric. It is the attitude of the Conseil that has stirred interest in the question of equality as a constitutional value, and has caused leading writers to revise some of their opinions. While drawing on the republican tradition, the Conseil is, in a real sense, engaged in the process of defining the substantive content of the Constitution.

This chapter will argue that the Conseil's notion of equality is quite limited. Rather than a value of general application, equality is mainly concerned with non-discrimination in relation to specific constitutional values. Here, and in the wider range of cases where it is invoked, the concern of the Conseil seems predominantly to ensure a rational relationship between the provisions of a *loi* and the constitutional or legislative objectives that they are intended to achieve. The conception of equality is also limited, in that, in keeping with the ideology of 1789, it is essentially formal.[5] Equality before the law is concerned with the rejection of irrelevant differences and with the recognition of relevant differences in the categories in which the law places citizens and in the rules that it applies to them, rather than with requiring the legislator to seek substantive equality of outcomes, resources, or opportunities. The existence of substantive inequalities may be accepted by the Conseil as an adequate reason for formal differences in treatment in a *loi*, but they are not used by the Conseil to require action by Parliament. Such a limited theory merits attention not merely because of the checks that it reveals on French constitutional review, but also for the light that it sheds on the value of equality in the liberal political tradition. That equality seems almost reducible to more specific constitutional values or to a requirement of rationality in legislation would give support to those political writers who reject egalitarianism as part of the liberal political ideal.[6] One must ask whether an appeal to equality really provides an additional factor for the legislator to bear in mind.

6.1. The Constitutional Sources for Equality

The renunciation of privileges by the nobles and clergy on 4 August 1789[7] emphasized that special status achieved simply by inheritance or traditional social position had no place in liberal society. The rejection of unjustified privilege explains much of the power that the concept of equality retains. Privilege was countered by a notion of equality that in the words of the 1795 version of the Declaration of the Rights of Man, 'consists in the fact that the law is the same for all'.[8] At the same time,

[5] See Rivero, *Libertés publiques*, i. 73.
[6] See, among others, J. Raz, *The Morality of Freedom* (Oxford, 1986), ch. 9.
[7] See J.-P. Hirsch, *La Nuit du 4 août* (Paris, 1978).
[8] Declaration of 5 fructidor, an III (22 Aug. 1795), art. 3.

the revolutionary tradition did show some concern for the material inequalities of citizens. Fraternity directed society's efforts towards helping the poor and the disadvantaged members of society. For example, the 24 June 1793 version of the Declaration states in article 21 that: 'Public assistance is a sacred debt. Society owes subsistence to disadvantaged citizens, either in procuring them work or in ensuring the means of existence to those who are unable to work.' The Girondin proposals for a Declaration of Rights of 26 April 1793 contained the statement that 'equality consists in everyone being able to enjoy the same rights'. The republican tradition, which unites the various legal and political sources of the present French Constitution, encompasses both these conflicting approaches of formal and substantive equality. Privilege as such is outlawed, but inequality in treatment by the law is not. All turns on the reasons why differences are made. As we shall see from the case-law, equality does not necessarily require that similar people be treated the same, or that different people be treated differently. For example, the same solidarity payment can be due from those in receipt of a pension who take up paid employment, whether they receive a full pension or only a partial one, and the same insurance scheme may apply to two different professions.[9] The concept of equality is used essentially to designate prohibited grounds for differentiating between citizens with regard to the provisions of the law.

Given the tension between conflicting values within the republican tradition, it is not surprising that the current, Fifth Republic, Constitution of 1958 treats equality in a fragmented way. We have already noted that this Constitution is not a single, coherent document, but a series of texts drawn from various periods and reflecting the republican tradition in all its forms. The 1958 Constitution has little to say on the subject, apart from the promise of equality before the law contained in article 2. Equality in voting (article 3) is the only other directly relevant provision, though some would see the indivisibility of France (article 2) as ensuring equal treatment throughout the Republic. The principal text invoked in recent years is the Declaration of 1789. While article 1 proclaims generally that 'Men are born and remain equal in their rights. Social distinctions may only be based on public utility,' other provisions are more specific. Article 6 proclaims equal treatment by the law, both in protection and punishment, as well as equal access to public offices. Article 13 affirms equality in paying for the forces of public order and the expenses of the administration (in short, equality

[9] See respectively, CC decision no. 85-200 DC of 16 Jan. 1986, *Cumulation of Pensions and Salaries, Rec.* 9, and CC decision no. 84-182 DC of 18 Jan. 1985, *Receivers and Liquidators, Rec.* 27.

before public burdens). The whole of the 1946 Preamble is concerned with equality, especially racial and religious equality (§1), equality between the sexes (§3), equality before public burdens and national disasters (§12), and equal access to education and culture (§13). If, since the *Nationalizations* case in 1982,[10] appeal has been made to equality without reference to one of the above texts, this does not mean that they have lost their value, merely that equality is recognized as outstripping the confines of these specific provisions.

The Conseil's decisions since 1973 have begun to build these fragmentary texts into a pattern. A Socialist Government committed to remedying social inequalities was bound to provoke serious consideration of the Constitution's requirements in the name of equality, and the litigation since 1981 has done much to clarify the Conseil's conception of equality as a constitutional value. There are those like Luchaire, a former member of the Conseil, and now adviser to its President, who do not consider equality to be a general constitutional value in a strict sense, but think that it creates specific constitutional and 'paraconstitutional' principles, which are values that bind the legislator to a greater or less extent, depending on the subject-matter.[11] For him, constitutional values have a greater obligatory force. Others are prepared to grant equality the status of constitutional principle,[12] and it certainly does provide a reason, based on constitutional texts, for invalidating *lois*. Whatever their precise status, the specific equality provisions have the effect of requiring the legislature to demonstrate a serious justification, related to genuine differences in the situation or activity of persons, whenever discrimination is imposed. Elsewhere, differences in treatment are less suspect, but they must still have a rational relationship to the policy of the legislation. The meaning of equality will therefore be examined by considering first the specific areas designated by the constitutional texts on which the Conseil has been required to pronounce, and then by considering the wider application of equality before the law.

6.2. Equality between Whom?

Although equality is a principle proclaimed in all the constitutional texts, its importance depends on the groups towards whom equality of treatment is required. In numerous cases the Conseil has stated that

[10] DECISION 29.

[11] Luchaire, 'Un Janus constitutionnel', 1234, and see the discussion of paraconstitutional principles in ch. 2.

[12] See C. Leben, 'Le Conseil constitutionnel et le principe d'égalité devant la loi', *RDP* 1982, 295; F. Miclo, 'Le Principe d'égalité et la constitutionnalité des lois', *AJDA* 1982, 115.

equality does not prevent different treatment being accorded to those in different situations, and may indeed require it. Between whom does equality of treatment apply, and when do situations become different? If one goes back to the Declaration of 1789, its original purpose was not straightforwardly egalitarian. It sought to remove the privileges of the aristocracy and the clergy, and its promulgation was connected to the abolition of these privileges on 4 August 1789. Difference in treatment is not justified by reference to birth or social status, but by reference to the various situations in which people find themselves. As Rivero pointed out, equality is sought through differentiation according to situation, rather than through generalization.[13] The idea is of a formal equality—the same rule for all in the same category—not an equality of outcomes.[14]

Even with that, there is the important question of whether all are included in this widened notion of equality. The dominant view of legal writers until 1990 was that equality as a constitutional value applied only to citizens.[15] Many of the constitutional provisions talk of citizens—article 6 of the 1789 Declaration, for example, and article 2 of the 1958 Constitution. All the same, some constitutional values clearly do apply to foreigners. In earlier decisions the Conseil had held that principles of individual liberty applied to citizens and foreigners alike,[16] and that immigrants had the constitutional right to a normal family life.[17] At the same time, it was agreed that foreigners did not have equal rights to vote in France. Indeed, the debate during the 1988 election campaign on giving immigrants the right to vote presupposed that this would require a change in the Constitution. There is thus a distinction between *human* rights and the rights of the *citizen*. Only citizens have equal rights to settle in France; to vote or to accede to public office. On the other hand, everyone has an equal right to human rights.

DECISION 33, *Miscellaneous Social Measures*, of 22 January 1990 develops this line of thought. It affirms that citizens are entitled to equality of treatment only indirectly. Paragraph 31 states that 'with regard to foreigners, the legislature may make specific provisions on condition that it respects international agreements to which France is a party and the fundamental rights and freedoms of constitutional value recognized to all who reside in the territory of the Republic.' Equality in human rights is therefore indirectly expressed as a constraint on the

[13] J. Rivero, 'Les Notions d'égalité et de discrimination en droit public français', in *Travaux Association H. Capitant*, xiv. 343 at 350–1.

[14] See e.g. DECISIONS 34, 35.

[15] See *GD* 271.

[16] CC decision no. 79-109 DC of 9 Jan. 1980, *Immigration Law*, *Rec.* 29. But see CC decision of 25 Feb. 1992, *Entry of Foreigners*, *Le Monde*, 27 Feb. 1992.

[17] CC decision no. 86-216 DC of 3 Sept. 1986, *Entry and Residence of Foreigners*, *Rec.* 135.

legislature's freedom of action. But what of other rights? The Conseil's reply is essentially that, even here, there must be some objective reason for discriminating against foreigners. Article 24 of the *loi* in question provided that the supplementary benefits from the National Solidarity Fund would only be paid to foreigners according to the requirements of EC law or reciprocal international treaties. This amendment to the rules of the Fund, created in 1956, was passed without a vote in Parliament in order to comply with the decision of the European Court of Justice in *Mme Giletti*,[18] which had considered that the benefit represented a social security payment within EEC Regulation 1408/71 on the free movement of workers and their families, and, as such, should apply even where the recipient did not reside in the territory of a Member State. The Conseil objected to the formulation on its own motion. The policy of the Fund was to provide elderly persons without adequate resources with a minimum amount for living. In the Conseil's view, the policy applied equally to citizens and foreigners alike, whether or not there was any international or Community obligation. In other words, where the policy of the law—here to help the indigent elderly—did not of itself require discrimination between foreigners and citizens, the legislature could not introduce such a distinction. By contrast, an earlier decision had held that to require a minimum period of residence for eligibility for benefit from the National Solidarity Fund was not unequal treatment.[19]

The judgment in DECISION 33 is important, and followed recent decisions of the administrative courts on the eligibility of foreigners for various social security or local benefits.[20] In essence, it applies a general requirement of relevance to any discrimination between foreigners and citizens. In the case of rights of the citizen, the relevance of discrimination is constitutionally enshrined. In the case of human rights, then it is specifically excluded. In the vast bulk of cases the Constitution has little specific to say. Here Parliament is free to adopt any policy it likes, but it must be applied without arbitrariness, and this demands that any discrimination must at least be required by the policy of the legislation, even if it is not required by some constitutional value. To this very important extent, what the Conseil refers to as 'the principle of equality' applies generally to citizens and foreigners alike.

Given the political debate about the rights to be accorded to foreigners in France, the decision is an important step forward. All the same, it must be recognized as limited. Only a fortnight before,[21] the

[18] Case no. 379/85 of 24 Feb. 1987.
[19] CC decision no. 86-225 DC of 23 Jan. 1987, *Séguin Amendment*, DECISION 16.
[20] See B. Genevois, *RFDA* 1990, 406 at 413.
[21] CC decision no. 89-266 DC of 9 Jan. 1990, *Entry and Residence of Foreigners*, *RFDC* 1990, 326.

Conseil had held that the legal framework for the entry and residence of foreigners placed them in a different situation from nationals, and that this alone justified different treatment (in this case, a more favourable outcome in relation to the suspension of administrative decisions). The equality that the Constitution requires is thus not an exact sameness in treatment; simply being a foreigner suffices in many cases to justify a difference. By contrast, equality between citizens in access to public offices does require that others should be accorded different treatment. Thus, access to the civil service could be extended to all EC citizens on the same terms as French nationals.[22]

6.3. Specific Constitutional Equalities

Given the political climate in France over the last few years, it is not surprising that references by deputies to the Conseil have been predominantly concerned with the equality provisions of the liberal-inspired Declaration of 1789, rather than with the more socialist Preamble of 1946. The Conseil has been anxious to define the limits that equality imposes on the legislator's attempts to discriminate between individuals in order to remedy perceived inequalities in the material conditions of people. Although some specific inequalities have not been the subject of litigation, particularly those drawn from the 1946 Preamble, a fairly full picture emerges of the Conseil's approach, particularly of the extent to which the legislator has freedom to evaluate differences and to adopt measures to deal with them.

Equality in Voting

Prior to 1973, French constitutional writers had very little to say on equality, because they did not believe that it was a principle that restricted the freedom of Parliament, whatever limits it might impose on the executive. However, the one area in which equality impinged at the constitutional level was equality in voting: universal suffrage seemed to represent the proper consequence of the natural equality of citizens.[23] Article 3 §3 of the 1958 Constitution affirms that voting 'shall always be universal, equal, and secret'. It is not specifically stated, but this principle applies to voting for all public authorities, not merely to the elections envisaged by the Constitution. Although, at first sight, equality might not seem to add anything to the requirement of universality in voting, questions of equality in the number of votes, and of their relative worth have attracted the attention of the Conseil.

[22] CC decision no. 91-293 DC of 23 July 1991, *RFDA* 1991, 918, D. 1991, 617, note Hamon.

[23] See J. Barthélemy and P. Duez, *Traité de droit constitutionnel* (Paris, 1933), 65; Rivero, Les Notions d'égalité', 345.

The leading early case on voting rights is *Conseil de prud'hommes*.[24] A *loi* proposed to base employers' votes for nominees to the Conseil de prud'hommes (industrial tribunals) on the number of their employees. The Conseil held that inequality in voting rights could only be justified by a relevant difference in situation related to the objectives of the *loi*. Since the text was concerned with the election of members to a court, and not to a representative body, the number of employees was irrelevant. Likewise, the legislature does enjoy a discretion to adjust the weight of votes, but it cannot create an absolute rule that agricultural shareholders will always be able to outvote others, whatever the composition of the shareholding in an agricultural credit society. Such a provision was held to exceed manifestly the protection necessary for the special position of farmers.[25]

The value of a vote depends in large part on the constituency within which the vote is exercised. A voter in a large constituency may have less influence on the outcome than one in a small constituency. Questions about the size and composition of constituencies have taxed the Conseil several times in recent years. Exact equality in constituencies is rarely possible or desirable. Thus, although the Conseil constitutionnel has held that 'the National Assembly, nominated by universal suffrage, must be elected on essentially demographic bases', it has gone on to state that this principle may be modified by the legislator to take account of public interest requirements.[26]

In the electoral boundary cases of 1986 the Conseil made it clear that such departures from basic principle could only take place to a limited extent. This affected the enabling law that the Chirac Government had put before Parliament in the summer of 1986. The Socialists had changed the system of a simple majority vote in two rounds, with single-member constituencies, in 1985, and had reintroduced proportional representation. This had produced substantial representation for the National Front and the Communists, and had reduced the majority that the right-wing coalition would otherwise have had. The Chirac Government wished to secure a return to the previous system of voting before any election could be called. Accordingly, it passed an enabling law to empower it to redraw the electoral boundaries during the summer recess, and to have them enacted by *ordonnance*. In its decision, *Electoral Enabling Law*,[27] the Conseil constitutionnel upheld the constitutional validity of the enabling law, subject to a number of 'strict

[24] CC decision no. 78-101 DC of 17 Jan. 1979, *Rec.* 23.

[25] CC decision no. 87-232 DC of 7 Jan. 1988, *Mutualization of the National Agricultural Credit Bank*, *Rec.* 17.

[26] CC decision no. 86-208 DC of 1, 2 July 1986, *Electoral Enabling Law*, *GD*, no. 42 §20, and CC decision no. 86-218 DC of 18 Nov. 1986, *Electoral Boundaries*, *GD*, no. 42 §7.

[27] CC decision no. 86-208 DC of 1, 2 July 1986.

reservations of interpretation'. While not criticizing the criteria that the *loi* laid down for departing from demographic principles in drawing up constituency boundaries, their use was restricted by the Conseil's interpretation to those situations mentioned in the *loi* itself—namely, where cantons did not have a continuous territory, or where the population exceeded 40,000 inhabitants. In addition, grounds permitted by the *loi* for departing from the 20 per cent variation on the average population could only be exceptional, and had to be justified by requirements of the public interest. In any case, there could be no arbitrary drawing of boundaries.

These specific injunctions provided the President with the ammunition to refuse to sign the *ordonnances* setting out the new boundaries of constituencies on 2 October. The Government then presented a bill to Parliament to enact the provisions of the *ordonnance* in question. Again the Conseil's decision in July provided the basis for the reference by the Socialist deputies. In November 1986 the Conseil's decision on *Electoral Boundaries*[28] rejected the complaints of arbitrariness, and found that there was no evidence of a manifest disregard for constitutional requirements. The allegations that the new single-member constituencies closely matched the notoriously unfair boundaries prior to 1985 did not provide adequate justification for ruling any provision of this new *loi* unconstitutional.

The Conseil's approach to equality in the second case on electoral boundaries is instructive. It stated that its function was not to ensure that boundaries were as fair as possible, nor was its role to suggest better boundaries, as the Conseil d'État might do. Equally, the scope of its control was not as extensive as that of an administrative court, which would look more fully into the manner and content of an administrative decision. The Conseil's control is really of manifest inequality, not the securing of equality. It is the political process that must achieve that.

All the same, the Conseil will intervene to invalidate legislation when the limits to legitimate discrimination have been exceeded, as was emphasized in DECISION 35, *Elections in New Caledonia*, in August 1985. As part of a process of eventual decolonization, election to the new territorial assembly was to be by proportional representation, and the islands were to be divided into four areas. The boundaries were fixed so that the non-white populations dominated in three areas, while the whites were in the majority in the highly populated south. The Government was in favour of the independence of the islands, and had drawn up the boundaries and distributed the seats so as to minimize the influence of the white settlers, who were resisting

[28] CC decision no. 86-218 DC.

independence. Accordingly, the northern region (population 21,602) had nine seats, the centre (population 23,248) also had nine seats, the Loyalty Islands (population 15,510) had seven seats, and the south (population 85,098) had eighteen seats. The opposition complained that the Government was creating the equivalent of the eighteenth-century English rotten boroughs, and in paragraph 16 the Conseil upheld the objection to the distribution of seats:

> in order to be representative of the territory and its inhabitants in accordance with article 3 of the Constitution, the Congress, . . . must be elected on an essentially demographic basis; that, even if it does not follow that representation must be necessarily proportional to the population of each region, nor that other requirements of the public interest cannot be taken into account, these considerations can only be brought in, however, to a limited extent, which, in this case, is manifestly exceeded . . .

Thus, although Parliament enjoys some discretion in the application of the principle of equality in voting, an exact equality is not required; the Conseil will intervene in those cases where there has been a manifest error in evaluating the relevant differences. The differences between the regions did not constitute a rational justification for the solution adopted, which seemed to owe more to a desire to limit the number of white settlers who would be elected.

Equality of treatment requires not merely the prohibition of irrelevant discrimination, but also a respect for relevant differences, which was illustrated in a case where university professors had to be given distinct representation in the government of the university.[29] Although this *University Professors* decision might seem to suggest that the Conseil is concerned with the practical outcome of legislation, and not merely formal categorization, this is not really the case. This is confirmed by a judgment given the day before the *University Professors*, where it was argued that to restrict membership of various governing boards to representatives of the unions negotiating for the personnel of the corporations was unequal as regards the management staff, since they were unlikely to be chosen as representatives.[30] The argument failed; the Conseil did not even look into the substantive likelihood of equal chances of representation. The nearest the Conseil has got to paying attention to practical equality came in its decision on *Election Expenses*.[31] Here the funding of political parties was to be based on the number of votes cast in the previous National Assembly election. While this was not considered unconstitutional in itself, the limitation

[29] See CC decision no. 83-165 DC of 20 Jan. 1984, *University Professors*, discussed above, in ch. 5. 3.

[30] CC decision no. 83-167 DC of 19 Jan. 1984, *Credit Corporations*, *Rec*. 23.

[31] CC decision no. 89-271 DC of 11 Jan. 1990, *JO*, 13 Jan. 1990: 573.

that the party should have received 5 per cent of the vote in each of at least seventy-five constituencies was held to be so. The 5 per cent limit was likely to hinder the development of new currents of opinion. On the whole, however, the concern in this case is still with the justification of formal differences between persons. If, in funding political parties, the policy of the legislator is to ensure that electors have a proper choice, then the exclusion of minority parties is difficult to justify. Even if this does represent a concern for real equality in voting, the principle does not seem to extend much beyond parliamentary and local elections.[32]

In the end, equality in voting seems primarily aimed at achieving a rational relationship between voting procedures and the right to vote. Given the importance of this right, differences are subjected to careful scrutiny, at least for political elections.

Equal Access to Public Office or Employment

The revolutionary slogan was that 'careers are to be open to talents', not to birth or privilege. Accordingly, article 6 of the 1789 Declaration provides that 'all citizens, being equal [in the eyes of the law], are equally eligible for all public dignities, positions, and employment according to their abilities, and without any distinction other than that of their virtues and talents.' This is not a simple equality provision; it sets out positively the relevant criteria for selection to public office, a pure meritocracy. Other reasons are excluded so that appeal to equality is a way of focusing attention on relevant, constitutionally entrenched reasons for selection.

There is an abundant case-law for the Conseil d'État's application of this principle to entry to the civil service. Although not formally invoked by the Conseil d'État itself until the *Barel* decision of 1954, it had been used in arguments by *commissaires du gouvernement* since 1912.[33] This enabled some advances to be made in securing equal treatment for women in public employment (1936), and the prohibition of discrimination on grounds of religious beliefs (1912), even before the 1946 Preamble was adopted.[34] Decisions of the Conseil constitutionnel have carried on this line of development.

This aspect of equality has mainly been used with regard to entry to

[32] See e.g. CC decision no. 82-148 DC of 14 Dec. 1982, *Social Security Organizations*, *Rec.* 73, and no. 83-162 DC of 19, 20 July 1983, *Democratization of the Public Sector*, *AJDA* 1983, 614 §3. Cf. CC decision no. 87-227 DC of 7 July 1987, *Paris-Lyons-Marseilles*, *Rec.* 41, where the principles established in relation to national elections were applied to local elections.

[33] CE Ass., 28 May 1954, *RDP* 1954, 505, *GA*, no. 90; CE, 10 May 1912, *Bouteyre*, D. 1914.3.74, *concl.* Heilbronner, *GA*, no. 28.

[34] See CE, 3 July 1936, *Dlle Bobard*, D. 1937.3.38, *concl.* Latournerie; CE, 10 May 1912, *Bouteyre*.

the civil service. In 1976 the Conseil had considered that article 6 was not breached by a promotion board consulting a candidate's dossier before making its decision, provided that this applied to all candidates in the same category.[35] This idea was tested further in relation to the Socialist reforms of 1983–4 designed to break the monopoly of the École nationale de l'administration (the national civil service college, called ENA for short) with regard to entry to the higher civil service. A *loi* permitted one in five posts in the higher civil service to go to applicants who were not drawn from the ENA, but from local government, the unions, or public bodies, whose recruitment procedure was to be different from that of the ENA. The Conseil upheld this difference in entry procedures, on the ground that it was a recognition of different kinds of talents.[36] Furthermore, the Conseil permitted managerial restrictions to be placed on the categories of person who could enter through the special procedure—for example, that no assistant to a mayor of a commune of under 10,000 inhabitants could apply. Such measures, designed to limit the number of potential candidates to a manageable size, were acceptable unless there was any manifest error in evaluation. The only inequality of this kind that the Conseil found was in the provisions on seniority. The *loi* proposed that the period prior to entry into the civil service was to count for the 'external' candidates, but not for the *Énarques* (as the products of the ENA are known colloquially), which amounted to an unjustified discrimination. Seniority should simply be determined by the length of office in public service. The same solution was adopted with regard to a *loi* that permitted ambassadors to be integrated into the diplomatic civil service and to retain their rank, irrespective of the duration of their public service.[37]

Equality in admission is only one aspect of equality in the public service. Equality in career progression is illustrated both in the 1976 decision and, more fully, in the decision on *Retirement Age in the Public Sector*.[38] Here the Conseil recognized that civil servants of the same category should receive equal treatment as regards the terms and conditions of employment, but this did not prevent them from having different retiring ages according to the nature of their duties. Thus, the members of the *Collège de France* and the Vice-President of the Conseil

[35] CC decision no. 76-67 DC of 15 July 1976, *Civil Service Statute*, *Rec*. 35; Nicholas, 167–8.

[36] CC decision no. 82-153 DC of 14 Jan. 1983, *Civil Service Entry*, *Rec*. 35.

[37] CC decision no. 85-204 DC of 16 Jan. 1986, *Miscellaneous Social Measures*, *Rec*. 18; cf. decision no. 84-178 DC of 30 Aug. 1984, *Status of the Territory of New Caledonia*, *Rec*. 69, where difference in background was treated as relevant in much the same way as in the *Civil Service Entry* case.

[38] CC decision no. 84-179 DC of 12 Sept. 1984, *Rec*. 73. See also CC decision no. 86-219 DC of 22 Dec. 1986, *Organic Law on Judges of the Cour de cassation*, *Rec*. 172, and CC

d'État could legitimately have a later retiring age than other civil servants. The equal treatment that is required here is of a formal kind, and the Conseil merely controls Parliament's decision to accord different treatment, ensuring that it is justified by objective differences in the virtues and talents of the applicant. Such a formal approach permits a degree of reverse discrimination by enabling special entry procedures for persons of different backgrounds. But the refusal to countenance a derogation from the rule that prior service to the State in another capacity does not count towards seniority shows an unwillingness to move far from formal equality.

In this light, it comes as no surprise that clear policies of reverse discrimination are not allowed. For instance, a parliamentary amendment proposed to modify the election rules for local authorities, requiring that lists of candidates comprise no more than 75 per cent of persons of the same sex, thus ensuring a minimum of 25 per cent of women on all lists. In DECISION 34, *Feminine Quotas*, the Conseil raised an ex-officio objection to this provision. Basing itself on article 3 of the 1958 Constitution and article 6 of the 1789 Declaration, it stated in paragraph 6 that 'from a comparison of these texts, it follows that the status of citizen itself gives rise to the right to vote and to be eligible for election, on identical terms to all who are not excluded by reason of age, incapacity, or nationality, or for any reason designed to protect the freedom of the voter or the independence of the person elected . . .' It is interesting to note that one of the members of the Conseil, Vedel, had written in 1979, before he was appointed, that 'a text that reserved a certain number of places for women . . . without doing the same for men . . . would be contrary to the principle of equality.'[39] Although the 1982 text was less obviously unequal, the philosophy of the Conseil seems close to that of Vedel. Substantive arguments in favour of reverse discrimination did not justify inequality. It is not equality of chances within the law that is favoured, but rather formal equality. This approach may well pose problems in the future with regard to policies for the advancement of ethic minorities that have involved measures of positive discrimination and the creation of quotas for such groups in other countries.

Equality before Public Burdens

The principle of equality before public burdens is usually related to article 13 of the Declaration of the Rights of Man, which states: 'For the

decision no. 86-220 DC of 22 Dec. 1986, *Retirement Ages and Recruitment of Civil Servants*, *Rec.* 174.

[39] Cited in J. T. S. Keeler and A. Stone, 'Judicial–Political Confrontation in Mitterrand's France', in G. Ross, S. Hoffmann, and S. Malzacher (eds.), *The Mitterrand Experiment* (Cambridge, 1987), 171.

upkeep of a police force and for the expenses of the administration, common taxation is indispensable. This should be shared equally among all citizens, according to their means.' In its original formulation this is concerned with equality in taxation, although the case-law of the administrative courts has extended it into a general principle covering matters such as public property and governmental liability. However, much of the Conseil d'État's case-law refers to equality before public burdens, without reference to a specific constitutional text.[40]

The Conseil first relied on the principle of equality before the law in DECISION 32, *Ex Officio Taxation*, where it struck down a provision permitting tax inspectors to assess the tax due ex officio if they considered that the returns of the taxpayer were insufficient. Although the taxpayer was in general allowed to rebut this assessment by showing that he had no illegal or hidden resources, this procedure was not available where the income came not from revenue, but from capital, or from capital gains, or from gifts. This discrimination among taxpayers was held to be contrary to equality before the law. Although the decision did not make explicit reference to article 13 of the 1789 Declaration, which requires equality before public burdens, this idea is at least implicit. The case-law of the Conseil constitutionnel tends to combine these two aspects of article 13 as part of the general concept of equality before the law, but it is very close in inspiration to the position of the Conseil d'État. Perhaps the clearest application of this principle is the decision on the *Cumulation of Pensions and Salaries*.[41] The *loi* proposed an increase of 40 per cent (from 10 per cent to 50) in the solidarity payment due from retired persons who, while in receipt of a pension, take up paid employment. This was to be paid by both employer and employee if the employee's combined income from his pension and salary was more than 2½ times the national minimum wage (the *salaire minimum interprofessionnel de croissance* (SMIC)). In an argument raised of its own initiative, the Conseil recalled that article 13 required the contribution to the common good to be according to the resources of individuals, and this did not prevent the legislature from imposing on one social group the burden of helping one or more other social groups, in this case the unemployed. All the same, the level of payment here amounted to 'a clear breach of the principle of the equality before public burdens of all citizens', and so the 50 per cent figure was declared unconstitutional. The burden of relieving social problems should not fall disproportionately on one section of society, even though there may be an obligation on that section to contribute to the cost.

[40] e.g. the leading decision on fiscal equality, CE, 1 Feb. 1944, *Guieysse*, *Leb*. 45; *RDP* 1944, 158 *concl*. Chenot, note Jèze.

[41] CC decision no. 85-200 DC of 16 Jan. 1986, *Rec*. 9.

The Conseil had an opportunity to consider this principle again in 1989, in connection with the French equivalent of local government taxation on citizens, the *taxe d'habitation*, or residence tax. Traditionally based on the rental value of property, this tax was to be modified by income-related scales and by local reductions in the rate of tax payable. A ceiling of 4 per cent of the income of the lowest category of taxpayers was introduced. Beyond this threshold, the tax was a progressive levy on the rental value of first and second homes, with a higher levy on second homes, and an even higher levy on expensive second homes. The opposition argued that the tax should be borne equally, according to citizens' means. But the Conseil held that the scale of taxation was not contrary to equality: 'despite the effects of thresholds inherent in the classifications adopted by the legislator, such provisions do not involve a clear breach of equality between taxpayers, having regard to the fact that the maximum applicable rate remains limited . . .'[42]

Whether a burden is called a 'solidarity payment' or a 'tax', the principles are the same. As the Conseil stated in another case, it is the nature of taxation measures to be unequal 'for reasons of the public interest that it is up to the legislature to evaluate'.[43] This applies both to fixing tax-rates and to determining the categories subjected to taxation. Thus, in its decision on the *Finance Law for 1982*[44] the Conseil did not object to property belonging to a spouse or to a child being included within the determination of a person's liability to wealth tax. The idea of a 'fiscal household' was perfectly acceptable, and did not infringe the principle of article 13 of the 1789 Declaration that taxation was to be based on a person's resources. Parliament has a similar discretion in determining the relevant differences for tax purposes. Thus, in the same case no objection was taken to the exclusion of professional property from the wealth tax if it amounted to at least 25 per cent of the total shareholding in the firm. The difference between major and minor shareholdings was acceptable, because major shareholders held them as part of their instruments of work. On neither of these provisions had there been a manifest error in evaluation. This approach of leaving a wide discretion to Parliament to determine the relevant differences for tax purposes can be seen in many other cases.[45]

[42] CC decision no. 89-268 DC of 29 Dec. 1989, *Finance Law for 1990*, *RFDA* 1990, 143, note Genevois; *RTDC* 1990, 122. See also CC decision no. 90-277 DC of 25 July 1990, *Local Direct Taxation*, *RFDA* 1991, 354.

[43] CC decision no. 86-209 DC of 3 July 1986, *Rectifying Finance Law for 1986*, *RDP* 1986, 1457, §17.

[44] CC decision no. 81-133 DC of 30 Dec. 1981, *Rec.* 41.

[45] e.g. the Conseil has approved the exclusion of electricity plants from an exemption from local taxation for newly established firms (CC decision no. 79-112 DC of 9 Jan. 1980, *Local Direct Taxation*, *Rec.* 32, D. 1980, 420, note Hamon), the exclusion from a special alcohol and tobacco tax of firms with an annual turnover of under 50 million F (CC

Thus, to take a recent example, even if a tax to the State is payable throughout the country, it is permissible for the legislator to allow local authorities to fund local exemptions or reductions in the rates paid by taxpayers. Thus, no breach of equality was found when such local reductions were allowed in relation to the *taxe d'habitation* on primary and secondary homes.[46]

The Conseil will ensure that differences, when made, are not inappropriate to the purpose of the legislation. For example, while permitting a tax amnesty for those who had illegally expatriated funds during a period of currency restrictions, it did declare unconstitutional a provision that reduced the period in respect of which the revenue authorities could go back over an individual's tax affairs. The provision related only to those whose income came exclusively from a salary, and did not apply to persons with even the smallest amount of investment income. The difference of category was too marked in relation to the purpose of relieving the tax burden on individuals, and was thus contrary to the principles of equality.[47]

Outside the area of taxes and charges, equality before public burdens provides the basis for governmental liability. At the end of the last century leading administrative lawyers disagreed as to whether the Declaration of 1789 or analogies with private law provided the principles for governmental liability.[48] In truth, there is real doubt about the extent to which the principle that an individual should be compensated for harm suffered for the benefit of the community can be drawn from the 1789 Declaration. As Mestre shows, the principle can be found in pre-revolutionary public law.[49] The liability of the community for defects in policing that led to damage from riots, or for the costs of billeting troops was recognized in the Middle Ages. The same applied to the occupation of private land and the extraction of materials for war, much as was stated in the English *Case of the King's Prerogative in Saltpetre* of 1605. As the eighteenth-century texts of Provence stated:

decision no. 82-152 DC of 14 Jan. 1983, *Social Security*, *Rec.* 31), differential rates of taxation for alcohols of different strengths (ibid.), an increase in tax allowances, where gifts are made to *fondations* or associations set up in the general interest that have a cultural function (CC decision no. 84-184 DC of 29 Dec. 1984, *Finance Law for 1985*, *Rec.* 94); and, equally, it has approved the refusal to distinguish between municipal and other water services as regards the application of VAT (CC decision no. 84-186 DC of 29 Dec. 1984, *Water Charges*, *Rec.* 107). See also CC decision no. 90-285 DC of 28 Dec. 1990, *Finance Law for 1991*, *RFDC* 1991, 136 (net-income basis for non-salaried taxpayers and gross-income basis for salaried are not unequal).

[46] CC decision no. 89-268 DC, *Finance Law for 1990*.

[47] CC decision no. 86-209 DC, *Rectifying Finance Law for 1986*.

[48] cf. J. Barthélemy, *Essai d'une théorie des droits subjectifs des administrés* (Paris, 1899), 150, and E. Lafferière, *Traité de la juridiction administrative* (2nd edn., Paris, 1896), 189–90.

[49] J.-L. Mestre, *Introduction historique au droit administratif français* (Paris, 1985), esp. no. 161.

'When an inhabitant suffers for the whole community, or provides something for this body, he should be indemnified. It is not the same where he only suffers from a burden imposed on every inhabitant.' Either the Declaration was merely a statement of best current practice (as is the case on the expropriation of property, mentioned in article 17), or it was simply a text to which longer-established ideas became attached for want of another specific peg on which to hang them.

Although it has been accepted in administrative law that the principle applies to losses caused by legislation, even in the absence of a specific provision, Errera points out that this is rarely invoked successfully before the administrative courts.[50] It is thus up to the legislature to make proper provision for compensation, and the Conseil constitutionnel has to ensure that it does indeed do so. The *Eiffel Tower Amendment* case[51] illustrates this. A Government amendment authorized newly created television corporations to instal and use transmitting equipment on private or public buildings. (In particular, the Government had in mind the Eiffel Tower, which belongs to the city of Paris). The Conseil held that the compensation provisions of the amendment were too narrow, since compensation was only paid for damage resulting from the installation of the equipment that occurred within a limitation period of two years. Neither subsequent damage, nor any interference with enjoyment as a result of the installation of the equipment would be recompensed, and this amounted to an inequality before public burdens. Equality here required that an individual should not bear an excessive burden for the collective good. The principle is one that is generally applied in administrative law, so the case is really an instance of inadequate justification for a derogation from the normal principles of law applicable to all citizens.

Equality before the Judicial Process

While the aspects of equality discussed above constitute fairly extended applications of equality before the law, equality before justice would seem to touch on a fundamental concept. It covers both equality before the judicial process and equality before legal penalties. The right to due process is proclaimed by article 7 of the 1789 Declaration, while the insistence that the law, including its penalties, is the same for all is

[50] R. Errera, 'The Scope and Meaning of No-Fault Liability in French Administrative Law' [1986] *Current Legal Problems* 157 at 160–1; CE, 14 Jan. 1938, *La Fleurette*, *Leb*. 25; *GA*, no. 58.

[51] CC decision no. 85-198 DC of 13 Dec. 1985, *AJDA* 1986, 171. The principle is limited to direct interference with property rights, and does not extend, for example, to legislative provisions varying the priority rights of creditors in bankruptcy: see decision no. 84-183 DC of 18 Jan. 1985, *Receivership of Companies*, *Rec*. 32.

contained in article 6. The two will be considered successively with regard to their provisions on equality.

DECISION 36, *Single Judge*, of 1975 demonstrated a fairly strict version of equality before the judicial process, by refusing to allow the President of a Tribunal de grande instance to decide whether a criminal case merited a single judge or a panel of judges. Such a choice would permit unequal treatment. However, the increasing burden of cases has led all French courts to develop more summary procedures for less important cases, and this movement seems to be reflected in a softening of the Conseil's position. Thus, in DECISION 18, *Security and Liberty*, allowing the *procureur de la République* to choose between sending a case to the investigating magistrate or directly to the Chambre d'accusation was considered to be constitutionally acceptable. The reasoning of the Conseil is interesting, in that it argued in paragraph 27:

> though, by virtue of article 7 of the Declaration of the Rights of Man and of the Citizen of 1789 and of article 34 of the Constitution, the rules of criminal procedure are fixed by *loi*, it is permissible for the legislature to provide different rules of criminal procedure according to the facts, circumstances, and persons to whom they apply, provided that these differences do not result from unjustified discrimination, and that equal safeguards for the accused are ensured . . .

This is far weaker in tone than the *Single Judge* decision, since it leaves the legislator with considerable flexibility. However, there is a limit imposed by the need to justify discrimination. For instance, in the same *Security and Liberty* case the Conseil struck down a provision that permitted a civil party to join in a criminal trial to claim damages either at the trial stage or on appeal. In some cases there would be two levels of decision, and in others only one—and this at the whim of the civil party. Such a distinction was prejudicial to the accused, and had no sufficient justification. Equally, in DECISION 19, *Terrorism Law*, the Conseil held that it was not contrary to the Constitution for there to be special procedures (including no juries) for terrorist offences, because of their distinctive character. But the extension of these measures to other serious crimes was not justified by this or any other special reason, and so the law was *pro tanto* unconstitutional. A similar approach was adopted with a provision that allowed tax authorities a more advantageous regime of time-limits than those that applied to other litigants, and the Conseil constitutionnel struck this down.[52] Given the nature of the interests at stake, it is not surprising that the Conseil looks more closely at the substantive equality of chances that

[52] CC decision no. 89-268 DC of 29 Dec. 1989, *Finance Law for 1990*, *RFDA* 1990, 143 §§61, 62.

litigants will have, and is keen to intervene where there is a potential disadvantage in the diversity of procedures.

Equality before Penalties

Article 6 of the 1789 Declaration proclaims that the law 'most be the same for all, whether it punishes or protects'. From this can be deduced the principle that the penalties imposed by the law should be the same for all offenders of the same kind, a consequence that would also follow from the objectives of retribution or deterrence in punishment. It is unnecessary to mention all the cases where this principle has been discussed, but it is useful to raise some significant points—namely, equality in the categorization of persons to be punished, equality in the penalties imposed, and equality in the application of those penalties. In addition, there is the vexed question of amnesty laws.

As in other aspects of equality, the choice of categories into which offences are placed has an element of arbitrariness, but the Conseil has been willing to accord a fairly wide margin of discretion to the legislature, particularly in non-criminal laws. Thus, in the *Press Law* decision[53] it upheld the choice of 15 per cent as the maximum market share to be held by a single individual, even if this was only a rough and ready figure and 12 or 20 per cent might equally well have been chosen instead.

The same indulgence is seen in relation to categorization by the legislator as a basis for civil liability. For example, in the *Trade-Union Immunity* case[54] the Conseil was prepared to accept that unions could be subject to a different regime of civil liability from others, but not that this could lead to their effective exemption from liability altogether. The Conseil merely requires a justifiable difference between the persons in each category.

Even in the case of criminal laws, the Conseil does seem loath to challenge the validity of the categorizations made by Parliament. Thus, in the *Suppression of Rape* decision[55] the Conseil refused to strike down a provision that punished indecent acts with a minor under 15 years of age, but punished indecent or unnatural acts with a minor of under 18 years of age if the parties were of the same sex. The Socialist opposition argued that this was treating homosexuals differently from heterosexuals. The Conseil rejected the claim, stating that the legislator was entitled to discriminate in those instances where the acts con-

[53] CC decision no. 84-181 DC of 10, 11 Oct. 1984, *AJDA* 1984, 684, note Bienvenu.

[54] DECISION 24. See also CC decision no. 89-260 DC of 28 July 1989, *Commission des opérations de Bourse*, *RFDA* 1989, 671, §§43–7.

[55] CC decision no. 80-125 DC of 19 Dec. 1980, *Rec.* 51.

cerned were different in kind, and so could distinguish between acts done by members of the same sex and those done by a member of the opposite sex. There was no unequal treatment in relation to the same kind of act.

Although the legislature is given a fairly free hand in the definition of an offence or a ground of civil liability, article 6 of the 1789 Declaration operates much more as a limitation on discrimination in the penalties imposed. Even though it has discretion to fix the appropriate level of penalty, the same penalties must apply for the same kinds of offence. For instance, in DECISION 18, *Security and Liberty*, the *loi* reduced the penalties to be imposed for certain offences, but this was only to apply to crimes committed after it came into force. Paragraph 71 of the Conseil's decision struck this down on the ground that equal treatment required that the lighter sentences should be applied to all cases judged after the *loi* came into force. This is a very strong case, and points to the view that discrimination in sentencing should come primarily from the individuation of the penalty by the judge. The limit on equal treatment imposed by this *loi* bore no real relationship to the goal of securing appropriate penalties, but was dictated purely by administrative convenience, a value that should have a low priority when individual liberty is at stake.

Although Parliament enjoys considerable discretion in fixing the appropriate level of penalties, this is limited by the requirement that there should be proportionality between the crime and the penalty, a requirement that the Conseil refers to by prohibiting any manifest error in evaluation. In most cases such a disproportion is not found to exist.[56] However, in the case of the *Finance Law for 1990*[57] it was proposed that a fine of 25 per cent of the total should be levied on non-businessmen who attempted to pay for goods and services of more than 150,000 F in value except by a crossed, non-negotiable cheque or by credit card. The idea behind the proposal was to exclude the possibility of tax evasion by payments in cash. All the same, the penalty was held to be disproportionate to the offence in question.

Once it comes to determining the regime of punishment for convicted persons, the Conseil is again happy to permit distinctions to be made, but within a framework of formal equality. Thus, in a decision of 1978[58] the Conseil agreed that the judge in charge of applying punishments should be given the discretion to decide whether a convicted person should be held in prison or in a more liberal establishment. As long as the possibility of a less strict regime was open to

[56] e.g. CC decision no. 86-215 DC of 3 Sept. 1986, *Criminality and Delinquency*, *Rec*. 130.
[57] CC decision no. 89-268 DC of 29 Dec. 1989, *RFDA* 1990, 143.
[58] CC decision no. 78-97 DC of 27 July 1978, *Criminal Procedure Reforms*, *Rec*. 31.

all prisoners accused of the same offence, and the measure was merely one of individuation, no unjustified discrimination was found.

Amnesty laws pose particular problems for equality before the law. Some will be punished for identical offences for which those amnestied will escape. At the same time, the public interest has frequently been served by amnesties to promote national reconciliation, for example, particularly on a change of regime, or, in due course, after the Liberation or the Algerian crisis. Under the Fifth Republic, it has become conventional for an amnesty to follow a presidential election—a sign of grace and favour to mark the beginning of a new era, just like the enthronement of a medieval monarch. Couched in general terms, there are few objections to such laws from a constitutional perspective, provided that they do not exempt particular individuals from the application of the general law of the land, which was the revolutionary complaint against James II in England.

The Conseil constitutionnel has generally left Parliament with a wide latitude to determine which offences should be punished and which should be amnestied. Amnesty falls within one of the specific categories of *loi* enumerated in article 34 of the Constitution. Parliament has considerable freedom of action in assessing what the public interest requires. This was made clear in the law reforming the local government of Corsica, where an amnesty for those agitating for local autonomy was granted. The Conseil held that equality did not prevent the amnesty from being limited to certain offences, provided that the categories of offence to be amnestied were defined in an objective way—in other words, not relying on the subjective discretion of the administration.[59]

More recently, the issue of amnesty has provoked more controversy. In the first set of cases the Conseil struck down the amnesty granted in successive laws to the Renault Ten. In the course of industrial disputes at Renault certain shop stewards had allegedly been involved in the use of violence and intimidation against non-strikers within the precincts of the factory itself. The ringleaders were dismissed by the management. The Communists supported the campaign for their reinstatement, and their inclusion within the *Amnesty Law of 1988*,[60] passed after the re-election of President Mitterrand, was one of the items in the bargaining to obtain Communist support for the minority Socialist Government of Michel Rocard. The provision in question sought to amnesty those who had been dismissed for offences com-

[59] CC decision no. 82-138 DC of 25 Feb. 1982, *Corsica*, *Rec.* 41; see also CC decision no. 89-265 DC of 9 Jan. 1990, *New Caledonia Amnesty Law*, *RFDC* 1990, 323.

[60] DECISION 3b, discussed in ch. 2. See also S. Wright, 'The French Constitutional Council: A Political Weapon in the Amnesty Arena' (1989–90) 14 *Holdsworth Law Review* 41.

mitted while acting as the representative of employees, unless they had been guilty of a serious fault amounting to assault or wounding, and entitling them to be reinstated. The Conseil did not take exception to the reinstatement nor to the amnestying of civil, as opposed to criminal or public, wrongs. However, the provision went too far in compromising the freedom both of employers to choose their employees and of the employees themselves, who might have to rub shoulders with those who had assaulted them. The *loi* placed undue burdens on the third parties and victims of the amnestied acts. Employees' representatives could legitimately be singled out for special treatment because of the difficulties of their functions, and this could justify the pardoning of some faults. But where serious fault that would amount to an abuse of functions was involved, the burden imposed on the employer, the victim of the abuse, exceeded manifestly the sacrifices that could be required of one person in the public interest. A further amnesty law was passed for the bicentenary of the Declaration of the Rights of Man in 1989. There was another attempt to obtain an amnesty for the Renault Ten. This time, mindful of the previous failure, provision was made for the courts not to apply the amnesty where this would create an undue burden on the employer. Again, in DECISION 4, *Amnesty Law for 1989*, the Conseil constitutionnel struck this down on the ground that the *loi* failed to take account of the burdens placed on employees who might have to work again with those who had assaulted them. The Conseil limits the right to grant an amnesty by reference to equality, but it is not so much a matter of equality of penalties, as equality before public burdens.

That equality before penalties is really not at stake here is seen with the new, infamous saga of the amnesty granted to electoral offences. Election irregularities had been noted in both the presidential and legislative elections of 1988, mainly concerning campaign payments by companies. A *loi* 'limiting electoral expenses and clarifying political activities' was passed by Parliament in December 1989. An amendment was proposed to the text, amnestying offences committed in relation to the financing of electoral campaigns and parties before 15 June 1989, except where there was personal profit, or where the offender was a member of the National Assembly, or where corruption of public officials was involved. There was no self-amnestying by members of the National Assembly. The Socialists voted for this, the centre-right UDF and UDC (Union démocratique du Centre) abstained, leaving the RPR and Communists voting against. In DECISION 37 the Conseil constitutionnel was presented with the argument that no consideration of the public interest or social appeasement justified this. The Conseil simply left the assessment of such a public interest to the legislator unless some manifest error was found. All the same, it did decide that it was unconstitutional to refuse to amnesty those who were members of the National Assembly at the time of the criminal acts. The policy of the legislation was claimed to be that no social appeasement would be achieved if members of Parliament were able to vote themselves an amnesty. The Conseil seized on this in paragraph 23, saying that this objective had no relevance to former members of Parliament, and their exclusion from the amnesty was unjustified on the principle of equal treatment.

Ironically, the limited exclusion of members of Parliament was the source of problems in the following spring. When the case of the false invoices returned in support of the pre-1989 election campaigners came to the Chambre d'accusation in Paris, of the thirty-six accused, only nine were sent for trial. Of the rest, thirteen were beneficiaries of the amnesty law, of whom four were former deputies. In addition, a former minister, M. Christian Nucci, was held to have no case to answer on various charges of fraudulent use of Government funds. Judges in the lower courts did consider that there was a case of unequal treatment before the law, and took the opportunity to pass low sentences on other criminals. Thus, on 26 April 1990 the Tribunal correctionnel of Vannes appealed to 'a recent development in the notion of public policy' to justify the imposition of a 30 F fine on a woman convicted of fiscal fraud of over 700,000 F.[61] Other judges followed suit by releasing a number of those accused of crimes against property.[62] In the words of *Le Monde* of 4 May 1990, it was a 'devastating amendment' in terms of its political consequences, which gave the impression of politicians corruptly defending their own backers and companions.

The Conseil's role in this case was very limited, and, if anything, compounded Parliament's policy options rather than called them into question. On the whole, the Conseil's concern seems to be with formal equality, not with the serious questioning of the policy options behind a text. Thus, when the amnesty for offences in overseas territories included those involving eighteen months' imprisonment, whereas the amnesty for the mainland was limited to sentences of twelve months, although the offences amnestied were otherwise identical, then the Conseil held there to be unequal treatment.[63]

6.4. Equality as a General Principle

The specific constitutional provisions on equality isolate areas in which unequal treatment at a formal level in the law will be viewed with

[61] *Le Monde*, 4 May 1990.
[62] Ibid. 8 May 1990.
[63] *Amnesty Law for 1988*, §§3–7 (not reproduced in DECISION 3b).

suspicion. Access to public office and to voting can only be governed by certain factors related to the talents of the applicant or to the nature of the representative relationship. Public burdens are borne in relation to resources, and penalties in relation to the acts committed and to the character of the offender. Procedure should be governed by the concern to do justice. Such reasons may justify or even require differences to be made in the legal categorization or treatment of individuals. Other reasons, such as public necessity, might justify discrimination, but such public interest reasons are rarely accepted. Equality is the appropriate treatment in those situations where the constitutionally permitted justifications do not argue in favour of discriminatory treatment. Given that article 6 of the 1789 Declaration, and article 2 of the 1958 Constitution speak of equality before the law in a general way, it might be thought that such a role is not limited to specific areas but is wider in its application, and this would seem to be borne out by decisions of the Conseil.

The principal authority on this point is Part II of DECISION 29, *Nationalizations*, in 1982. The *loi* in question proposed to nationalize the major private banks in French hands, as well as other important sectors of the economy. A number of its provisions were challenged on the ground of breach of equality before the law, under article 2 of the 1958 Constitution. In its statement of principle the Conseil fully accepted that different rules could be applied to different bodies 'where the non-identity is justified by a difference in situation and is not incompatible with the purpose of the *loi*'. Thus, Parliament could legitimately exclude foreign-owned banks, because to nationalize them would entail problems both in international law and in France's relations with other states. Equally legitimate were a *de minimis* provision excluding small banks with less than 1 milliard F. in deposits, and the choice of the top five commercial companies to be taken into public ownership along with the banks. The decision on the extent to which public necessity required nationalization was a matter for Parliament, subject to review by the Conseil only if there was a manifest error in evaluation. However, the decision to exclude co-operative and mutualist banks, which included such major banks as the Crédit Agricole, was an unjustified discrimination. If the purpose of the *loi* was to control the banking sector, then the exclusion of some major banks from nationalization constituted unequal treatment in the implementation of that purpose. Although this decision does not prevent Parliament from balancing several purposes within the same *loi*, it does require a degree of sincerity in the statement and pursuit of the purposes of the *loi*, and that these purposes justify any discrimination that the law creates. Given that the Conseil can draw on the parliamentary debates and on an unpublished memorandum of observations sent to it by the

Secretary-General of the Government, it should have no difficulty in discovering whether there is a plausible justification for any discriminatory provision. Only where the justification is absent or manifestly inadequate in relation to the purpose of the legislation, will unequal treatment be found.

A similar approach can be seen when a *loi* seeks to derogate from some existing, generally applicable principle or rule of law in relation to a specific group. As the decisions in the *Eiffel Tower Amendment* and *Trade-Union Immunity* have demonstrated, such a derogation requires justification, and may be struck down if this is inadequate.[64]

Despite these decisions, it cannot be said that discrimination in a *loi* is, in itself, suspect. Provided that Parliament offers a justification related to the policy of the *loi*, the Conseil will nearly always defer to its assessment. Thus, in the *Credit Corporations* decision[65] the Conseil upheld an authorization for some, though not all, credit corporations to compete with banks in the provision of services, since those authorized, such as the Bank of France and the Posts and Telecommunications, offered stronger guarantees of security for investors. The reasons justified a difference of treatment within the activity of credit provision. By contrast, separate but similar professions could be subjected to the same insurance scheme if Parliament thought it appropriate.[66] It is not so much inequality that is the concern of the Conseil, as the rationality of the relationship between the measures proposed and the purpose of the *loi* or of the constitutionally protected values that they are designed to secure. As long as Parliament can justify the measures in terms of these objectives, and does not appear to have acted disproportionately, then the solution adopted will be accepted by the Conseil whether it involves equal treatment or not. Only where there is no rational relationship between the discrimination and the purpose of the *loi*, will a provision be invalid. For example, where different appointment procedures were used to engage doctors as heads of service and heads of department, the Conseil found no public interest reason to justify this discrimination, especially when the doctors were performing tasks in the same public service.[67]

In the *Conseil de Prud'hommes* decision of 1979, discussed above in Chapter 6.3, the Conseil constitutionnel invoked article 2 of the 1958 Constitution, and noted that the principle of equality did not prevent the application of different rules to those in different situations pro-

[64] See above, pp. 215, 217.

[65] CC decision no. 83-167 DC of 19 Jan. 1984, *Rec.* 23.

[66] CC decision no. 82-182 DC of 18 January 1985, *Law on Receivers and Liquidators, Rec.* 27; for a similar decision before the Conseil d'État, see CE, 26 June 1959, *Syndicat général des ingénieurs-conseils, Rec.* 394; *GA*, no. 96.

[67] CC decision of 29 July 1991, *Hospital Reforms, Le Monde,* 31 July 1991; see DECISION 33.

vided that the difference of situation justifies this discrimination, and that the non-identity of rules is compatible with the purpose of the *loi*. Subsequent decisions confirm that this is basically a negative approach: the legislature has significant freedom in assessing relevant differences, and compatibility with the purposes of the *loi* is not very strictly examined. The suggestion by Leben that equality amounts to a principle of 'for similar cases, similar rules; for different cases, different rules', is too wide.[68] As Luchaire suggests,[69] the general application of equality would seem to have more to do with an overall prohibition on arbitrariness by Parliament or by those to whom it entrusts power. Where the Conseil is not persuaded that legislation entails or may entail arbitrariness, then it has no objection to unequal treatment.

6.5. Equality in French Constitutional Law

In 1965 Jean Rivero wrote that:

> the equality to which French public law is attached is a legal equality, not a factual equality. The law has fulfilled its duty towards equality when it is the same for all. That the factual situation in which each subject of the law is placed makes it beneficial to one and ruinous for another, is something of which the law cannot take account: it is none of its making.[70]

Such he considered to be a continuation of the 1789 search for equality through the generality of the law. This study of the case-law of the Conseil constitutionnel would seem to suggest that the formal conception of equality through generality remains dominant. Where the *loi* treats citizens on the basis of formal equality, the Conseil seems only to intervene to impose a difference in treatment where this is required to promote a constitutionally protected value, such as the academic freedom of teachers. Where the *loi* treats citizens on the basis of formal inequality, then the Conseil seems more wary. It requires that the discrimination thus effectuated should be justified by a difference in situation determined in relation to a constitutionally protected value or in relation to the purpose of the *loi*. Given the pre-eminent position of the legislator in assessing what is required to promote the purpose of the *loi* or even to protect constitutional values, the Conseil will intervene only where the discrimination is not based on 'objective and rational criteria';[71] or is otherwise manifestly unjustified. In some areas, such as equality in voting, or equality before the judicial process,

[68] 'Le Conseil constitutionnel', 310.
[69] 'Un Janus constitutionnel'.
[70] 'Les Notions d'égalité', 349.
[71] A phrase first used in CC decision no. 83-164 DC of 29 Dec. 1983, *Finance Law for 1984*, *Rec*. 67 at 69.

or in the criminal penalties imposed, the Conseil is stricter in its control of the legislator's justifications. For the most part, the Conseil is content to ensure the genuineness of any differences in situation that the legislator is using to justify unequal treatment, and, since it only intervenes for *manifest* errors of evaluation, the investigation of the situations need not be very deep or detailed. Only where constitutional values are affected is there much attention to the consequences in legislative practice, but that serves more to confirm that formal inequality is indeed unequal treatment, as in the cases of legal procedure, than to reveal latent inequality in apparently justified equal treatment.

Such an approach does not mean that factual inequalities are irrelevant. First, there are specific constitutional values to be promoted, values that define individual rights, and may require, or at least justify, unequal treatment in legislation. Freedom of academics, and the right to strike could be given as illustrations of this. Secondly, the legislator is free to identify factual differences as the reason for introducing discrimination into legislation, provided that these are rationally related to the purposes of the legislation. The limited nature of a constitutional review operated through the principle of equality gives plenty of scope for the legislator to pursue what Rivero called 'equality through differentiation',[72] the application of different rules to different starting-points. What is perhaps more important here is the range of constitutionally permissible unequal treatment, rather than a constitutionally imposed requirement of equality. As Miclo argues,[73] equality really has the role of a residual principle of review.

The Conseil constitutionnel's conception of equality is a far cry from the notions of equality of resources, of welfare, or of opportunity, proposed by liberal theorists. For the Conseil, it remains a weapon of assault against a privilege (or disadvantage) conferred by law that is not justified by common utility or by another constitutionally recognized value. But in the modern Welfare State such a formal conception has limited attraction. Respect for others as equals requires attention to their concrete situations and inequalities. However, the Conseil uses equality to remind Parliament of the limited reasons that may justify interference with certain specific and constitutionally protected rights, and of the general requirement of rationality in lawmaking. Given that Parliament is pre-eminent in deciding what common utility requires, this is as far as the limited nature of constitutional review can push the Conseil. As a result, most of the arguments against *lois* that are based on the principle of equality fail before the Conseil.

72 'Les Notions d'égalité', 351.
73 'Le Principe d'égalité'.

As de Gaulle wryly remarked: 'the desire for privilege and the taste for equality [are] dominant and contradictory passions for the French of all periods.' The role of the Conseil constitutionnel is not to enforce equality, but to control the justifications for inequality. Its theory of equality ensures that inequalities are adequately justified by the legislator in terms of constitutionally acceptable reasons. Of course, it could be said that Parliament is already constitutionally bound to act for those reasons if it chooses to legislate in a particular way, and that the requirement of equality changes nothing in the process of reasoning or justification. However, the Conseil's approach of emphasizing the limited reasons that justify departure from certain constitutionally specified equalities, and the need for serious and rational justifications in other cases requires the legislator to take a hard look at the pattern of burdens and benefits that a *loi* creates. It demands especially careful justification for particular aspects of a policy. In that way, it has a special place in legislative reasoning.

The study of equality shows the way in which the Conseil constitutionnel has been 'fleshing out' the Constitution. It has taken specific values from the constitutional texts, and required the legislator to respect them. Such texts have been developed to acquire a significantly substantive content that reflects the republican tradition and the principles of public law. Thus, they represent important limits on what the legislator can do. Beyond this, the Conseil has more recently been moving towards preventing the abuse of legislative power by the adoption of rules that appear arbitrary or disproportionate to the objectives to be achieved. This opens up the spectre of legislative arbitrariness being replaced by that of the Conseil.[74] However, as yet, and despite the delicate nature of French politics, the Conseil has adopted a modest and prudent role, predominantly reaffirming traditional values. It is unlikely that equality will acquire the same place in constitutional litigation as it has in the United States, and that reflects as much on the role of constitutional review in French Government as on French conceptions of equality.

[74] See Luchaire, 'Un Janus constitutionnel', 1261; Leben, 'Le Conseil constitutionnal', 331.

7

CONCLUSION

THE Conseil constitutionnel has introduced a new dimension to both law and politics in France. It has an important role in setting the framework for what both law and politics can achieve, and its operation seems to attract widespread public support. But has this revolution in constitutional practice and attitudes produced a beneficient political system? Three criticisms are typically made of constitutional review: (1) that it leads to government by judges; (2) that it politicizes the judiciary; and (3) that it leads to legalistic politics. In addition, it may be asked whether the Conseil has operated beneficially in (4) defining the roles of Government and Parliament; and (5) in definding fundamental rights. Finally, it might be asked whether (6) there has really been much of a revolution; and (7) whether there are lessons from the French experience for other jurisdictions.

7.1. GOVERNMENT BY JUDGES

As Teitgen suggested in 1958,[1] constitutional review has always been criticized on the ground that it will lead to effective political decisions being taken by judges, not politicians. This fear has been supported by the analysis of some political scientists, but has been rebutted by most French lawyers. As the leader of the latter, Louis Favoreu, argues: 'government by judges' is a myth that does not correspond to political reality.[2] Judges do not take the effective decisions, but deal with questions of legality.

The argument of the political scientists has been put clearly by John Keeler and Alec Stone.[3] In the first place, a constitutional court is necessarily a *contre-pouvoir*, a centre of power that lies outside the control of the executive and the legislature. When a group fails to find redress through the latter two institutions of government, it looks to the Conseil constitutionnel as a third branch. Of course, the rules on

[1] *Avis et débats*, 77.

[2] See esp. 'Le Mythe du gouvernement des juges', Colloquium paper, Oxford, 16–17 Oct. 1987.

[3] J. T. S. Keeler and . Stone, 'Judicial–Political Confrontation in Mitterrand's France', in G. Ross, S. Hoffmann, and S. Malzacher (eds.), *The Mitterrand Experiment* (Cambridge, 1987), ch. 9.

who may make references to the Conseil are restrictive, and limit its role. All the same, it does act as a counterweight among the institutions of government. Secondly, and consequently, the Conseil can block policies that elected Governments wish to pursue. The Keeler and Stone analysis concentrates on the Socialist Government of 1981–6, but Stone takes this further, into the cohabitation period of 1986–8.[4] A number of important policies, such as those on nationalization, the press and the media, feminine quotas in local elections, and the electoral boundaries for New Caledonia, all came in for at least partial annulment on significant aspects. Indeed, the 1981–6 Socialist Government saw 51.5 per cent of the *lois* submitted to the Conseil annulled either in whole or part. They argue that the formal annulments by the Conseil are only the tip of the iceberg. Stone examines the procedure for the preparation of *lois* within the Government during the cohabitation period, and confirms his conclusion reached with Keeler, that Governments engage in a substantial amount of self-limitation in anticipation of reactions from the Conseil constitutionnel. In the *Nationalizations* case the Government had already increased the amount of compensation paid to the shareholders of nationalized companies as a result of advice from the Conseil d'État. As a result of DECISION 29 of the Conseil constitutionnel, the Government had to increase the compensation yet further. The lois Pasqua on criminal law and public order of August and September 1986 had to be toned down in significant respects, as did some aspects of privatization. Indeed, the proposal to create private prisons had to be dropped altogether. The Conseil is thus an obstacle to the implementation of certain important Government policies. Thirdly, Harrison notes[5] that amendment to the Constitution is a cumbersome process. It failed in 1984, over the proposal to allow the President to submit some *lois* to referendum, and in 1990, over the right of the Conseil d'État and the Cour de cassation to refer matters to the Conseil constitutionnel. In effect, the decisions of the Conseil constitutionnel are practically irreversible. Where does it get its legitimacy from? Its present functions, particularly on fundamental rights, were not approved in 1958; the Conseil simply arrogated this power to itself in 1971, and has continued to define its own role ever since. It may be that the Conseil constitutionnel now receives approval from 84 per cent of the population in opinion polls,[6] but does this really mean that people accept it as legitimate that the Conseil should strike down a *loi* proposed by their party?

[4] A. Stone, 'In the Shadow of the Constitutional Council: The "Juridicisation" of the Legislative Process in France' (1989) 12 *West European Politics* 12.

[5] M. Harrison, 'The French Constitutional Council: A Study in Institutional Change' (1990) 38 *Political Studies* 603 at 617

[6] See D. Maus, *RFDC* 1990, 470.

Favoreu, by contrast, takes the view that the Conseil constitutionnel is simply performing a normal judicial task of interpreting legal texts and applying them, albeit in abstract, to specific situations. Indeed, he sees the decisions of the Conseil as giving the seal of authenticity to changes in the law. Far from preventing changes in Government policy, the role of the Conseil in 1981–2 and 1986–7 was to secure and facilitate political change.[7]

The argument of some French lawyers stresses that the Conseil constitutionnel operates very much as a court. Given that, even in the recent past, other French lawyers were saying that it was not a court,[8] this defensiveness is not surprising on the part of those who wish to establish the study of the decisions of the Conseil constitutionnel as a truly 'legal' subject. First, they draw attention to the legal issues that the Conseil addresses, and the legalistic approach of its decisions. The arguments addressed to the Conseil, and to which its reasons reply, are legalistic in that they involve the interpretation of texts in a lawyerly fashion, seeking to draw rules from them without offering substantive reasons based on political argument. Bockel suggests that, while in its early years the main concern of the Conseil was to maintain a certain political balance between the institutions of government, the approach adopted today is very much a legalistic one, whatever the political implications of a decision may be.[9] Secondly, they put forward the 'transformation' thesis. They suggest that the need to frame arguments in a legal form for submission to the Conseil transforms the political debate into a legal one. A successful argument is one that provides a coherent and consistent reading of the legally binding texts. Indeed, like any other court, the authority and impartiality of the Conseil constitutionnel depends on its willingness to entertain arguments from all sides, and to respond to them without arbitrariness. Legal argumentation provides such a framework for the Conseil's neutrality, and, since the passion of political debate gives way to legal argument, the debate is transformed.[10] Thirdly, they point to the way

[7] 'Le Conseil constitutionnel et l'alternance' (1984) 34 *Revue française de science politique* 1002 at 1014–15; F. Luchaire, 'Le Conseil constitutionnel et le gouvernement des juges', Colloquium paper, Oxford, 16–17 Oct. 1987, §36, argues that the protection of fundamental rights also secures alternation between political groups, since it guarantees their position when in a political minority.

[8] See e.g. D. Tallon, 'The Constitution and the Courts in France' (1979) 27 *American Journal of Comparative Law* 567. Cf. L. Favoreu, 'Le Droit constitutionnel: droit de la Constitution et constitution du droit', *RFDC* 1990, 71.

[9] A. Bockel, 'Le Pouvoir discrétionnaire du législateur', in G. Conac *et al.* (eds.), *Itinéraires* (Paris, 1982), 43 at 48.

[10] J.-M. Garrigou-Lagrange, 'Les Partenaires du Conseil constitutionnel ou de la fonction interpellatrice des juges', *RDP* 1986, 647; L. Favoreu, *La Politique saisie par le droit* (Paris, 1988), 81; F. Morton, 'Judicial Review in France: A Comparative Analysis' (1988) 36 *American Journal of Comparative Law* 89 at 104.

in which decisions of the Conseil constitutionnel are increasingly treated as authoritative by ordinary courts and lawyers, not simply for the precise *ratio decidendi* of the decision, but also for the doctrinal theory of constitutional principles that it expounds. Indeed, it could now be said that the Conseil constitutionnel takes the lead among French courts in establishing principles on the operation of governmental institutions and on the protection of fundamental rights, to the extent that other institutions, such as the Conseil d'État, definitely play second fiddle. Fourthly, they note that the Conseil constitutionnel is very limited as a political actor. It has no power of initiative. It merely reacts to the issues that are submitted to it. In such a context, it does not enjoy anything like the political freedom of the Government or Parliament in shaping policy.[11]

Commenting on this debate, Davis rightly suggests that the French writers are frequently engaged in formulating a rather exaggerated contrast between law and politics, starting from a rather rigid conception of the separation of powers.[12] On the other hand, the (mainly) American commentators on the Conseil constitutionnel come from a tradition where law and politics are very closely tied together in the role of the Supreme Court.

Two issues are central to a proper analysis of the relationship between law and politics in the functioning of the Conseil constitutionnel. First, as much jurisprudential literature recognizes, interpretation is not simply an act of cognition, it is an act of will. The interpreter has to offer a version of the law that best fits the institutional morality of the constitutional system.[13] The interpretation offered for a decision fits an express or implicit constitutional theory about the proper role of the various institutions of government, the content of fundamental rights, and the proper form for their protection. To this extent, to adopt a particular interpretation of the Constitution is to take up a stance in the political arena that may well be controversial.

This is particularly so in France, because of the incomplete character of the formal constitutional texts. In numerous areas, such as parliamentary procedure, delegations of legislative power under article 38, and the 'fundamental principles recognized by the laws of the Republic', the Conseil has had to develop constitutional norms from fragmentary provisions. With particular reference to the protection of fundamental rights, Luchaire, a former member of the Conseil con-

[11] Bockel, 'Le Pouvoir discrétionnaire', 49.

[12] M. H. Davis, 'The Law/Politics Distinction: The French Conseil constitutionnel and the US Supreme Court' (1982) 34 *American Journal of Comparative Law* 45 at 88.

[13] R. M. Dworkin, *Law's Empire*, (London, 1986) 254–7, ch. 9; M. Troper, 'Justice constitutionnelle et démocratie', *RFDC* 1990, 31.

stitutionnel, has stated that, 'in ways that the Constitution of 1958 did not envisage, the Conseil participates in the construction of the constitutional edifice'.[14] But he goes on to show that, through 'reservations of interpretation' about *lois* of which the Conseil does approve, it effectively shares in the task of the legislature in drafting the rules to be applied. Thus, although limited by the constraints inherent in the nature of constitutional arguments, the Conseil constitutionnel, as interpreter and administrator of the Constitution, is an active participant in the political process. It sets out its interpretation as a framework within which the other political actors, the Government and Parliament, and their legislation can operate.

Secondly, a current member of the Conseil has argued that it is not a third chamber of Parliament, and is not, in that way, an integral part of the legislative process. Its ambition is merely 'to act in the wings before and after the vote in order to ensure respect for the rules'.[15] But as a body to whom appeal can be made by those who feel that they are losing out in the political process, usually the opposition, it is an integral part of that process. It may only entertain references for breach of the Constitution, but these do have an important bearing on the outcome of political action. As Keeler and Stone point out, unlike the US Supreme Court, the Conseil makes its decisions very soon after the vote on the political merits of a *loi*. It receives references as part of the overall pattern of political action available to the opposition and the Government, and it is expected to decide whether certain reforms are constitutionally permissible or not. The Conseil has frequently stated that it does not enjoy a 'general power of judgment and decision identical to that of Parliament' (see DECISION 18 §8). But Luchaire[16] points out that this does not mean that the Conseil never looks at the merits of what Parliament has passed. The notion of 'manifest error in evaluation' gives the Conseil the power to intervene where Parliament has made a gross error of judgment, as in DECISIONS 35 and 37. As was seen in Chapter 6, the Conseil will also satisfy itself that there are good reasons for unequal treatment, and that discrimination is neither arbitrary nor unjustified. Finally, in adapting the law to overseas territories under article 73 of the Constitution, or in nationalization or expropriation under article 17 of the Declaration of the Rights of Man Parliament must demonstrate the 'necessity' for its actions. Its involvement in judging the merits of political policies may therefore be rather limited and typically negative, but this still makes the Conseil an important political actor.

By its rulings on what is constitutionally permissible, the Conseil

[14] 'Le Conseil constitutionnel et le gouvernement des juges', §§8–16.
[15] J. Robert, cited in *Le Monde*, 6 Feb. 1991.
[16] 'Le Conseil constitutionnel et le gouvernement des juges', §§17–21.

does sometimes prevent politicians from doing what they want: schoolteachers must respect the religious character of the school in which they teach (DECISION 22); the Renault Ten could not be amnestied (DECISION 3, 4); and there must be legislative guarantees of pluralism in the media (DECISION 26). All are obligations because the Conseil refused to allow previous *lois* to be repealed or amended. Constitutional review does mean that some political policies are off limits. But it does not simply act negatively in ruling out certain options, it also defines objectives and values that Parliament must pursue in some way, as in the case of the media.[17] If the members of the Conseil are responsible for drawing those boundaries—often, as in the media case, without a clear warrant from established constitutional texts—then the judges are responsible in part for the design of the political system that is allowed to operate.

In the end, the legal and the political are more closely connected than much of the French debate might suggest. 'Political' activity may best be described as an exercise of power giving direction to society.[18] Under such a description, the French Conseil constitutionnel is very much taking part in politics. But this is not to say that the debate in which it participates has the same partisan character as political debate in Parliament.

7.2. A Politicized Judiciary

From this situation, Stone rightly draws the conclusion that members of the Conseil constitutionnel cannot help but be political actors. The abstract review of the whole of a text, isolated from any dispute, immerses the Conseil further in the political process, since its decisions come in the heat of the political debate. Since it has to develop the rules of the political game in such a process, it performs a political function.

As was seen in Chapter 1, political controversy has occasionally been raised by appointments to the Conseil. More often, controversy focuses on its particular decisions, especially when they interfere with cherished political reforms, such as the *Nationalizations* decision of 1982 (DECISION 29), or the *Séguin Amendment* of 1987 (DECISION 16). Given the timing of decisions of the Conseil constitutionnel, it is inevitable that they will be the subject of partisan political comment, and will be seen

[17] See above, ch. 5.5. Similarly, C. Graham and T. Prosser argue that, while the Conseil constitutionnel did not prevent changes of policy in the cases of nationalization and privatization, it did have an important effect on the way in which they were implemented: *Privatizing Public Enterprises* (Oxford, 1991), 245–8.

[18] J. Bell, *Policy Arguments in Judicial Decisions* (Oxford, 1983), 6; R. Wassermann, *Der politische Richter* (Munich, 1972), 18.

by the public as victories or defeats for the Government of the day. The legitimacy of its decisions is necessarily questionable when, as justifications, the Conseil offers 'interpretations' of the Constitution that are by no means settled in legal doctrine, and may even have been constructed out of fragments of constitutional texts or 'fundamental principles recognized by the laws of the Republic'.

At the same time, Favoreu is right to point out that the Conseil tries to develop legally applicable principles and to apply them in a legalistic fashion. It acts very much like a court receiving a reference on a point of law. Its reasons have to be grounded in the interpretation of legal texts. In that way, it does not set out to attract political criticism.

These two views of the Conseil are not incompatible. It is a judge, but with an important role as an on-going drafter of the Constitution, however much it likes to dress up its decisions as 'interpretations' of texts. All the same, its members are not partisan politicians serving party political interest. It is fair to accept that most of its views on basic constitutional principles are unanimous, or at least do not follow party lines. As Luchaire puts it: '[unlike Parliament] in the Conseil constitutionnel, composed of courteous men who know each other well and respect each other, a certain consensus is sought quite understandably'.[19] Like a group of Privy Counsellors, they are representatives of the national political tradition, not of party political advantage, and this colours their approach. They will owe more allegiance to the Conseil than to their ex-colleagues in Parliament. After all, in recent times it is more like a political graveyard than the launching pad for a prominent political career. The role is political, but not in quite the way that Favoreu fears. The general public acceptance of the Conseil constitutionnel's role demonstrates that it is not widely viewed as a partisan political institution.

7.3. Legalistic Politics?

Favoreu sees the great achievement of the Conseil as introducing a kind of political convergence through the rule of law. The legal discipline of constitutional principles provides a common set of assumptions within which partisan politics must work, and political debate turns around the interpretation of legal texts.[20]

Stone's study of the cohabitation period does show how legalistic arguments are raised in Parliament by way of a preliminary point of order (the *exception d'inconstitutionnalité*), and bills are drafted to meet the anticipated legal requirements of the Conseil constitutionnel, often

[19] 'Le Conseil constitutionnel et le gouvernement, des juges', §30.

[20] See also Garrigou-Lagrange, 'Les Partenaires du Censeil constitutionnel', 651.

as interpreted by the Conseil d'État acting as the Government's legal adviser. As one minister put it: 'the legislator legislates under the shadow cast by the Conseil constitutionnel.'[21] Legalistic arguments about the constitutional implications of texts form a greater part of political activity and debate than in previous Republics. Indeed, the process of making references to the Conseil constitutionnel thrusts lawyers into the activity of political parties.[22]

But it would be an exaggeration to suggest that politics has become 'legalistic'. Yes, politicians are now able to add a new string to their bow. The preliminary point of order and the reference to the Conseil are new political strategies. Constitutional legal considerations also provide an additional hurdle for the Government in getting its reforms passed. But these additional features do not represent the introduction of a great legalism into politics. Politics continues to be about strategies and policies, about the direction to be given to society. Constitutional review, however stringent, operates on the margins of this process. For example, a finance bill of 130 articles might have two that cause serious problems for the Conseil. The main political debate about budgetary strategy rages relatively unhampered by the potential intervention of the Conseil.

The real difference today is that the final sanction for failure to comply with constitutional values is legal, not political, as it was in the Fourth Republic and still is in the United Kingdom. To that extent, law has entered the political process and given it a legalistic dimension, but it is not one that seriously alters traditional political activity.

7.4. Protector of the Government or Protector of Parliament?

The original function of the Conseil constitutionnel was to police the separation of legislative power established by articles 34 and 37 of the Constitution, both in relation to individual measures and also, in advance, by approving parliamentary standing orders. Breaches of procedural rules are still the main reason for the quashing of provisions in *lois* submitted to the scrutiny of the Conseil.[23]

It is the areas of *loi* and *règlement*, and in relation to finance law procedure, that the Conseil has produced its most complex and developed case-law. Nearly every finance law is referred to the Conseil, and the procedure of declassification under article 37 §2 ensures that the issue

[21] F. Léotard, cited in Stone, 'In the Shadow of the Constitutional Council', 19.

[22] See M. Charasse, 'Saisir le Conseil constitutionnel' (1986) 13 *Pouvoirs* 87.

[23] L. Favoreu, 'Les Cents Premières annulations prononcées par le Conseil constitutionnel, *RDP* 1987, 443 at 451, states that they accounted for 60% of the provisions quashed between 1959 and 1987.

of the boundary between *loi* and *règlement* is regularly brought to the Conseil. By contrast, the Conseil's decisions on other issues, such as fundamental rights, are sporadic.

The initial duty of the Conseil was to preserve the freedom of the Government to govern—to stay in power and implement its policies. Its role in approving parliamentary standing orders has confirmed this, in that, from DECISION 5 onwards, it has refused Parliament the power to address injunctions to the Government. It has refused to limit the use of the major steamroller powers to require Parliament to vote on a text as a block under article 44 §3, and to make an issue a matter of confidence under article 49 §3 (DECISION 15). Even though it does require a statement of the purposes for which the power to pass *ordonnances* is delegated, it does not set significant limits on when such delegations can occur. The Conseil's rather limited control seems to reinforce the major powers of Government.

All the same, within the new sphere of its operation, Parliament is protected against abuse of authority by the Government, most notably in the use of the power of amendment, as in DECISION 16, and in attempts to introduce budgetary riders.[24] To assess the role of the Conseil here, one has to look not at what a pre-1958 Parliament could have done, but at the powers given to the 'rationalized Parliament' in 1958. On the whole, as was seen in Chapters 3 and 4, the Conseil's interpretations have favoured an enhanced role for Parliament compared with what might have been anticipated in 1958, and they have allowed its role to overlap with that of the Government in legislative activity, as long as the latter does not object. Here the Conseil has undoubtedly been influenced by the more parliamentarian approaches, compared with their predecessors, of the Giscard d'Estaing and Mitterrand presidencies.

As Neuborne suggests,[25] much constitutional law can be seen as requiring special majorities in order to pass certain kinds of legislation. Indeed, constitutional theorists often see such an approach as necessary if judicial review is to have constitutional legitimacy.[26] As long as a special majority is reached, or, as in the case of France, as long as special procedures are followed, almost any reform can succeed. The Conseil constitutionnel merely charts the route that a reform must follow. In this way, the Conseil's role in procedural questions is central to its activity and to its legitimacy. All the same, it has to take decisions on procedure in relation to specific areas of social life. In some

[24] See above, p. 133.
[25] 'Judicial Review and Separation of Powers in France and the United States' (1982) 57 *New York University Law Review* 369 at 368–9.
[26] S. Freeman, 'Constitutional Democracy and the Legitimacy of Judicial Review' (1990) 9 *Law and Philosophy* 327 at 364.

areas, such as fundamental rights, the Conseil's interpretations of the Constitution make it more difficult for reform to occur than in others. Indeed, where it notes that a constitutional amendment is required to pass a reform, this will typically put an end to the practical possibilities of achieving the change, since the amendment procedure is rarely successful. Thus, the mere fact that, in the ultimate analysis, the Conseil's role may be predominantly as the guardian of procedural rules does not prevent it from having an important impact on substantive policies.

7.5. Protecting Fundamental Freedoms

Various justifications might be offered for a judicial role in protecting freedoms. The first might be that judges can protect citizens against elective dictatorship. The second would be that they can protect minorities from the adverse consequences of majority rule. The third is that they are concerned to curb the excessive zeal of the administration. The function of the Conseil involves all three of these objectives, though the greatest prominence is given to the latter.

Preventing Elective Dictatorship

Lord Hailsham's description of modern government as 'elective dictatorship'[27] is apt to describe the current French system of government in the Fifth Republic. A Government, even without a parliamentary majority in the National Assembly, can steamroller *lois* through Parliament by requiring the text to be voted upon as a block under article 44 §3, and by making the vote an issue of confidence under article 49 §3. The range of the matters covered by the 1986 enabling laws, especially privatization (DECISION 30), shows that a Government with a suitable majority can obtain the delegation of wide powers to legislate by *ordonnance* in areas that the Constitution reserves to Parliament, thereby dispensing with Parliament itself. The Conseil constitutionnel offered limited resistance to that decision, pointing merely to the obligation of the Government to respect constitutional values.

Under article 16, the Conseil constitutionnel has to give an opinion to the President before a state of emergency is declared, but the President is not bound by that advice, and it is clear that de Gaulle used the power in 1961–2 for longer than the Conseil thought necessary.[28] In some areas it has very much stood in the way of what the Government wished to do. DECISION 1, *Associations Law*, of 1971 is a clear

[27] *The Dilemmas of Democracy* (London, 1978), ch. 26.
[28] J. Boudéant, 'Le Président du Conseil constitutionnel', *RDP* 1987, 589 at 627–9.

example here. The same could be said of DECISION 35, where the Government wanted to give the Kanaks special treatment in the electoral boundaries of New Caledonia, DECISION 23, where it wanted to repeal provisions in the *loi Guermeur* requiring teachers to respect the special character of religious schools, DECISION 26, where it wanted to repeal the limits on press ownership, DECISIONS 3 and 4, where it wanted to amnesty serious wrongs committed during trade-union activities, and DECISION 24, where it wanted to grant an absolute immunity to unions during industrial action, and where it wanted to impose a 50 per cent tax on those who took up employment after they were in receipt of a retirement pension.[29] There is a hint in many of these cases of the Government trying to benefit its friends, or at least to harm its enemies, rather than to serve the national interest. The harming of left-wing organizations in the *Associations Law* case, and the benefiting of political allies in the union and press cases might be evidence of this, as would numerous administrative law cases concerning mayors of communes preventing rivals from being buried, getting planning permission, or holding religious processions.[30] In addition, the Conseil may strike down clearly oppressive provisions of a *loi* as constituting manifest errors in evaluation, but instances of this are rare. On the whole, however, preventing elective dictatorship is not a major feature of the Conseil's role.

Protecting Minorities

As was seen in Chapter 6, the French have a very formal ideal of equality: in form, the law must be the same for all, and this does not have regard to substantial outcomes. All that it really requires is that there must be come non-arbitrary reason for making a discrimination. Thus, in the *Suppression of Rape* decision[31] the Conseil constitutionnel refused to strike down a provision that discriminated between homosexual and heterosexual acts in the criminal law. Reverse discrimination for women in DECISION 34, *Feminine Quotas*, and for Kanaks in New Caledonia in DECISION 35 were not permitted. Substantive considerations in favour of reverse discrimination seem to have had little impact.

Leading French constitutional lawyers do not see protecting a social minority against the majority as the role of the Conseil constitutionnel. Majority rule as a conception of democracy is too entrenched to enter the Dworkinian empire of law, in which a constitutional court defends

[29] CC decision no. 85-200 DC of 16 Jan. 1986, *Cumulation of Pensions and Salaries*.

[30] See J. Bell, 'The Expansion of Judicial Review over Discretionary Powers in France' [1986] *PL* 99 at 103.

[31] CC decision no. 80-125 DC of 19 Dec. 1980, *Rec.* 51.

minorities against the excesses of majority voting power.[32] They are more commonly concerned to ensure that the political minority of the moment is not unduly disadvantaged in the political process. Loïc Philip suggests that the role of the Conseil ensures some protection against the tyranny of the majority:

> The current control over the constitutionality of *lois* is well suited to the situation of French political life. The separation of the majority and the opposition into two clearly antagonistic groups exacerbates political divisions. The power to make references on constitutionality that the opposition possesses gives it the guarantee that the majority of the moment cannot abuse its power beyond certain limits, and avoids it feeling completely disarmed in the legislative process.[33]

But the concern here is to ensure a reasonable sharing of political power over time, rather than the typical American concern of protecting entrenched social minorities, which Dworkin exemplifies.

Curbing Excessive Zeal

In many ways, the curbing of executive zeal is the most important feature of the French protection of civil liberties. The Government is not typically conceived as malevolent or oppressive of minorities. It simply is blinkered in its pursuit of goals. The police force is there to catch criminals, the tax inspectors to detect fraud. They may act in the interests of the State, but they may well unjustifiably infringe individual rights. The courts exist to oppose and sanction this, and the Conseil constitutionnel tries to prevent the opportunities for infringement coming into the hands of administrators. Thus, in DECISION 17, *Vehicle Searches*, the Conseil constitutionnel was concerned to ensure that the grounds and procedural safeguards for exercising police search powers were adequately defined. As a result, subsequent *lois* have been more precise, and have included such provisions as a right to a medical examination for a suspect held for more than twenty-four hours.

A similar restriction on zeal came in DECISION 32, when the Conseil constitutionnel annulled a provision in a bill that authorized the tax authorities to substitute their own view of how much tax a citizen owed and did not allow the citizen to rebut this in those instances where income came from capital gains or gifts. Similarly, the Conseil struck down the 50 per cent solidarity payment imposed on pensioners

[32] See R. M. Dworkin, *A Matter of Principle* (Cambridge, Mass., 1985), 197–8.

[33] 'Bilan et effets de la saisine du Conseil constitutionnel' (1984) 34 *Revue française de science politique* 988 at 998. See also Luchaire, 'Le Conseil constitutionnel et le gouvernement des juges', §31.

and their employers, and the penalty levied on consumers who paid for goods and services above the value of 150,000 F in cash.[34] The Conseil effectively stated here that a person's freedom of choice could not be so greatly restricted, however laudable the goal that Parliament and the administration were pursuing.

Commentators such as Vroom have compared the protection of fundamental rights in France unfavourably with that in the United States.[35] It would be fairer to say that the ambitions of the Conseil constitutionnel are more modest than those of the Supreme Court. In the first instance, Parliament remains the guardian of fundamental rights, and the Senate sees this as its particular mission. The Conseil receives a reference only at the end of the legislative process, when all else has failed, and references are not systematic. For the most part, the Conseil is not concerned with violation of fundamental liberties, but with violation of constitutional procedures (especially in finance laws).[36] Where it and the ordinary courts do intervene, it is typically to set standards for the strict scrutiny of decisions and for the interpretation of texts in favour of fundamental rights, much as might happen in the United Kingdom. The collective welfare in public order, or the continuity of public services are allowed to restrict the scope of freedoms. A Government that wishes to carry out its programme can, by and large, do so, as has been seen with nationalization, privatization, media reform, police powers, and immigration. Failure to balance competing interests in a rational way, or the downright abuse of power may well be sanctioned. In more normal cases, the protection and definition of rights is very much a matter for Parliament, from whom the Conseil frequently takes its lead for the content of rights. The French are concerned to create institutional procedures for intervention prior to the event (for example, the Commission nationale de l'informatique et des libertés, the Conseil d'État, or the investigating magistrate), rather than to control *ex post facto* by some judicial review. It is in such a relative 'hands on' way that some effective control of the administration can be operated.

7.6. Continuity and Convergence in Constitutional Principle

The Constitution of 1958 did not arise out of nowhere. Most of its rules and procedures had been gleaned from provisions of previous Con-

[34] CC decision nos. 85-200 DC of 16 Jan. 1986, *Cumulation of Persions and Salaries*, and 89-268 DC of 29 Dec. 1989, *Finance Law for 1990*.

[35] C. Vroom, 'Constitutional Protection of Civil Liberties in France' (1988) 63 *Tulane Law Review* 266.

[36] Favoreu, 'Les Cents Premières Annulations', 453, suggests that breaches of fundamental rights account for only a third of provisions quashed in the 1981–6 period.

stitutions, suitably adapted and improved. Further, the 'French republican constitutional tradition' has shaped not merely its content, but also the way in which it has been operated and interpreted. In the areas of *loi* and *règlement*, and fundamental rights, especially through notions such as the 'fundamental principles recognized by the laws of the Republic', there has been a significant continuity with the past—at least at the level of detail, even if there has been discontinuity at the level of constitutional theory—especially in relation to parliamentary sovereignty. This is not to say that important changes have not taken place through the 1958 Constitution. A rationalized Parliament, presidentialism, and constitutional review are notable departures from the past. Jurists closely connected with the Conseil have been keen to stress the continuity between the case-law of the Conseil constitutionnel and the traditional rules and principles of public law, especially those developed by the Conseil d'État. Georges Vedel, a distinguished professor of public law and a former member of the Conseil constitutionnel, argues that, since 1789, social and psychological unity and continuity have been easier to secure in France than institutional continuity, and this is reflected in the content of law, where changes have more of the character of 'topological transformation' than wholesale 'rupture' with the past.[37]

More specifically, in recent years there has been a substantial unity in public law, and a continuity with principles developed under the Third and Fourth Republics, although the separation of powers now operates in a new context. The current Secretary-General of the Conseil constitutionnel, Bruno Genevois, a member of the Conseil d'État, sees overlap and convergence between the case-law of the Conseil constitutionnel and the Conseil d'État.[38] For a long time only the Conseil d'État ensured the continuity of public law principles through different political regimes, and its case-law served as an important inspiration for the positions adopted by the Conseil constitutionnel. But now the latter has its own views, and has become very much the dominant partner, if only because it can have the final say. The Conseil constitutionnel is influenced by private lawyers, especially those among its members, and a number of issues such as media law and criminal procedure lie outside the traditional expertise of the Conseil d'État. The Conseil constitutionnel has been selective in its choice of legal principles from public and private law, and now acts as a bridge between

[37] 'La Continuité constitutionnelle en France de 1789 à 1989', *RFDC* 1990, 5 at 13.

[38] See esp. 'Continuité et convergence des jurisprudences constitutionnelle et administrative', *RFDA* 1990, 143; also G. Vedel, 'Discontinuité de droit constitutionnel et continuité du droit administratif: Le rôle du juge', in *Mélanges Waline* (Paris, 1974), 777; L. Favoreu, 'Dualité ou unité d'ordre juridique?', in *Conseil constitutionnel et Conseil d'État* (Paris, 1988), 145.

them. At the same time, it has been an innovator in areas such as media law, parliamentary procedure, the protection of property, and education law. It looks forward to the kind of political community that France should become, and not simply back at the heritage of the past.

A very large amount of legal continuity has been achieved, just as the institutions of the Fifth Republic owe much to previous regimes. The Conseil has contributed to this, especially in the relations between Government and Parliament, but also in fundamental rights. But the situation in which it has to operate today is significantly different from that of 1958, and the Conseil has helped to adapt the Constitution to this new field of operation.

7.7. Lessons for Other Jurisdictions?

A number of lessons might be drawn by other countries from the French experience of constitutional review since 1958. In the first place, that experience demonstrates that a country can radically change its attitude to the legitimacy and practice of constitutional review, and very rapidly reach a new constitutional consensus. It was less than fifteen years between the first constitutional review of *lois* to protect fundamental rights and the firm establishment and political acceptance of this process. Indeed, it took place gradually and cautiously, such that the Conseil could have retreated if there was too much criticism. It may be that this gradual approach helped to convince those who would have rejected it when it was proposed in 1958, and this may show the advantages of not making a sudden and dramatic break with the past.

The practice of the Conseil shows that constitutional review need not involve any major erosion of the importance of the political and parliamentary process; indeed, it may enhance it. It was the new constitutional settlement of 1958 that established the revised powers of Parliament and Government. Constitutional review has served to support this system and to ensure that Parliament is not abused by the power of the Government, reinforced as it is by the new structure of political parties to which the Fifth Republic has given rise.

One of the major functions of constitutional review in France is the enforcement of constitutional rules of parliamentary procedure. The Conseil thereby protects the proper operation of the political process. It is in this low-profile area that the Conseil makes perhaps its most significant contribution to the enhancement of democracy. And it is here that political debate in other countries, especially in the United Kingdom, has been weakest.

The Conseil also has an important role in protecting fundamental rights. But its approach demonstrates an alternative to an American view that rights are anti-majoritarian. The Conseil's approach is basically limited to requiring rationality and proportionality in legislative decisions. It would, however, be an exaggeration to suggest that this has improved significantly the public justification of the legislature's decisions, which some would see as a rationale for constitutional review.[39] Some rights are entrenched, and there is undoubtedly strict scrutiny of the reasons for interfering with them. But, given the Conseil's approach, it is never likely that it would become a forum for political debate comparable to the US Supreme Court. Rights are predominantly shaped in Parliament, with the Conseil operating as a supervisor of last resort when a major infringement of rights is likely to occur. The contrast between DECISION 20, *Abortion Law*, and decisions of the Supreme Court provides a marker for the difference in approach between the two institutions.

The success of the Conseil constitutionnel has been the subject of very varying assessments. A leading lawyer, Loïc Philip, argues that French constitutional review of *lois* has become 'one of the widest in the world'.[40] By contrast, a leading political scientist is more cautious: 'The gradual extension of the role of the Conseil should not be misunderstood. Its decisions remain less numerous and more cautious than those of its foreign counterparts.'[41] It has not been as adventurous as the German or US Constitutional Courts, or even, more recently, Canada and India. But it has made a contribution to the development of the Constitution and the structuring of law and politics that has not been unimportant in ensuring the stability and durability of the Fifth Republic.

[39] S. Freeman, 'Constitutional Democracy', 365.
[40] 'Le Développement du contrôle de constitutionnalité et l'accroissement des pouvoirs du juge constitutionnel, *RDP* 1983, 401 at 415.
[41] M. Duvergier, *La Cohabitation des français* (Paris, 1987), 131.

PART TWO

Constitutional Texts, Materials, and Decisions

CONSTITUTIONAL TEXTS

I. The Constitution of 1958

Preamble

The French people solemnly proclaims its attachment to the rights of man and to the principles of national sovereignty such as are defined by the Declaration of 1789, confirmed and completed by the Preamble to the 1946 Constitution.

In virtue of these principles and that of the self-determination of peoples, the Republic offers to overseas territories that express the desire to join them new institutions based on the common ideal of liberty, equality, and fraternity, and conceived with a view to their democratic evolution.

TITLE 1: SOVEREIGNTY

Article 1

The Republic and the peoples of the overseas territories who, by an act of free determination, adopt this present Constitution thereby institute a Community.

The Community is founded upon equality and solidarity among the peoples who compose it.

Article 2

France is an indivisible, lay, democratic, and social Republic. It ensures equality before the law for all citizens, without distinction as to origin, race, or religion. It respects all beliefs. The language of the Republic is French.

The national emblem is the tricolour flag of blue, white, and red.

The national anthem is 'La Marseillaise'.

The motto of the Republic is 'Liberty, Equality, Fraternity'.

Its principle is: government of the people, by the people, and for the people.

Article 3

National sovereignty belongs to the people, which shall exercise it by its representatives and by means of referendum.

No section of the people, nor any individual, may arrogate its exercise to itself.

Suffrage may be direct or indirect, under the conditions provided by the Constitution. It shall always be universal, equal, and secret.

Within the terms settled by *loi*, all adult French nationals of both sexes, enjoying civil and political rights, are voters.

Article 4

Parties and political groups contribute to the expression of suffrage. They may form and exercise their activity freely. They must respect the principles of national sovereignty and democracy.

TITLE 2: THE PRESIDENT OF THE REPUBLIC

Article 5

The President of the Republic shall supervise respect for the Constitution. By his arbitrament [*arbitrage*], he shall ensure the proper functioning of public authorities, as well as the continuity of the State.

He is the guarantor of the nation's independence, of the integrity of its territory, and of respect for Community agreements and for treaties.

Article 6 (as amended by the Referendum Law of 6 November 1962)

The President of the Republic is elected for seven years by direct universal suffrage.

The procedures for implementing this article shall be determined by an organic law.

Article 7 (as amended by the Constitutional Law of 18 June 1976)

The President of the Republic is elected by an absolute majority of the votes cast. If this is not obtained on the first ballot, a second ballot shall be held on the second Sunday thereafter. Only those two candidates may present themselves for this who have obtained the highest number of votes on the first ballot, having taken account of any withdrawal by candidates who obtained even more votes.

The ballot is held on orders from the Government.

The election of a new President takes place at least twenty days, and at most thirty-five days, before the end of the mandate of the President in office.

Where the post of President of the Republic is vacant for whatever reason, or where the Conseil constitutionnel decides, on a reference from the Government and by an absolute majority of its members, that he is incapable, the functions of President of the Republic, except for those set out in articles 11 and 12, shall be provisionally exercised by the President of the Senate and, if the latter is unable to exercise these functions, by the Government.

In the case of a vacancy or an incapacity declared permanent by the Conseil constitutionnel, the ballot for the election of a new President takes place at least twenty days, and at most thirty-five days, from the occurrence of the vacancy or the declaration of permanent incapacity, unless the Conseil constitutionnel finds that a supervening event prevents this.

If, within the seven days preceding the last date for receipt of nominations, one of those persons who, at least thirty days before that date, had announced publicly his decision to be a candidate dies or becomes incapacitated, the Conseil constitutionnel may decide to postpone the election.

If one of the candidates dies or becomes incapacitated before the first ballot, the Conseil constitutionnel shall postpone the election.

In the case of death or incapacity of one of the two candidates receiving the most votes on the first ballot, taking account of any withdrawals, the Conseil constitutionnel shall declare that the whole election process must be repeated; the same shall occur in the case of the death or incapacity of one of the two candidates standing in the second ballot.

All cases shall be referred to the Conseil constitutionnel under the provisions of paragraph 2 of article 61 below, or under those established by the organic law mentioned in article 6 above, for the presentation of candidates.

The Conseil constitutionnel may extend the time-limits set out in paragraphs 3 and 5 above, provided that the ballot does not take place more than thirty-five days after the decision of the Conseil constitutionnel. If the application of the provisions of this paragraph have the effect of delaying the election until after the expiry of the mandate of the President in office, he shall remain in post until the proclamation of his successor.

While the post of President of the Republic is vacant, or during the period between the declaration of his permanent incapacity and the election of his successor, no application may be made of articles 49 and 50, nor of article 89.

Article 8

The President of the Republic appoints the Prime Minister. He terminates his functions when the latter presents the resignation of the Government.

On the recommendation of the Prime Minister, he appoints and dismisses the members of the Government.

Article 9

The President of the Republic presides over the Council of Ministers.

Article 10

The President of the Republic shall promulgate *lois* within fifteen days of the transmission to the Government of the *loi* as definitively adopted.

Article 11

On the recommendation of the Government during sessions [of Parliament], or on a joint recommendation of both chambers published in the *Journal officiel*, the President of the Republic may submit to a referendum any bill concerning the organization of public authorities, or requiring the approval of a Community agreement, or providing for authorization to ratify a treaty that, without being contrary to the Constitution, would affect the functioning of its institutions.

Where a referendum has decided in favour of the bill, the President of the Republic shall promulgate it within the time-limit set out in the preceding article.

Article 12

After consulting the Prime Minister and the Presidents of the chambers, the President of the Republic may dissolve the National Assembly.

General elections shall be held at least thirty days, and at most forty days, after the dissolution.

The National Assembly shall convene as of right on the second Thursday following its election. If this meeting takes place outside the periods provided for ordinary sessions, a session shall be held, as of right, for a period of fifteen days.

There may not be a further dissolution within a year of these elections.

Article 13

The President of the Republic signs *ordonnances* and decrees deliberated upon in the Council of Ministers.

He makes appointments to the State's civil and military services.

Conseillers d'État, the Grand Chancellor of the Legion of Honour, ambassadors and extraordinary envoys, the *conseillers-maîtres* of the Cour des comptes, prefects, Government representatives in the overseas territories, general officers, rectors of academies, *directeurs* of central government departments are appointed in the Council of Ministers.

An organic law shall determine those other posts to be filled in the Council of Ministers, as well as the terms under which the powers of appointment of the President of the Republic may be delegated by him or exercised on his behalf.

Article 14

[Accreditation of ambassadors.]

Article 15

The President of the Republic is the head of the army. He shall preside over the councils and higher committees of national defence.

Article 16

When the institutions of the Republic, the nation's independence, the integrity of its territory, or the implementation of international agreements are threatened seriously and immediately, and the proper functioning of constitutional public authorities is interrupted, the President of the Republic shall take such measures as these circumstances require, having consulted officially the Prime Minister, the Presidents of the chambers, and also the Conseil constitutionnel.

He shall inform the nation by message.

These measures must be prompted by a desire to ensure for public constitutional authorities the means of fulfilling their functions within the shortest possible time. The Conseil constitutionnel shall be consulted about them.

Parliament shall convene as of right.

The National Assembly cannot be dissolved during the exercise of these exceptional powers.

Article 17

The President of the Republic has the right of pardon.

Article 18

The President of the Republic shall communicate with the two chambers by messages that he shall cause to be read out and that shall not be subjected to debate.

Article 19

[Countersignature of decisions.]

TITLE 3: THE GOVERNMENT

Article 20

The Government shall determine and conduct national policy.

It shall have the administration and the army at its disposal.

It shall answer to Parliament under the terms of, and following the procedures set out in, articles 49 and 50.

Article 21

The Prime Minister directs the operation of the Government. He is responsible for national defence. He ensures the implementation of laws. Subject to the provisions of article 13, he exercises the power to make regulations, and makes appointments to the civil and military services.

He may delegate certain powers to ministers.

Should the occasion arise, he may deputize for the President of the Republic in presiding over the councils and committees mentioned in article 15.

Exceptionally, he may deputize for him in presiding over the Council of Ministers by virtue of an express delegation and for a specified agenda.

Article 22

[Countersignature of decisions.]

Article 23

The functions of members of the Government are incompatible with the exercise of any parliamentary mandate, any role of representing a profession at national level, and any public employment or professional activity.

An organic law shall determine the conditions under which the holders of such mandates, roles, or employments are replaced.

TITLE 4: PARLIAMENT

Article 24

Parliament comprises the National Assembly and the Senate.

Deputies to the National Assembly are elected by direct suffrage.

The Senate is elected by indirect suffrage. It ensures the representation of the local authorities of the Republic. French citizens living outside France are represented in the Senate.

Article 25

An organic law shall determine the duration of the powers of each chamber, the number of its members, their salaries, their conditions of elegibility, and the system of inelegibilities and incompatibilities.

It shall also determine the terms under which persons are elected to replace deputies or senators when seats fall vacant, until the next general or partial election of the chamber to which they belong.

Article 26

No member of Parliament may be prosecuted, pursued, arrested, detained, or tried for opinions or votes expressed by him in the course of his functions . . .

[Other provisions on parliamentary immunity.]

Article 27

Any binding instruction [from an outside body] is void.

The right to vote of members of Parliament is personal.

In exceptional circumstances an organic law may authorize the delegation of voting. In such a case, no one can be delegated more than one proxy vote.

Article 28 (as amended by the Constitutional Law of 30 December 1963)

Parliament shall meet as of right for two ordinary sessions a year.

The first session shall begin on 2 October and shall last for eighty days.

The second session shall begin on 2 April and shall not last more than ninety days.

If either 2 October or 2 April is a public holiday, the session shall begin on the first working day thereafter.

Article 29

Parliament shall meet for an extraordinary session, for a specified agenda, at the request of the Prime Minister or of a majority of the members of the National Assembly.

When an extraordinary session is held at the request of members of the National Assembly, the decree closing the session shall be made as soon as Parliament has exhausted the agenda on which it was convened, or, at the latest, twelve days from the opening of the session.

Only the Prime Minister may request another session within a month of the decree closing the session.

Article 30

Except where Parliament meets as of right, extraordinary sessions are begun and closed by a decree of the President of the Republic.

Article 31

Members of the Government shall have access to the two chambers. They shall be heard on request.

They may be assisted by Government representatives.

Article 32

The President of the National Assembly is elected for the duration of the legislature. The President of the Senate is elected after each partial election.

Article 33

The sittings of the two chambers shall be public. The full minutes of debates shall be published in the *Journal officiel*.

Each chamber may sit in secret session at the request of the Prime Minister or of one-tenth of its members.

TITLE 5: RELATIONS BETWEEN PARLIAMENT AND THE GOVERNMENT

Article 34

Loi [law] is passed by Parliament.

Loi shall determine the rules concerning:

- civic rights and the fundamental safeguards granted to citizens for the exercise of civil liberties; the burdens imposed on citizens in their persons and on their property for the purposes of national defence;
- the nationality, status, and capacity of persons, matrimonial regimes, succession, and gifts;
- the determination of *crimes* and *délits*, as well as the penalties applied to them; criminal procedure; amnesties; the creation of new types of court, and the status of judges (*magistrats*);
- the basis of assessment, rates, and means of recovery of taxes of all kinds; the system for the issuing of currency.

Loi shall also determine the rules concerning:

- the electoral system for parliamentary and local assemblies;
- the creation of categories of public corporation;
- the fundamental guarantees granted to the civil and military servants of the State;
- the nationalization of undertakings, and the transfer of undertakings from the public to the private sector.

Loi shall lay down the fundamental principles:

- of the general organization of national defence;
- of the free administration of local authorities, their powers, and their resources;
- of education;
- of the regime for property, property rights, and civil and commercial obligations;
- of labour law, trade-union law, and social security.

Finance laws specify the resources and charges upon the State under the terms of, and with the reservations specified in, an organic law.

Programme laws specify the objectives of the economic and social action of the State.

The provisions of the present article may be clarified and complemented by an organic law.

Article 35

A declaration of war shall be authorized by Parliament.

Article 36

A state of siege is decreed in the Council of Ministers.

Its extension beyond a period of twelve days may only be authorized by Parliament.

Article 37

Matters other than those within the province of *loi* have a regulatory character.

Texts in the form of a *loi* passed on these matters can be modified by decree following the advice of the Conseil d'État. Those texts that are passed after the coming into force of the present Constitution can only be modified by decree if the Conseil constitutionnel has declared that they are of a regulatory character by virtue of the preceding paragraph.

Article 38

For the implementation of its programme, the Government may ask Parliament for authority to take measures that are normally within the province of *loi* by *ordonnance* [ordinance] for a limited period.

The *ordonnances* are made by the Council of Ministers on the advice of the Conseil d'État. They come into force once they are published, but they lapse if a bill to ratify them is not laid before Parliament by the date fixed by the enabling law.

On the expiry of the period mentioned in the first paragraph of this article, the *ordonnances* may no longer be amended except by *loi* with respect to those matters which are within the province of *loi*.

Article 39

The right to propose *lois* belongs concurrently to the Prime Minister and to members of Parliament.

Government bills [*projects de loi*] are considered by the Council of Ministers after the advice of the Conseil d'État, and are tabled in one of the chambers. Finance bills are submitted first to the National Assembly.

Article 40

Private members' bills [*propositions de loi*] and amendments drafted by members of Parliament are not admissible if their adoption would have the consequence of reducing public resources or of creating or increasing a public charge.

Article 41

In the course of the legislative procedure, if it appears that a private Member's bill or an amendment is not within the province of *loi*, or is contrary to a delegation of authority granted by virtue of article 38, the Government can oppose it as inadmissible.

In case of disagreement between the Government and the President of the relevant chamber, and at the request of one or other of them, the Conseil constitutionnel shall give a ruling within a period of eight days.

Article 42

In the first chamber in which it is tabled, debate on Government bills shall take place on the text submitted by the Government.

When referred a text from the other chamber, a chamber shall consider the text transmitted to it.

Article 43

At the request of the Government or of the chamber in which it is tabled, Government bills and private members' bills are sent for scrutiny by committees specifically appointed for the purpose.

Government bills and private members' bills in relation to which no such request has been made are sent to one of the permanent committees, which shall not number more than six for each chamber.

Article 44

The members of Parliament and the Government have the right of amendment.

After a debate has begun, the Government may oppose discussion of any amendment that has not previously been submitted to the committee.

If the Government so requests, the chamber shall decide by a single vote on all or part of the text under discussion, including only those amendments proposed or accepted by the Government.

Article 45

Any bill must be debated successively by the two chambers with a view to the adoption of a single text.

If, following a disagreement between the two chambers, a bill has not been adopted after two readings by each chamber, or, if the Government has declared it urgent, after a single reading by each of them, the Prime Minister has the power to call a meeting of a joint committee to propose a text on the provisions remaining to be discussed.

The text drafted by the joint committee can be submitted by the Government for approval by the two chambers. No amendment can be accepted without the Government's agreement.

If the joint committee does not reach agreement on a common text, or if the text is not adopted under the provisions of the previous paragraph, then, after a further reading by the National Assembly and by the Senate, the Government may ask the National Assembly to make a final decision. In this case, the National Assembly may reconsider either the text produced by the joint committee or the last text passed by itself, modified as necessary by one or more amendments adopted by the Senate.

Article 46

Lois on which the Constitution confers the character of organic laws shall be passed and amended in the following way.

The bill is only submitted for discussion and voting in the first chamber if fifteen days have elapsed after it was tabled.

The procedure set out in article 45 may be applied. All the same, where the two chambers disagree, the text may only be adopted by the National Assembly after a final reading by an absolute majority of its members.

Organic laws relating to the Senate must be passed in identical terms by both chambers.

Organic laws may only be promulgated after the Conseil constitutionnel has ruled that they are consistent with the Constitution.

Article 47

Parliament shall pass finance bills under the conditions set out in an organic law.

If the National Assembly has not voted on the first reading within forty days of the tabling of a bill, the Government shall refer it to the Senate, which must decide on it within fifteen days. Thereafter, the procedure set out in article 45 shall be followed.

If Parliament does not decide within seventy days, the provisions of the bill may be brought into force by *ordonnance*.

If the finance bill determining the resources and charges for the financial year has not been tabled in time for its promulgation before the beginning of that year, the Government shall request, as a matter of urgency, the authorization from Parliament to collect taxes and to make credits available by decree for appropriations already approved.

The time-limits set out in this article are suspended when Parliament is not in session.

The Cour des comptes shall assist Parliament and the Government in monitoring the implementation of finance laws.

Article 48

As a priority, and in the order determined by the Government, the agenda of the chambers shall include discussion of bills tabled by the Government and private members' bills accepted by it.

One sitting a week is reserved as a priority for questions by members of Parliament and Government replies.

Article 49

After consideration by the Council of Ministers, the Prime Minister may make the Government's programme or a statement of general policy, as the case may be, an issue of confidence in the National Assembly.

The National Assembly may call the confidence of the Government into question by means of a censure motion. Such a motion is only admissible if it is signed by at least one-tenth of the members of the National Assembly. A vote may only take place forty-eight hours after it is tabled. Only votes in favour of the censure motion are counted, and it can only be passed by a majority of the members of the Assembly. If a motion of censure is rejected, its signatories may not propose another one during the same session, except in the situation described in the following paragraph.

After consideration by the Council of Ministers, the Prime Minister may make a vote on a text an issue of confidence in the National Assembly. In this case, the text is treated as having been passed unless a motion of censure, tabled within twenty-four hours, is endorsed under the terms set out in the previous paragraph.

The Prime Minister can request approval from the Senate on a statement of general policy.

Article 50

When the National Assembly passes a motion of censure or disapproves the programme or the statement of general policy of the Government, the Prime Minister must tender the resignation of the Government to the President of the Republic.

Article 51

Where necessary, the closure of ordinary or extraordinary sessions shall be delayed to enable the application of the provisions of article 49.

TITLE 6: TREATIES AND INTERNATIONAL AGREEMENTS

Article 52

The President of the Republic negotiates and ratifies treaties.

He is informed of all negotiations leading to the signing of an international agreement not submitted for ratification.

Article 53

Peace treaties, commercial treaties, treaties relating to international organizations, those which commit the finances of the State, those relating to the status of persons, and those which involve the transfer, exchange, or addition of territory may only be ratified or approved by a *loi*.

They only take effect after they are ratified or approved.

No transfer, exchange, or addition of territory is valid without the consent of the populations concerned.

Article 54

If, on a reference from the President of the Republic, the Prime Minister, or the President of either chamber, or 60 deputies or senators, the Conseil constitutionnel declares that an international agreement includes a clause contrary to the Constitution, authorization to ratify or approve it may only be given after a revision of the Constitution.

Article 55

From their publication, duly ratified or approved treaties or agreements have a higher authority than *lois*, subject, for each treaty or agreement, to its implementation by the other party.

TITLE 7: THE CONSEIL CONSTITUTIONNEL

Article 56

The Conseil constitutionnel is composed of nine members, whose term of office lasts for nine years and is not renewable. A third of the Conseil constitutionnel is replaced every three years. Three of its members are appointed by the President of the Republic, three by the President of the National Assembly, and three by the President of the Senate.

In addition to the nine members mentioned above, former Presidents of the Republic are members of the Conseil constitutionnel for life.

The President is appointed by the President of the Republic. He has the casting vote in the case of a tie.

Article 57

The office of member of the Conseil constitutionnel is incompatible with that of a minister or a member of Parliament. Other incompatibilities shall be determined by an organic law.

Article 58

The Conseil constitutionnel shall monitor the validity of the election of the President of the Republic.

It shall investigate complaints, and declare the results of the ballot.

Article 59

In cases of dispute, the Conseil constitutionnel shall rule on the validity of the election of deputies and senators.

Article 60

The Conseil constitutionnel shall monitor the validity of referendum procedures, and declare the results.

Article 61 (as amended by the Constitutional Law of 29 October 1974)

Organic laws, before they are promulgated, and the standing orders of the parliamentary chambers, before they are brought into force, must be submitted to the Conseil constitutionnel, which shall decide on their compatibility with the Constitution.

For the same purpose, *lois* may be referred to the Conseil constitutionnel, before they are promulgated, by the President of the Republic, the Prime Minister, the President of the National Assembly, the President of the Senate, or sixty deputies or sixty senators.

In the cases set out in the previous two paragraphs, the Conseil constitutionnel must reach a decision within a month. Nevertheless, at the request of the Government, if there is urgency, this time-limit is reduced to eight days.

In these same cases, a reference to the Conseil constitutionnel suspends the time-limit for promulgation.

Article 62

A provision that has been declared unconstitutional may neither be promulgated nor applied.

Decisions of the Conseil constitutionnel bind public powers and all administrative and judicial authorities.

Article 63

An organic law shall determine the rules of organization and functioning of the Conseil constitutionnel, the procedure that it is to follow, and, especially, the time-limits for referring disputes to it.

TITLE 8: THE JUDICIARY

Article 64

The President of the Republic is the guardian of the independence of the judiciary.

He is assisted by the Conseil supérieur de la magistrature.

An organic law shall establish the status of judges.
Judges (*magistrats de siège*) are irremovable.

Article 65

The Conseil supérieur de la magistrature is presided over by the President of the Republic. The Minister of Justice is its Vice-President ex officio.

In addition, the Conseil supérieur comprises nine members appointed by the President of the Republic under terms set out in an organic law.

The Conseil supérieur de la magistrature makes proposals for the appointment of judges to the Cour de cassation and for the posts of First President of the Courts of Appeal. In circumstances determined by the organic law, it gives its opinion on the proposals of the Minister of Justice relating to the appointment of other judges. It is consulted on pardons in circumstances determined by an organic law.

The Conseil supérieur de la magistrature makes rulings as a disciplinary committee for judges. In this case, it is chaired by the First President of the Cour de cassation.

Article 66

No one may be detained arbitrarily.

The judiciary, the guardian of individual liberty, ensures respect for this principle in circumstances provided for by *loi*.

TITLE 9: THE HIGH COURT OF JUSTICE

TITLE 10: THE ECONOMIC AND SOCIAL COUNCIL

TITLE 11: LOCAL AUTHORITIES[*]

Article 72

The local authorities of the Republic are the communes, the departments, and the overseas territories. Every local authority is created by *loi*.

These authorities administer themselves freely through elected councils, under conditions set out by *loi*.

In departments and overseas territories the Government representative is responsible for national interests, administrative supervision, and respect for the law.

Article 73

The legislative regime and organization of overseas departments may be adapted by measures made necessary by their special situation.

Article 74

The overseas territories of the Republic shall have a special organization taking account of their specific interests among the totality of interests in the Republic. . . .

TITLE 14: EUROPEAN COMMUNITIES AND EUROPEAN UNION

Article 88–1

The Republic shall participate in the European Communities and in the European Union, composed of states which have chosen freely to exercise certain of their competences in common in accordance with the treaties which created them.

Article 88–2

Subject to reciprocity and according to the terms laid down by the treaty on European union signed on 7 February 1992, France consents to the transfer of competences necessary for the creation of European economic and monetary union as well as for the determination of the rules relating to the crossing of the external frontiers of member states of the European Community.

Article 88–3

Subject to reciprocity and according to the terms laid down by the treaty on European union signed on 7 February 1992, the right to vote and to be elected in municipal elections may be granted only to citizens of the Union resident in France. These citizens may not hold the office of mayor or deputy mayor, nor participate in the nomination of senatorial electors or in the election of senators. An organic law voted in the same terms by the two chambers shall determine the provisions for implementing this article.

Article 88–4

The Government shall submit to the National Assembly and the Senate drafts of community decisions which include provisions having legislative effect, as soon as they are transmitted to the Council of the Communities. During sessions [of Parliament] or outside them, resolutions may be voted under this article according to the terms fixed by the standing orders of each chamber.

TITLE 15: REVISION

Article 89

The initiative for revision of the Constitution belongs concurrently to the President of the Republic on the proposal of the Prime Minister and to members of Parliament.

The bill for revision must be passed by the two chambers in identical terms. The revision becomes definitive after it has been approved by referendum.

Nevertheless, the revision bill is not submitted to a referendum when the President of the Republic decides to submit it to Parliament convened in Congress; in this case, the revision bill is only approved if it obtains a majority of three-fifths of votes cast. The office of the Congress is that of the National Assembly.

No procedure for revision may be undertaken or pursued when it would infringe the integrity of [national] territory.

The republican form of government may not be the subject of revision.

TITLE 16: TRANSITIONAL PROVISIONS

II. THE DECLARATION OF THE RIGHTS OF MAN AND OF THE CITIZEN, 1789

Article 1

All men are born and remain equal in their rights. Social distinctions may only be based on public utility.

Article 2

The ultimate purpose of every political institution is the preservation of the natural and imprescriptible rights of man. These rights are to liberty, property, security, and resistance to oppression.

Article 3

The source of all sovereignty lies ultimately in the nation. No body or any individual can exercise any authority that is not expressly derived from it.

Article 4

Liberty consists in the power to do anything that does not harm another. Therefore, the only limits on the exercise of the natural rights of each man shall be those that ensure the enjoyment of the same rights by other members of society. Such limits may only be established by law (*loi*).

Article 5

The law may only prohibit actions harmful to society. Anything that is not prohibited by law cannot be prevented, and no one may be forced to do anything that it does not require.

Article 6

Loi is the expression of the general will. All citizens have the right to participate in its creation, either personally or through their representatives. It must be the same for all, whether it punishes or protects. All citizens, being equal in its eyes, are equally eligible for all public dignities, positions, and employment according to their abilities, and without distinction other than that of their virtues and talents.

Article 7

No individual may be accused, arrested, or detained except where the law (*loi*) so prescribes, and in accordance with the procedures it has laid down. Those who solicit, further, execute, or arrange for the execution of arbitrary orders must be punished; but any citizen charged or detained by virtue of a *loi* must obey it immediately; resistance renders him culpable.

Article 8

The law (*loi*) may only create penalties that are strictly and evidently necessary. No one may be punished except according to a *loi* passed and promulgated prior to the offence, and lawfully applied.

Article 9

Since a man is presumed innocent until he has been declared guilty, if it is judged indispensable to arrest him, any force that is not necessary to secure his person should be severely punished by the law.

Article 10

No one may be troubled on account of his opinions or religion, provided that their expression does not infringe public policy as established by *loi*.

Article 11

The free communication of thoughts and opinions is one of the most precious rights of man; hence, every citizen may speak, write, and publish freely, save that he must answer for any abuse of such freedom in cases specified by *loi*.

Article 12

The safeguarding of the rights of man and of the citizen requires a police force; such a force is thus created for the benefit of all, and not for the private advantage of those to whom it is entrusted.

Article 13

For the upkeep of a police force and for the expenses of the administration, common taxation is indispensable. This should be shared equally among all citizens, according to their means.

Article 14

All citizens have the right to satisfy themselves, either personally or through their representatives, that a public tax is necessary, to consent to it freely, to monitor its spending, and to determine its amount, its basis of assessment, its collection, and its duration.

Article 15

Society has the right to demand an account of administration from any public official.

Article 16

Any society in which the safeguarding of rights is not assured, and the separation of powers is not established, has no constitution.

Article 17

Property, being an inviolable and sacred right, none can be deprived of it, except when public necessity, legally ascertained, evidently requires it, and on condition of a just and prior indemnity.

III. PREAMBLE TO THE 1946 CONSTITUTION

Paragraph 1

On the morrow of the victory won by free peoples over regimes that tried to enslave and degrade the human person, the French people proclaims anew that any human being possesses inalienable and sacred rights, without distinction as to race, religion, or beliefs. It solemnly reaffirms the rights and liberties of man and of the citizen consecrated by the Declaration of Rights of 1789, and the fundamental principles recognized by the laws of the Republic.

Paragraph 2

In addition, it proclaims the following political, economic, and social principles as particularly necessary for our times.

Paragraph 3

The law shall guarantee to women, in all spheres, equal rights to men.

Paragraph 4

Any person persecuted for his activities on behalf of freedom has the right of asylum in the territories of the Republic.

Paragraph 5

Everyone has the duty to work and the right to obtain a job. No one may be harmed in his work or employment on account of his origins, opinions, or beliefs.

Paragraph 6

Every individual may defend his rights and his interests by union action, and belong to the union of his choice.

Paragraph 7

The right to strike shall be exercised within the framework of *lois* that regulate it.

Paragraph 8

Every worker shall participate, through his representatives, in the collective determination of conditions of work, and also in the running of businesses.

Paragraph 9

Any property or business whose exploitation has or acquires the character of a national public service or a *de facto* monopoly should become the property of the community.

Paragraph 10

The nation shall ensure to the individual and to the family the conditions necessary for their development.

Paragraph 11

It guarantees to all, especially to the child, the mother, and aged workers, the protection of health, material security, rest, and leisure. Any human being who, by reason of his age, physical or mental health, or economic situation, is unable to work, has the right to obtain appropriate means of subsistence from the community.

Paragraph 12

The nation proclaims the solidarity and equality of all French men and women in the face of the burdens that result from national calamities.

Paragraph 13

The nation guarantees the equal access of children and adults to instruction, to professional training, and to culture. The organization of free and secular public education at all levels is a duty of the State.

Paragraph 14

Faithful to its traditions, the French Republic shall comply with public international law. It shall never undertake any war of conquest, and never deploy its forces against the freedom of any people.

Paragraph 15

Subject to reciprocity, France shall consent to those limitations on sovereignty necessary for the organization and defence of peace . . .

Paragraph 16

[The French Union.]

Paragraph 17

[Colonies.]

IV. *Ordonnance* no. 58-1067 of 7 November 1958 (Organic Law on the Conseil constitutionnel)

TITLE 1: ORGANIZATION OF THE CONSEIL CONSTITUTIONNEL

Article 1

The members of the Conseil constitutionnel, other than ex-officio members, are nominated by decisions of the President of the Republic, the President of the National Assembly, and the President of the Senate.

The President of the Conseil constitutionnel is nominated by a decision of the President of the Republic. He is chosen from among either the nominated or ex-officio members of the Conseil.

The above decisions shall be published in the *Journal officiel*.

Article 2

[Provisions about the terms of office of the initial members of the Conseil.]

Article 3

Before they take up office, members of the Conseil constitutionnel swear an oath before the President of the Republic.

They swear to fulfil their functions well and faithfully, to exercise them impartially, respecting the Constitution, to keep secret the deliberations and votes, and not to take a position in public, or to give any consultancy, on matters within the jurisdiction of the Conseil.

A formal record is taken of the oath.

Article 4

The functions of members of the Conseil constitutionnel are incompatible with those of being a member of the Government, or of Parliament, or of the Economic and Social Council.

Members of the Government, or of Parliament, or of the Economic and Social Council nominated to the Conseil constitutionnel are deemed to have opted for these latter functions unless they express their intention to the contrary within a week of the publication of their nomination.

Members of the Conseil constitutionnel nominated to governmental or elective functions in one of the two chambers of Parliament, or appointed a member of the Economic and Social Council, are replaced.

Article 5

During their term of office, members of the Conseil constitutionnel cannot be appointed to any public employment nor, if they are civil servants, can they obtain a promotion by choice.

Article 6

The President and the members of the Conseil constitutionnel receive respectively an honorarium equal to the salary paid to the two highest classes of State employment outside the [ordinary] scale.

The honorarium is reduced by half for members of the Conseil constitutionnel who continue with an activity compatible with their functions.

Article 7

A decree made in the Council of Ministers on a proposal by the Conseil constitutionnel shall define the duties imposed on members of the Conseil constitutionnel in order to safeguard the independence and dignity of their functions. The duties must include, in particular, a ban on members of the Conseil constitutionnel taking a position in public on matters that have been or are capable of being the subject of a decision on the part of the Conseil, or of giving a consultancy on such matters during their term of office.

Article 8

The replacement of members of the Conseil shall be ensured at least a week before the end of their term of office.

Article 9

A member of the Conseil constitutionnel may resign by a letter addressed to the Conseil. A replacement shall be appointed at the latest within a month of his resignation. The resignation takes effect from the nomination of the replacement.

Article 10

Should the case arise, the Conseil constitutionnel shall verify the resignation from office of any of its members who is carrying on an activity, or has accepted a function or elective mandate, incompatible with his status as a member of the Conseil, or who no longer enjoys civil and political rights.

He is then replaced within a week.

Article 11

The rules laid down in article 10 above apply to members of the Conseil constitutionnel who are permanently prevented by a physical incapacity from carrying out their duties.

Article 12

Members of the Conseil constitutionnel nominated to replace those who cease to hold office before the expiry of their normal term complete the mandate of those whom they replace. At the expiry of this mandate, they may be nominated as a member of the Conseil constitutionnel if they have performed the duties of a replacement for less than three years.

TITLE II: OPERATION OF THE CONSEIL CONSTITUTIONNEL

Chapter 1: Common Provisions

Article 13

The Conseil constitutionnel shall meet upon a summons by its President or, where he is incapacitated, upon summons by the oldest of its members.

Article 14

The decisions and opinions of the Conseil constitutionnel are given by at least seven members, except in the case of a supervening event [*force majeure*] duly noted in the minutes.

Article 15

A decree made in the Council of Ministers on a proposal by the Conseil constitutionnel shall set out the organization of the secretariat.

Article 16

The credits necessary for the operation of the Conseil constitutionnel shall form part of the general budget. The President shall authorize expenditure.

Chapter 2: Declarations of Compatibility with the Constitution

Article 17

Organic laws passed by Parliament are communicated to the Conseil constitutionnel by the Prime Minister. The letter of communication shall indicate, if the case arises, that it is a matter of urgency.

The standing orders and amendments to standing orders passed by one or other of the chambers are communicated to the Conseil constitutionnel by the President of the chamber.

Article 18

When a *loi* is referred to the Conseil constitutionnel on the initiative of members of Parliament, the Conseil is seised of it by one or more letters containing signatures of a total of at least sixty deputies or sixty senators.

The Conseil constitutionnel, seised in accordance with articles 54 or 61 §2 of the Constitution, shall immediately notify the President of the Republic, the Prime Minister, and the Presidents of the National Assembly and the Senate. These latter shall inform the members of the chambers.

Article 19

The assessment of whether a *loi* is compatible with the Constitution is made following a report of one member of the Conseil within the time-limits fixed by article 61 §3 of the Constitution.

Article 20

The pronouncement of the Conseil constitutionnel shall be reasoned. It shall be published in the *Journal officiel*.

Article 21

The publication of a pronouncement by the Conseil constitutionnel that a provision is not incompatible with the Constitution puts an end to the suspension of the time-limit for its promulgation.

Article 22

In cases where the Conseil constitutionnel declares that the *loi* referred to it contains a provision that is contrary to the Constitution and is inseparable from the rest of this *loi*, the latter may not be promulgated.

Article 23

In cases where the Conseil constitutionnel declares that the *loi* referred to it contains a provision that is contrary to the Constitution without also finding that it is inseparable from the rest of this *loi*, the President of the Republic may either promulgate the *loi* except for this provision, or request the chambers to give it another reading.

In cases where the Conseil constitutionnel declares that the parliamentary standing order communicated to it contains a provision that is contrary to the Constitution, this provision may not be put into practice by the chamber that voted it.

Chapter 3: Examination of Texts in Legislative Form

Article 24

In the cases provided for in article 37 §2 of the Constitution, the Conseil constitutionnel is seised by the Prime Minister.

Article 25

The Conseil constitutionnel shall decide within a period of one month. This period is reduced to a week when the Government declares it to be urgent.

Article 26

The Conseil constitutionnel shall determine, by a reasoned pronouncement, whether the provisions submitted to it fall within the province of *loi* or *règlement*.

Chapter 4: Examination of Motions of Inadmissibility (fins de non-recevoir)

Article 27

In the cases provided for by article 41 §2 of the Constitution, the discussion of a private member's bill or of an amendment against which the Government raises a motion of inadmissibility, is immediately suspended.

The authority that seises the Conseil constitutionnel shall immediately notify the authority that also has the power to make such a reference under article 41 of the Constitution.

Article 28

The Conseil shall decide within a week, by way of a reasoned pronouncement.

Article 29

The pronouncement is sent to the President of the chamber in question, and to the Prime Minister . . .

CONSTITUTIONAL MATERIALS AND DECISIONS OF THE CONSEIL CONSTITUTIONNEL

I. Sources of the Constitution

1. LAW AND CONVENTION

MATERIAL 1: Extracts from the message of President Mitterrand to the National Assembly, 8 April 1986.

Background: This statement was made at the beginning of the period of 'cohabitation' between the Socialist President of the Republic and the right-wing Government that had just won the elections to the National Assembly. Since the Constitution did not provide specific rules on what should be done, the President and Prime Minister each worked out a series of conventions about their respective roles. This message of the President contains his understanding of those conventions.

TEXT

Our institutions are to be tested by events. From 1958 until now, the President of the Republic has been able to carry out his functions by relying on a majority and a Government that appealed to the same opinions as himself. No one can ignore the fact that the situation resulting from the recent legislative elections is totally different.

For the first time the parliamentary majority is made up of different political tendencies from those which were found in the presidential election, and this is what the composition of the Government expresses, as it should.

Faced with such a state of affairs, which they have, moreover, desired, many of our fellow citizens ask how the public authorities will operate. To this question, I only know one reply, the only one possible, the only reasonable one, the only one consistent with the interests of the nation: the Constitution, nothing but the Constitution, the whole Constitution.

Whatever view one has of it—and I personally do not forget my initial rejection, nor the reforms that I previously proposed on behalf of a vast current of opinion, and which I continue to believe desirable—it is the fundamental law. There is, in this matter, no other source of law. Let us adhere to this rule.

The circumstances that accompanied the birth of the Fifth Republic, the 1962 reform of the election of the head of State by universal suffrage, and an enduring identity of views between the parliamentary majority and the President of the Republic have created and developed usages that, beyond the texts, have added to the role of the former in public affairs. The change that has just occurred requires a new practice from all sides.

I will not delay you here with a declaration of the current competencies, of which I presume you are aware. I will recall merely that the Constitution attributes powers to the head of State that cannot be affected by a consultation of the electorate, in which his position is not at issue.

The proper functioning of public authorities, the continuity of the State, the nation's independence, the integrity of its territory, respect for treaties—article 5 designates these as areas where he exercises his authority and his arbitrament, and the provisions that result from it make this more precise. To these should be added his obligation to guarantee the independence of the administration of justice, and to safeguard the rights and liberties defined by the Declaration of 1789 and the Preamble of the 1946 Constitution.

For its part, the Government has the task, under the terms of article 20, to determine and conduct national policy. Subject to the prerogatives of the President of the Republic and to the confidence of the Assembly, it undertakes to implement the pledges to which it has bound itself before the French people. This responsibility is its own.

That being clearly established, President and Government have to seek, in all cases, the means that will enable them to serve the great interests of the country to their best ability and with a common accord.

But, Ladies and Gentlemen, what about Parliament? The legislative power retains, and should retain, the plentitude of its rights. To be sure, article 38 authorizes *ordonnances*, and the majority of Governments, including recently, have had recourse to this procedure. Thus, I did not believe that I should refuse this possibility to the present Government, having reminded it, however, that the great reforms of the preceding legislature, such as decentralization, nationalization, the rights of workers, the new liberties, the arrangement of working hours, had followed the normal legislative route.

I consider, therefore, that the *ordonnances*, which I have already said could not go back on social acquisitions (*les acquis sociaux*), should be few in number, and the enabling laws should be sufficiently precise for Parliament and the Conseil constitutionnel to make decisions with full information. I made known this observation to the Prime Minister when he presented to me the two bills that will be studied by the Council of Ministers tomorrow. It seemed to me all the more necessary since the combination of the *ordonnances* and article 49 §3 of the Constitution would, at the end of the day, risk reducing the deliberations of the Assembly excessively.

2. FUNDAMENTAL PRINCIPLES RECOGNIZED BY THE LAWS OF THE REPUBLIC

DECISION 1: CC decision no. 71-44 DC of 16 July 1971, *Associations Law*, *Rec.* 29, *GD*, no. 19; *SB* 64; J. Rivero, *Le Conseil constitutionnel et les libertés* (Aix, 1984), part I, ch. 1; J. Beardsley, (1972) 20 *American Journal of Comparative Law* 431; Nicholas, 87–92.

Background: This is the first decision of the Conseil constitutionnel that struck down a provision of a *loi* for breach of fundamental rights. Its justification appealed to the Preamble of the 1958 Constitution and to a fundamental principle recognized by the laws of the Republic, to be found in the *loi* of 1 July 1901 on associations. That *loi* provides that, before an association may be recognized as having legal status, it must file certain particulars with the prefect, who must then issue a certificate of registration.

In this case the National Assembly sought, against the opposition of the Senate, to pass a *loi* that would empower the prefect to refuse registration pending a reference to the courts over the legality of the objectives of a proposed association. The President of the Senate referred the *loi* to the Conseil. The principal issue was the constitutionality of prior restraint of the freedom of association.

DECISION

In the light of the Constitution and notably of its Preamble;

In the light of the *ordonnance* of 7 November 1958 creating the organic law on the Conseil constitutionnel, especially chapter 2 of title II of the said *ordonnance;*

In the light of the *loi* of 1 July 1901 (as amended) relating to associations;

In the light of the *loi* of 10 January 1936 relating to combat groups and private militias;

1. Considering that the *loi* referred for scrutiny by the Conseil constitutionnel was put to the vote in both chambers, following one of the procedures provided for in the Constitution, during the parliamentary session beginning on 2 April 1971;

2. Considering that, among the fundamental principles recognized by the laws of the Republic and solemnly reaffirmed by the Constitu-

tion, is to be found the freedom of association; that this principle underlies the general provisions of the *loi* of 1 July 1901; that, by virtue of this principle, associations may be formed freely and can be registered simply on condition of the deposition of a prior declaration; that, thus, with the exception of measures that may be taken against certain types of association, the validity of the creation of an association cannot be subordinated to the prior intervention of an administrative or judicial authority, even where the association appears to be invalid or to have an illegal purpose;

3. Considering that, even if they change nothing in respect of the creation of undeclared associations, the provisions of article 3 of the *loi*, the text of which is referred to the Conseil before its promulgation for scrutiny as to its compatibility with the Constitution, is intended to create a procedure whereby the acquisition of legal capacity by declared associations could be subordinated to a prior review by a court as to its compliance with the law;

4. Considering that, therefore, the provisions of article 3 of the *loi* are declared not to be compatible with the Constitution . . .

3. THE HIERARCHY BETWEEN THE DECLARATION OF THE RIGHTS OF MAN AND OF THE CITIZEN OF 1789, AND THE PREAMBLE TO THE 1946 CONSTITUTION

DECISION 2: CC decision no. 81-132 DC of 16 January 1982, *Nationalizations*, *Rec.* 18; *GD*, no. 33; *SB* 67; L. Favoreu, *Nationalisations et Constitution* (Aix, 1982); J. Rivero, *Le Conseil constitutionnel et les libertés*, part I, ch. 8. (The substance of this decision is reproduced in DECISION 29.)

Background: The Socialist Government elected in May–June 1981 sought to nationalize a number of strategic companies as well as the major banks. In its 1979 *programme commun* with the Communists it was envisaged that these nationalizations would require a constitutional amendment. Since the Socialists were in a minority in the Senate, a *loi* to amend the Constitution could not be passed. In any case, this would have taken some time, and thus impeded progress on implementing the policy. The Government adopted the course of having an ordinary *loi* passed by Parliament, with the National Assembly imposing its will over the Senate. The *loi* was referred to the Conseil constitutionnel by opposition deputies and senators. Among other questions, the issue was raised of whether such a reform could be undertaken within the existing Constitution, or whether an amendment was necessary. The Conseil d'État had advised that an amendment was necessary. The Conseil constitutionnel found that it was not, but it did impose strict limits on the power to nationalize by giving priority to the Declaration of 1789 over the Preamble to the 1946 Constitution.

DECISION

On the principle of nationalizations

13. Considering that article 2 of the Declaration of the Rights of Man and of the Citizen of 1789 proclaims that 'The ultimate purpose of every political institution is the preservation of the natural and imprescripible rights of man. These rights are to liberty, property, security, and resistance to oppression', and that article 17 of the same Declaration also proclaims that 'Property, being an inviolable and sacred right, none can be deprived of it, except when public necessity, legally ascertained, evidently requires it and on condition of a just and prior indemnity';

14. Considering that the French people, by the referendum of 5 May 1946, rejected a draft Constitution that would have preceded the provisions on the institutions of the Republic with a new Declaration of the Rights of Man, including, notably, a statement of principles differing from those proclaimed in 1789 by the above-mentioned articles 2 and 17;

15. Considering that, by contrast, by the referendums of 13 October 1946 and 28 September 1958, the French people have approved texts conferring constitutional value on the principles and rights proclaimed in 1789; that, in fact, the Preamble to the 1946 Constitution 'solemnly reaffirms the rights and liberties of man and of the citizen consecrated by the Declaration of Rights of 1789', and aims simply to complete them by the formulation of 'political, economic, and social principles as particularly necessary for our times'; that, in the terms of the Preamble to the Constitution of 1958, 'the French people solemnly proclaims its attachment to the rights of man and to the principles of national sovereignty such as are defined by the Declaration of 1789, confirmed and completed by the Preamble to the 1946 Constitution';

16. Considering that, if, since 1789 until today, the objectives and the conditions for the exercise of the right of property have undergone an evolution characterized both by a significant extension of its sphere of application to particular new areas and by limitations required in the name of the public interest, the same principles proclaimed by the Declaration of the Rights of Man retain full constitutional value both in so far as they concern the fundamental character of the right of property, whose preservation constitutes one of the purposes of political society, and which is placed on the same level as liberty, security, and resistance to oppression, and in so far as they concern the safeguards given to the holders of this right and the prerogatives of public authorities; that the freedom which, in the terms of article 4 of the Declaration, consists in the power to do anything that does not

cause harm to another itself cannot be preserved if arbitrary or abusive restrictions are imposed on the freedom of enterprise;

17. Considering that article 9 of the Preamble to the 1946 Constitution provides that: 'Any property or business whose exploitation has or acquires the character of a national public service or a *de facto* monopoly should become the property of the community'; that this provision has neither the purpose nor the effect of rendering inapplicable the principles of the Declaration of 1789 recalled above to the operations of nationalizations;

18. Considering that, if article 34 of the Constitution places within the province of *loi* 'the nationalization of undertakings, and the transfer of undertakings from the public to the private sector', this provision, just like the one confiding to *loi* the determination of the fundamental principles of the regime for property, does not dispense the legislature, in the exercise of its powers, from respecting the principles and rules of constitutional value that bind all organs of the State;

19. Considering that it appears from the preparatory materials for the *loi* submitted for scrutiny by the Conseil constitutionnel that the legislature intended to justify the nationalizations effected by the said *loi* by claiming that they are necessary to give public authorities the means to deal with the economic crisis, to promote growth, and to combat unemployment, and that they therefore arise from a public necessity within the meaning of article 17;

20. Considering that the legislature's judgment of the necessity of the nationalizations decided upon by the *loi* submitted for scrutiny by the Conseil constitutionnel should not be called into question by the latter, in the absence of any manifest error of evaluation, so long as it is not established that the transfers of property and businesses currently effected would restrict the area of private property and the freedom of enterprise to such an extent as to violate the said provisions of the Declaration of 1789 . . .

4. THE REPUBLICAN TRADITION

DECISION 3a: CC decision no. 88-244 DC of 20 July 1988, *Amnesty Law of 1988*, *Rec.* 119, *AJDA* 1988, 752, note Wachsmann; *SB* 75.

Background: This decision related to the amnesty law passed after the presidential elections of May 1988. *Inter alia*, the *loi* was challenged on its scope. The particular provision in question attempted to give an amnesty to the Renault Ten. These were employees' representatives sacked by Renault in 1986 for 'gross fault'. They had been involved in organizing demonstrations in July and August 1986 at the factory at Billancourt, on the outskirts of Paris, during which offices were ransacked and administrative personnel were attacked and threatened. These union officials were the subject of a campaign by the Communist party and the Communist trade union, the CGT. In an effort to achieve social harmony, the minority Socialist Government sought to include the amnestying of their breaches of contract within the amnesty law passed at the beginning of President Mitterrand's new term of office.

DECISION

10. Considering that the deputies, authors of the first reference, claim that article 15 exceeds the scope of an amnesty law, in that it applies not only to 'the criminal and paracriminal area', but means to govern 'matters occurring within the framework of a contract of employment between two private persons'; that this breaches both the republican tradition and the will of the drafters of the Constitution;

Concerning the republican tradition

11. Considering that the republican tradition can be validly invoked to argue that the text of a *loi* that contradicts it is contrary to the Constitution only in so far as this tradition has given rise to a fundamental principle recognized by the laws of the Republic;

12. Considering that, even if a large majority of the texts enacted on amnesty matters in republican legislation passed before the coming into force of the Preamble of the 1946 Constitution do not include provisions on the relationships arising from private law contracts of employment, apart from any criminal proceedings to which they may have given rise, it is nevertheless the case that the amnesty law of 12 July 1937 departed from this tradition; that, therefore, the tradition invoked by the authors of the reference could not be regarded, in any case, as having given rise to a fundamental principle recognized by the laws of the Republic in the sense of the first paragraph of the Preamble to the 1946 Constitution;

Concerning the will of the drafters of the Constitution

13. Considering that, according to article 34 of the Constitution: '*loi* shall determine the rules concerning . . . the determination of *crimes* and *délits*, as well as the penalties applied to them; criminal procedure; amnesties; the creation of new types of court, and the status of judges';

14. Considering that it cannot be concluded from the terms of these provisions, which are not only concerned with criminal law and the place of amnesty within it, that the Constitution has limited the competence of the legislature in amnesty matters to the area of *crimes* and *délits*, and, more generally, to offences that have been punished criminally;

15. Considering that, therefore the legislature could, without infringing any principle or rule of constitutional value, extend the scope of

the application of the amnesty law to disciplinary or professional sanctions, with the objective of establishing political or social peace . . .

5. DECISIONS OF THE CONSEIL CONSTITUTIONNEL

DECISION 3b: CC decision no. 88-244 DC of 20 July 1988, *Amnesty Law of 1988*, *Rec.* 119.

Background: By a decision of 22 October 1982 (DECISION 24), the Conseil constitutionnel had ruled that a *loi* could not create an immunity from liability in civil law for all actions committed in furtherance of a trade dispute, at least where they involved serious fault. It was argued that this decision prevented Parliament from passing an amnesty in the terms proposed for the Renault Ten.

DECISION

Concerning infringement of decisions of the Conseil constitutionnel

16. Considering that, according to the senators, authors of the second reference, the provisions of article 15-II infringe decision no. 82-144 DC of 22 October 1982, by which the Conseil constitutionnel declared that provisions forbidding all actions against employees, elected or appointed representatives, or organizations of workers for compensation for losses caused by a criminal offence, and for losses caused by actions manifestly incapable of being connected to the right to strike or trade-union rights, were contrary to the Constitution;

17. Considering that, by virtue of the second paragraph of article 62 of the Constitution, the decisions of the Conseil constitutionnel 'bind public powers and all administrative and judicial authorities';

18. Considering that the authority of *res judicata* that applies to the decision of the Conseil constitutionnel of 22 October 1982 is limited to the declaration of unconstitutionality concerning certain provisions of the *loi* that was then submitted to it; that it cannot be validly invoked against another *loi* couched, moreover, in different terms . . .

22. Considering that the provisions of article 15 risk threatening the freedom of enterprise of the employer, who, as the person responsible for the business, should be able, in consequence, to choose his collaborators; that, in certain cases, they may also infringe the personal freedom of the employer and employees of a business, by imposing on them the presence in the work-place of those who have committed acts of which they have been the victims;

23. Considering that respect for the rights and freedoms of third parties to the amnestied acts, and, *a fortiori*, of those who have suffered the consequences through no fault of their own, imposes restrictions

on the exercise of the power conferred on the legislature in amnesty matters . . .

25. Considering that, thus, taking account of the necessary reconciliation that has to be made between the rights and liberties of an individual and the rights and liberties of others, the amnesty law might validly provide that an employees' representative or a union official has the right to reinstatement in circumstances specified by the *loi*, namely, where he has committed a wrong in the course of his difficult duties that does not amount to gross fault; that the restrictions following from this reinstatement do not exceed in scope the burdens that society may impose on its members in the public interest, and are not manifestly disproportionate in relation to this public interest objective;

26. Considering that, by contrast, the right to reinstatement should not be extended to employees' representatives or union officials dismissed for gross fault; that, in such a case, there has been a clear abuse of the protected offices or mandates; that, in addition, the restriction that such a reinstatement would impose on an employer who has been the victim of this abuse, or who is at least not responsible for it, would manifestly exceed the personal or property sacrifices that can be required of individuals in the public interest; that, in particular, reinstatement has to be excluded in those situations where the serious fault justifying the dismissal has harmed employees of the business, who could, moreover, themselves be employees' representatives or union officials;

27. Considering that . . . the cited provisions of paragraph II of article 15 of the *loi* only remove the right of reinstatement from employees' representatives or union officials dismissed for gross fault amounting to an assault (*coups et blessures*), [which is not amnestied by article 7 of the *loi*]; that it follows that reinstatement is imposed in cases where an assault may have been really serious in character; that, moreover, reinstatement is granted as of right in all cases where the gross fault consisted of an offence other than assault; that such provisions manifestly exceed the limits that respect for the Constitution imposes on the legislature in amnesty matters . . . [Article 15-II was struck down to the extent that it restricted gross fault to assaults not amnestied by article 7 of the *loi*.]

DECISION 4: CC decision no. 89-258 DC of 8 July 1989, *Amnesty Law of 1989*, *Le Monde*, 11 July 1989; *SB* 80.

Background: The Renault Ten had been dismissed for 'gross fault' following incidents of violence during an industrial dispute. In order to meet the aspirations of unions and the Communists, the Government sought to amnesty the Ten after the labour courts and the Cour d'appel had upheld the dismissals in April 1989 (Versailles, 26 Apr. 1989, D. 1989, 386), and new industrial unrest was imminent. The amnesty law for the bicentenary of the Revolution of 1789 was drafted in such a way as to escape the restrictions imposed by DECISION 3b.

DECISION

1. Considering that in its decision no. 88-244 DC of 20 July 1988 the Conseil constitutionnel declared that, within article 15-II of the amnesty law relating to the right to reinstatement, the words, 'having consisted of assaults punished by a condemnation not referred to in article 7 of the present *loi*', were contrary to the Constitution; that it is clear from the reasons given for this decision that the right of reintegration could not be extended to employees' representatives or shop stewards dismissed for reasons of grave fault; that, indeed, the decision of 20 July 1988 points out that this situation would represent 'a clear abuse of protected offices or mandates' and, furthermore, that 'the restriction that such a reinstatement would impose on the employer, who has been the victim of this abuse, or who is at least not responsible for it, would manifestly exceed the personal or property sacrifices that can be required of individuals in the public interest; that, in particular, reinstatement has to be excluded in those situations where the gross fault justifying the dismissal had harmed employees of the business, who could, moreover, themselves be employees' representatives or union officials';

2. Considering that article 3 of the amnesty law currently under scrutiny intends to complement the first paragraph of article 15-II of *loi* no. 88-828 of 20 July 1988 by the following sentence: 'These provisions apply in the case of gross fault, except where the reinstatement would impose on the employer an excessive personal or property sacrifice'; that this recognizes a right to reinstatement in the business, distinct from the amnesty already granted, to employees' representatives or shop stewards dismissed for gross fault;

3. Considering that article 3 creates an exception in those cases where reinstatement would 'impose on the employer an excessive personal or property sacrifice';

4. Considering that the tempering thus effected leaves in place the general rule laid down by this article, which recognizes a right to reinstatement in the case of serious fault; that, in particular, it does not take into consideration the case where the victims of gross fault are employees of the business, who could themselves be employees' representatives or union officials;

5. Considering that such a provision violates the authority that attaches, by virtue of article 62 of the Constitution, to the decision of the Conseil constitutionnel of 20 July 1988; that it therefore follows that article 3 of the *loi* has to be declared incompatible with the Constitution . . .

II. *LOI* AND *RÈGLEMENT*

THE DIVISION BETWEEN *LOI* AND *RÈGLEMENT* PRIOR TO THE FIFTH REPUBLIC

MATERIAL 2: *Conclusions* of Romieu, CE, 4 May 1906, *Babin*, S. 1908. 3.110.

Background: These *conclusions* offer a classic statement of the respective legislative roles of the executive and the legislature as they were understood in the Third Republic. Many of the principles stated here form the basis of the separation of legislative functions in the Fifth Republic, and their interpretation by the Conseil constitutionnel and the Conseil d'État.

TEXT

There are no written rules of law concerning the separation of powers, a principle so important from a constitutional point of view. It is for case-law to fix them, drawing upon the terms in which the legislature's intervention in these various matters has been manifested until now. Sometimes this intervention is general, complete, and so the matter is said to belong by nature to the legislative sphere; sometimes this intervention is rare, accidental, and limited to specific points, and the matter can be considered as a whole to belong to the executive power, except for the spheres that have been separated by the legislature for its own purposes. . . .

Generally, one can say that to the legislative power, by their very nature, belong all those matters that, directly or indirectly, concern obligations to be imposed on citizens by authority, without any contractual relationship (for example, everything relating to the right to command and restrain, the organization of the police force and the courts, the seizure of private property, the voting of taxes and the public expenditure that gives rise to them, etc.). The legislature can obviously, as it often does, delegate its powers to other bodies, and invest them with the right to regulate these matters in its place; but, in the absence of a general or specific delegation, whether express or implied, from the legislature, the executive power is absolutely without competence, and it is for the legislative power alone to make decisions.

Conversely, it is, in principle, the executive power that regulates the internal organization of public services, and the conditions for their operation that do not affect the rights of third parties. In particular, it determines the rules of the contract between the administration and its agents, their recruitment, promotion, discipline, dismissal, etc. . . . [T]he executive power has full authority to decide on, and can freely determine, the terms of this contract, except with regard to those matters that the legislature has, by way of exception, made its own by regulating them itself.

MATERIAL 3: *Loi* no. 48-1268 of 17 August 1948.

Background: This *loi* was passed in an attempt to restrict the scope of delegation of legislative powers to the Government, and to define the appropriate areas in which Parliament should focus its legislative activity.

ARTICLE 7

The matters belonging to the regulatory sphere are:

- in the *public service*: the organization, abolition, transformation, fusion, operational rules, and control of services of the State, or those whose cost is primarily borne either by it or by State public bodies; the limitation and abolition of public employments; the retirement ages of civil servants and military personnel;
- on *public bodies*: the organization, transformation, fusion, operational rules, and control of public bodies of an industrial and commercial character, nationalized industries, and other bodies in receipt of public funds or taxes;
- on *social security*: the rules of functioning, funding, and financial and technical control over the various regimes of social assistance, social security, and family allowances, as well as over the organizations providing the service;
- in the *public finance* sector: the issue of Treasury bonds and loans, the administration of the public share portfolio; currency operations, exchange rates, price controls, and the functioning of economic control;
- on *social resources*: the control and use of energy, and the rationing of raw materials and industrial products. . . .

ARTICLE 8

[Exempted the status of radio, television, and the press, which remained within the sphere of *loi*.]

MATERIAL 4: Opinion of the Conseil d'État, April 1953,
RDP 1953, 170.

Question: What is the definition and exact scope of the prohibition contained in article 13 of the Constitution? How far may the Government, as expressly authorized by a *loi*, exercise regulatory power in legislative areas, and, consequently, repeal, modify, or replace provisions in *lois* by way of *règlements*?

OPINION

Whereas, on the one hand, the legislature may sovereignly determine the competence of the regulatory power, and may, to this end, decide that certain matters belonging to the competence of the legislative power shall become part of the competence of the regulatory power. The decrees made in these areas may modify, repeal, or replace provisions of *lois;* and these may themselves be modified by further decrees until the legislature reassumes [authority for] these matters in terms excluding henceforth the competence of the regulatory power.

Whereas, nevertheless, certain matters are reserved to *loi,* either by reason of provisions of the Constitution, or by the republican constitutional tradition resulting especially from the Preamble to the Constitution and from the Declaration of the Rights of Man of 1789, whose principles have been reaffirmed by the Preamble. The legislature cannot, therefore, extend the competence of the regulatory power to these areas; but it can limit itself to stating the essential rules, leaving to the Government the task of completing them . . .

Whereas, furthermore, the extension of the competence of the regulatory power would be contrary to article 13 if, by its generality and its imprecision, it demonstrated the will of the National Assembly to abandon to the Government the exercise of national sovereignty.

INTERPRETATIONS GIVEN TO THE 1958 CONSTITUTION

1. *The Power of Parliament in the Province of règlement*

DECISION 5: CC decision no. 59-1 FNR of 27 November 1959, *Price of Agricultural Leases, Rec.* 71; *GD,* no. 6; Nicholas, '*Loi, règlement* and Judicial Review in the Fifth Republic', 261, 271-2.

Background: Early in 1959 the Government decided to alter the method of determining the premium for agricultural leases, then fixed in accordance with article 812 of the Rural Code (created by the *loi* of 23 March 1953). By a decree of 7 January 1959 (the *loi* was passed prior to the 1958 Constitution, and so was capable of amendment by decree), instead of having the premium fixed by reference to the market price of wheat, one or both of the parties could stipulate any other foodstuff as the basis for calculation and index-linking. This provoked fierce opposition from farmers, and two senators, MM Bajeux and Boulanger, introduced a bill to abolish the decree. The Prime Minister invoked article 41 of the Constitution, claiming that the bill was inadmissible. The President of the Senate disagreed, and referred the matter to the Conseil constitutionnel on 19 November 1959.

DECISION

1. Considering that the provisions of article 34, §4 of the Constitution reserve to *loi* the determination of fundamental principles concerning the matters listed by that text; that it follows from the very terms

of these provisions, and from the comparison that has to be made between them and paragraphs 2 and 3 of the same article, that the Constitution has not included in the province of *loi* the determination of the rules necessary for the implementation of these fundamental principles on the matters in question; that, by virtue of the provisions of article 37, it belongs only to the authority invested with regulatory power to promulgate these rules whilst respecting the fundamental principles;

2. Considering that the bill, the admissibility of which is currently under discussion, aims to repeal the decree of 7 January 1959 relating to the price of agricultural leases, on the ground that it is *ultra vires* the regulatory power, and that the decree has as its essential purpose to allow either or both of the parties to request, at the expiry of each of the first two three-year periods of the lease, the partial substitution for the price of wheat of one or other of the foodstuffs mentioned in paragraph 1 of article 812 of the Rural Code, where the amount of an agricultural premium, stipulated to be payable in money, is fixed totally by reference to the price of wheat;

3. Considering that, in opposing the motion of inadmissibility to the aforesaid bill raised by the Prime Minister, who claimed that the regulatory power alone was competent on the matter of agricultural leases, the President of the Senate invoked the breaches that had been made by the decree of 7 January 1959 against the fundamental principles of the system of property and of civil obligations;

4. Considering that those of these principles which are involved here—that is to say, the free disposal by any owner of his property, the freedom of the will of contracting parties, and the unalterability of contracts—have to be understood within the framework of the generally applicable restrictions that have been imposed on them by previous legislation to permit certain interventions judged necessary for public authorities in the contractual relationships between individuals;

5. Considering that, since this concerns the issue of agricultural leases in particular, public authorities could therefore, without calling into question the existence of the principles recalled above, limit the scope for the free expression of wills of lessors and lessees by imposing certain terms on the implementation of their agreements, especially those concerning the methods of calculating and revising the amount of the premiums;

6. That the provisions of the decree of 7 January 1959, which merely confines itself to modifying the provisions contained in previous *lois*, could not, therefore, be regarded as involving a change in the fundamental principles on the subject;

7. That it thus follows that these provisions have a regulatory character, and that the Prime Minister was entitled to oppose the bill

mentioned above that aimed at repealing them as inadmissible within article 41 of the Constitution . . .

DECISION 6: CC decision no. 82-143 DC of 30 July 1982, *Freeze on Prices and Salaries*, *Rec*. 57; *RDP* 1983, 350; *GD*, no. 37; *SB* 93.

Background: Part of the counter-inflation measures that the Mauroy Government was obliged to introduce within a year of coming to office, this *loi* sought to co-ordinate the Government *règlements* and *lois* passed by Parliament. The Socialists were keen to present the full package of measures to Parliament, yet the opposition claimed that the *loi* voted by the majority, among other things, failed to respect the competence of the executive. While the objection might appear to have little merit, the decision marks the confirmation of an overlap between the competencies of the executive and Parliament, at least when the Government does not avail itself of the power given in article 41 of the Constitution to object to a bill as infringing its prerogatives.

The part of the *loi* considered here concerns the limitation on the payments of dividents to shareholders.

DECISION

Concerning article 3 §5 of the loi

9. Considering that, by the terms of this paragraph, 'Companies that contravene the provisions of this article are liable to a fine of between 20 and 50 F per share'; that, according to the deputies making the reference, this provision is objectionable on two grounds, in that it breaches the principle of criminal law whereby only human beings can be subjected to criminal penalties, and that it is not within the province of *loi* to create a fine at the level for a *contravention*;

10. Considering that, on the first point, there is no principle of constitutional value that prevents a fine from being imposed on a legal person;

11. Considering that, on the second point, if articles 34 and 37 §1 of the Constitution establish a separation between the province of *loi* and that of *règlement*, the meaning of these provisions has to be understood whilst taking account of that of articles 37 §2 and 41; that the procedure of article 41 permits the Government, during the parliamentary process and by way of a motion of inadmissibility, to oppose the inclusion of a regulatory provision in a *loi*, whilst that of article 37 §2 has the effect, after the promulgation of the *loi* and by means of the declassification procedure, of restoring to the Government the exercise of its regulatory power, and of giving it the right to amend such a provision by decree; that each of these procedures is optional; that it appears, thus, that the Constitution did not intend, through articles 34 and 37 §1, to make unconstitutional a provision of a regulatory character contained

in a *loi*, but had wanted, alongside the province reserved to *loi*, to grant the regulatory power an inherent province, and to confer on the Government, by the implementation of the special procedures of articles 37 §2 and 41, the power to ensure that it is protected against possible encroachment by *loi*; that, in these circumstances, the deputies making the reference cannot rely on the fact that the legislature has intervened in the province of *règlement* to argue that the challenged provision is contrary to the Constitution. . . .

2. Restrictions on Article 37 over and above Article 34: General Principles of Law

DECISION 7: CC decision no. 69-55 L of 26 June 1969, *Protection of Beauty Spots and Monuments, Rec.* 27; *GD*, no. 18.

Background: Following the presidential elections of 1969, in which environmental issues had been important, the Government sought to determine the status of certain provisions of three *lois* concerning the period of notice to be provided by the owners of beauty spots and monuments when seeking planning permission to carry out works on their sites.

DECISION

Concerning the provisions of article 4 §4, of the loi *of 2 May 1930, as amended by article 3 of the* loi *no. 67-1174 of 28 November 1967*

1. Considering that the provisions thus submitted for scrutiny by the Conseil constitutionnel fix a period of four months within which, as provided by article 4, the owners of sites or natural monuments registered on a departmental list have to notify the administration of their intention to undertake works on the lands that are included among the restrictions set out in the order creating the list in question, namely, for rural lands, works other than normal exploitation, and for buildings, works other normal maintenance;

2. Considering that, according to the terms of article 21 of the *loi* of 2 May 1930, as amended by article 8 of *loi* no. 67-1174 of 28 November 1967, breaches of the provisions of article 4 §4, mentioned above, are punished by a penalty on the correctional scale; that, thus, the failure to observe the notice period fixed in that paragraph is one of the constituent elements of a *délit*;

3. Considering that article 34 of the Constitution reserves to *loi* the determination of *crimes* and *délits*; that, therefore, the fixing, as well as any amendment, of the notice period provided by the above-mentioned provision, submitted for scrutiny by the Conseil constitutionnel, belongs to the competence of the Legislature; . . .

Concerning the provisions of article 9 §1, and of article 12 of the loi *of 2 May 1930, as amended respectively by articles 6 and 7 of* loi *no. 68-1174 of 28 December 1967*

4. Considering that the provisions in question are submitted for scrutiny by the Conseil constitutionnel only in so far as they can be regarded as conferring an express character on the special authorization that the owners of listed beauty spots or monuments, or those being listed, must obtain from the Minister of Cultural Affairs in order to proceed with alterations to the state or appearance of these places or monuments;

5. Considering that, according to a general principle of our law, silence maintained by the administration is equivalent to a decision to reject an application, and that, in this case, it cannot be derogated from except by a legislative decision.

Concerning the provisions of article 98-1 §2, of the Planning Code, as amended by article 44 of loi *no. 67-1253 of 30 December 1967 and article 3 of* loi *no. 69-9 of 3 January 1969*

9. Considering that, even if the principle lies within the competence of the legislature, whereby the implicit granting of a certificate stating that works comply with planning requirements can follow from the absence of any explicit expression of disapproval on the part of the administration during the period available for this purpose, the methods of applying this principle do not call into question any of the fundamental principles or any of the rules that article 34 of the Constitution places within the province of *loi*; that the provisions in question belong thus to the competence of the regulatory power . . .

3. *Rules and Principles: An Example*

DECISION 8: CC decision no. 61-17 L of 22 December 1961, *Ante-Natal and Maternity Allowances*, *Rec.* 43; S. 1963, 280, note Hamon.

Background: This case concerns the long-standing French governmental policy of attempting to increase the birth-rate. The allowances in question, part of the Social Security Code, were aimed at encouraging women to have children at a young age, within a few years of marriage. In order to improve the policy, the Government wished to widen some of the existing rules, and sought to have various articles contained in the *ordonnance* of 30 December 1958 (the finance law for 1959) declassified. Since *loi* only has to lay down the fundamental principles of social security, the case illustrates the wide scope left to *règlement* in settling the content of social security law.

DECISION

On article 13-1 of the ordonnace *of 30 December 1958*

1. Considering that, even if, in relation to the special regime of antenatal allowances, the existence of such allowances is itself among the 'fundamental principles of social security' belonging to the province of *loi* by virtue of article 34 of the Constitution, it is up to the regulatory power to fix the level of the said allowances, and, consequently, to make the amendments to which those levels may possibly be subject;

2. Considering that article 13-1 of the *ordonnance* of 30 December 1958, codified as article 518 §2, of the Social Security Code, is limited to amending the level of the monthly allowances in question; that this provision is thus of a regulatory character;

On article 14-1 §2, of the ordonnance *of 30 December 1958*

3. Considering that, even if, among the fundamental principles of social security specific to the system of maternity allowances, which, as such, belong within the province of *loi*, one must place the actual existence of these allowances, as well as the nature of the conditions required for them to be granted, it belongs within the province of the regulatory power to specify their content and especially those conditions relating to the age of the mother at the time of the birth, or to the rhythm of successive births, provided there is no distortion of the aforementioned conditions;

4. Considering that it follows that, in so far as it is limited to fixing the content of the conditions indicated above, relating to the maximum age of the mother at the birth, as well as the periods between the birth of children, the date of the marriage, or the date of previous births, the provision submitted for scrutiny by the Conseil constitutionnel is of a regulatory character.

4. *Inherent Regulatory Powers: Policing Powers*

MATERIAL 5: CE, 8 August 1919, *Labonne*, *Leb*. 737; *GA*, no. 39.

Background: Driving-licences were created by presidential decree in 1899. Having committed two traffic offences within a year, M. Labonne had his driving-licence withdrawn by the local prefect, as was authorized by the presidential decree. M. Labonne sought to argue that the removal of his licence was unlawful, and that the system of driving-licences was, in any case, invalid.[1]

[1] In this and other texts the French word 'police' has been translated as 'policing', but it has a much wider sense than in English.

DECISION

Considering that, in his application to quash the prefectoral decision withdrawing his driving-licence, the applicant limits himself to challenging the legality of the decree of 10 March 1899, of which the decision is a mere application; that he claims that the aforementioned decree is *ultra vires* with respect to articles 11, 12, and 32, by which the licence is created and provision is made for the possibility of withdrawing it;

Considering that, since departmental and municipal authorities are authorized by statute, notably by the *loi* of 22 December 1789–8 January 1790 and that of 5 April 1884, to oversee the maintenance of public highways and the safety of traffic, the head of State is entitled, without any delegation by a *loi* and by virtue of his inherent powers, to specify policing measures that should, in any case, apply to the whole of the territory, provided, of course, that the above-mentioned authorities retain, each in its own area, full and complete competence to add to the general regulations promulgated by the head of State any additional regulatory prescriptions that the public interest may require in its locality;

Considering that, therefore, because of the dangers that automobile traffic presents, the decree of 10 March 1899 could validly require that all motor-car drivers should carry a driving-licence, issued in the form of a certificate of competence; that the power to grant this licence, given to the administrative authorities by that decree, carried with it necessarily the right of the same authority to withdraw the licence in the case of a serious failure to conform to traffic regulations; that it follows, thus, that the decree of 10 March 1899 and the prefectoral decision of 4 December 1913 are not illegal . . .

The claim was rejected.

DECISION 9: CC decision no. 87-149 L of 20 February 1987, *Protection of the Environment*, *Rec*. 22.

Background: The Prime Minister made a reference under article 37 §2 of the Constitution to declassify certain provisions of the Rural Code and other related texts on environmental protection.

DECISION

Concerning articles 366 bis *I, 366* bis *II, 366* bis *III §1, 373 §4 no. 4, and §§8 and final of the Rural Code, article 22-I of* loi *no. 74–1114 of 27 December 1974, article 3 §2 of* loi *no. 68–918 of 24 October 1968, as modified by article 13 of* loi *no. 75–347 of 14 May 1975, articles 19 §1, 21, 22 §3, 27 §§2, 4 of* loi *no. 76–629 of 10 July 1976, as modified by article 58 of* loi *no. 83–663 of 22 July 1983*

1. Considering that, in so far as they allocate competence, the above-mentioned provisions designate the administrative authority empowered on behalf of the State to exercise the functions that, by virtue of the *loi*, lie within the competence of the executive power; that they are, therefore, of a regulatory character;

Concerning article 373 §§5 and final of the Rural Code, insofar as they fix rules of procedure

2. Considering that the above-mentioned provisions have as their objective, on the one hand, to prescribe that the plan for the hunting of large game is created at a departmental level, 'on the recommendation of the prefect, presented at the joint request of the keeper of lakes and forests and of the departmental president of hunters', and, on the other hand, to subordinate the drafting of a plan for the hunting of large game in mountainous zones to the prior opinion 'of the president of the departmental federation of hunters and of the communes affected';

3. Considering that, to the extent that they determine the functions of State officials or of officials of an association performing a public service in the exercise of a power conferred upon the State by *loi*, the provisions referred to do not call into question fundamental principles of the system of property, nor any other fundamental principles or rules that article 34 of the Constitution places within the province of *loi*;

4. Considering that, having regard to the purpose and effects of the creation of a plan for hunting large game in mountainous zones, the obligation to obtain the prior opinion of the affected communes before it is drawn up cannot be considered as affecting the fundamental principles of the free administration of local authorities, nor any other principles or rules that fall within the province of the legislature;

5. Considering that, therefore, the provisions submitted for examination by the Conseil constitutionnel are of a regulatory character;

Concerning article 384 §1 of the Rural Code

6. Considering that *loi* no. 67–468 of 17 June 1967 re-enacted, as article 384 §1 of the Rural Code, provisions whose origin goes back to the *loi* of 23 February 1926, and the terms of which state: 'The Government shall exercise supervision and policing of hunting in the public interest';

7. Considering that, even if article 34 of the Constitution has not withdrawn from the head of the Government the general policing functions that he used to exercise before, by virtue of his inherent powers and outside any specific legislative authorization, the creation of a special policing regime for hunting calls into question fundamental principles of property; that it follows that, to the extent that they confer

the task of policing hunting on a State body, the provisions submitted for consideration by the Conseil d'État are within the province of *loi*;

8. Considering, nevertheless, that the distribution of functions between State administrative bodies falls within the province of *règlement*; that, in consequence, and in so far as they have the effect of designating among those bodies the one that will exercise the policing of hunting, the above-mentioned provisions have a regulatory character . . .

Concerning article 6 §1 of loi *no. 64–696 of 10 July 1964*

14. Considering that, by the terms of article 6 §1 of the *loi* of 10 July 1964, 'communal associations have to be constituted within a period of one year from the publication of ministerial or prefectoral orders establishing or completing the list of departments or communes mentioned in article 2';

15. Considering that, even if the period thus provided is not prescribed on pain of the nullity of the constitution of the association, it none the less imposes an obligation on the administration that, in the light of the conditions for creating associations and for their activities, affects 'fundamental principles of the system of property' and belongs to the competence of the legislature; that, in consequence and to that extent, the cited provisions fall within the province of *loi*; that, however, the specification of the administrative bodies entrusted with the performance on behalf of the State of the functions provided by the *loi* belong to the competence of the regulatory power . . .

Concerning article 14-VIII §1 of loi *no. 68–1172 of 27 December 1968*

16. Considering that the provisions of article of 14-VIII §1 of the *loi* of 27 December 1968 have the purpose of submitting litigation relating to compensation for damage caused to crops either by wild boars or by large game to 'the jurisdiction of the Tribunal d'instance, which decides in first and last instance, within the limits of its jurisdiction of first and last resort on personal and property matters, and by way of appeal, on whatever value the claim may amount to';

17. Considering that, even if, in applying the provisions of article 34 of the Constitution, by virtue of which *loi* determines the rules concerning the fundamental safeguards granted to citizens for the exercise of civil liberties, it is the role of the legislature, whilst respecting principles of constitutional value, to determine the competence of administrative and private law courts, by contrast, the designation of the competent court within the private law court system, in an area outside criminal procedure, does not affect fundamental principles nor any of the rules that are within the province of *loi*; that, therefore, the provisions submitted to the Conseil constitutionnel, in so far as

they specify, within the system of private law courts, the Tribunal d'instance as competent, fall within the province of *règlement* . . .

5. *Criminal Penalties*

MATERIAL 6: CE, 12 February 1960, *Société Eky*, *Leb*. 101; Nicholas, '*Loi, règlement* and Judicial Review in the Fifth Republic', 266.

Background: The company operated a radio station just over the border from France, but broadcasting into French territory. As part of its advertising, it ran competitions, the prizes for which were vouchers redeemable against commercialized goods. New legislation on the reform of the currency made it an offence to use and accept methods of payment for goods and services other than in French currency. The company considered that this would adversely affect its operations, and made two applications on 24 February 1959 to quash the new articles creating the offences, introduced into the Penal Code by article 13 of the *ordonnance* of 23 December 1958, and article 2 of decree no. 58–1303 of 23 December 1958. The first was held to be unchallengeable on the ground that it was a legislative act made under the transitional powers of article 92 of the Constitution. The second application concerning the decree is considered here.

DECISION

On the arguments based on the breach of article 8 of the Declaration of the Rights of Man, and of article 34 of the Constitution

2. Considering that, even if article 8 of the Declaration of the Rights of Man of 1789, to which the Preamble of the Constitution refers, lays down the principle that 'no one may be punished except according to a *loi* passed and promulgated prior to the offence', article 34 of the Constitution, which lists the matters falling within the province of *loi*, provides that *loi* shall determine 'the rules concerning the determination of *crimes* and *délits*, as well as the penalties applied to them'; that neither this article nor any other provision of the Constitution provides that the area of *contraventions* belongs to the province of *loi*; that, thus, it follows from the Constitution as a whole, especially from the quoted terms of article 34, that its authors have excluded from that province the determination of *contraventions* and the penalties attached to them, and, consequently, have intended specifically to derogate on this point from the general principle expressed by article 8 of the Declaration of the Rights of Man; that, therefore, the area of *contraventions* falls within the province of *règlement* according to article 37 of the Constitution;

3. Considering that, according to article 1 of the Penal Code, an offence that is punished by penalties on the police scale is a *contravention*; that it follows from articles 464, 465, and 466 of the same Code, that penalties on the police scale are imprisonment for a period not exceeding two months, a fine of up to a maximum of 200,000 F,

and the confiscation of certain articles seized; that the challenged provisions of articles R30 *et seq.* of the Penal Code punish those who have accepted, held, or used methods of payment with the purpose of supplementing or replacing monies of legal tender with a fine of 2,000 to 4,000 F, and with imprisonment for up to three days, and eight days in the case of persistent offenders; that these offences, being punished by penalties on the police scale, constitute *contraventions*; that therefore it is a correct application of the Constitution that the Government, by *règlement*, has defined and fixed the penalties applicable to them . . . [The company likewise failed in its other arguments.]

DECISION 10: CC decision no. 73-80 L of 28 November 1973, *Criminal Penalties (Rural Code), Rec.* 45; D. 1974, 269; *SB* 95; Nicholas, 92–7.

Background: The reference under article 37 §2, was made as part of the reform of the Rural Code, submitting a number of provisions contained in *lois* for declassification. Only one interests us here, the occasion on which the Council included an observation in 'a little phrase' (Touffait) that went against the established view expressed in the *Société Eky* decision.

DECISION

Concerning the provisions of article 188–9 §1 of the Rural Code, submitted for scrutiny by the Conseil constitutionnel

1. Considering that the provisions mentioned above have the purpose of punishing, with a fine of 500 to 2,000 F, the failure to submit an application for prior authorization, or a prior declaration in the case of the concurrent holding or joinder of agricultural holdings or of certain other concurrent holdings;

2. Considering that it follows from the provisions of paragraphs 3 and 5 of article 34 and of article 66 of the Constitution, combined with the Preamble, that the determination of *contraventions* and the penalties applicable to them are within the province of *règlement* when those penalties do not involve measures depriving a person of their liberty;

3. Considering that it emerges from the combined provisions of articles 1 and 466 of the Penal Code that the penalties of a fine, which may not exceed 2,000 F, constitute penalties on the police scale punishing *contraventions*; that, therefore, the above-mentioned provisions of article 188–9 of the Rural Code, which only provide for a penalty of a fine not exceeding 2,000 F, belong to the competence of *règlement* . . .

6. Specific Delegations: The Use of Article 38

DECISION 11: CC decision no. 76-72 DC of 12 January 1977, *Djibouti Elections*, D. 1980, 233, note Hamon.

Background: Soon after coming into power, the Barre Government sought parliamentary approval for its programme. Subsequently, as part of the independence measures being settled for Djibouti, it proposed to fix the boundaries of constituencies for the new elections by way of *ordonnance*. The opposition deputies challenged the validity of this authorization.

DECISION

1. Considering that, according to the terms of article 38 §1 of the Constitution: 'For the implementation of its programme, the Government may ask Parliament for authority to take measures that are normally within the province of *loi* by *ordonnance* for a limited period';

2. Considering that, even if the wording of article 38 §1 of the Constitution specifies that it is for the implementation of its programme that the Government is given the possibility of asking Parliament for power to legislate by *ordonnance* for a limited period, this text should be understood as requiring the Government to indicate with precision to Parliament, when the enabling bill is presented and in order to justify it, the objective of the measures that it proposes to take;

3. Considering that, therefore, it is necessary to exclude any other interpretation, and especially that drawn by analogy with the provisions of article 49 §1; that the latter, in reality, tends towards giving a meaning to 'programme' that is analogous to the expression, 'declaration of general policy', which, on the one hand, would leave no room among potential justifications for recourse to the provisions of article 38 for notions of unforeseen circumstances or for situations requiring urgent measures, and, on the other hand, because of its generality, would result in an extension, without defined limits, of the field of application of the enabling procedure provided in article 38, to the detriment of respect for the prerogatives of Parliament;

4. Considering that, in this case, the detail required by reason of article 38 §1 of the Constitution has been duly provided by the Government in support of its request for the authority to amend by *ordonnance* the constituencies for the election of members to the Chamber of Deputies in the French territory of the Afars and the Issas; [The law was thus declared to be consistent with the Constitution.]

DECISION 12: CC decision no. 86-207 DC of 25, 26 June 1986, *Privatizations*, *Rec*. 61; *GD*, no. 41; *SB* 82. [The substance of this decision is set out further in DECISION 30.]

Background: Following elections in March 1986, the right-wing coalition headed by Chirac formed the Government, whilst the Socialist President Mitterrand remained at the Elysée Palace. There was no guarantee that this period of 'cohabitation' between right and left would last for a long time. As a result, the new Government asked Parliament to pass two enabling laws to implement the main parts of its election platform as soon as possible. The first of these laws sought to liberalize the economy by reducing State controls and by 'letting the public sector breathe' (*la respiration du secteur public*), a programme that involved privatizing not merely the companies nationalized by the Socialists in 1982, but also those nationalized by earlier administrations as far back as the Liberation of 1944. A total of sixty-five companies were to be privatized over a period of five years.

REFERENCE BY DEPUTIES: JO *27 JUNE 1986: 7984–5*

Mr President, Messrs, Counsellors: In accordance with paragraph 2 of article 61 of the Constitution, we have pleasure in referring to the Conseil constitutionnel the text of the *loi* enabling the Government to take miscellaneous economic and social measures, in the form that has been finally adopted by Parliament . . .

Concerning article 1 of the loi

Since, in the first place, it concerns competition law, the delegation of power granted obviously exceeds the limits permitted by article 38. In its decision no. 76–72 DC of 12 January 1977, the Conseil constitutionnel emphasized the obligation imposed on the Government to indicate with precision to Parliament, when the enabling bill is presented, the objective of the measures that it proposes to take. Naturally, this is not a simple act of courtesy with regard to Parliament, no more is it an exclusively political obligation, but rather it is a condition on which the constitutionality of the enabling bill depends. Moreover, it is this requirement of precision on the objectives that makes all the difference between an enabling law and a law granting full power [to do whatever the Government wants]. The Constitution of the Fifth Republic only permits the first, and excludes the second.

Now, this condition is obviously not satisfied in this case. Moreover, implicitly, but inevitably, the text of the *loi* recognizes this, not without a degree of candour. In fact, the use of the verb 'define' suffices to demonstrate that it is impossible for Parliament to know and assess the objectives of the measures that the Government proposes to take in the area of competition law.

In vain is the second paragraph of article 1 [of the *loi*] cited. This cannot be seriously considered as providing the indispensable details (and, in any case, such details are the opposite of the notion of a definition in the future). At most, it provides superfluous verbiage in a measure in which the prescriptions that it lays down are necessary in any case, because they follow from constitutional requirements, and their inclusion in the *loi* does not add anything to their binding force, just as their omission would not detract from that force.

Thus, in so far as a definition of a new competition law is concerned, Parliament has not been in position to know with precision the objectives of the measures that the Government proposes to take.

But there is more. We would not bring an action against the Government for having wanted to hide its intentions. The reality is more prosaic and more surprising. If the Government has not indicated what it intended to do with the delegation, it is simply because it did not know itself. For proof, one could cite the mission given to a high-ranking civil servant, Mr Donnedieu de Vabres, to produce proposals for a new definition of competition law (see the statement of the Minister of State for the Economy, Finance, and Privatization, *Journal officiel*, 'Débats parlementaires', AN 208).

Once again the requirements of logic, like those of the Constitution are simultaneously breached. In truth, logic and, moreover, article 38 of the Constitution as it is interpreted by the Conseil constitutionnel would have wanted the Government to confide its mission to Mr Donnedieu de Vabres, then, in the light of his findings and if it seemed useful, it could have requested the necessary delegation from Parliament for the implementation of those proposals which the Government had chosen to adopt, and on which it could then, and only then, provide the necessary details.

The Government has chosen to proceed in reverse order, an attitude that goes against good sense, which is regrettable, but, above all, it breaches constitutional requirements, which is unacceptable. That is why the Conseil constitutionnel will be driven to declare the provisions authorizing the Government to define by *ordonnance* 'a new competition law' inconsistent with the Constitution.

Since, in second place, it concerns 'the freedom of operation of businesses', article 1 pursues the objective of the repeal of *ordonnance* no. 45–1483 of 30 June 1945 relating to prices. It is not simply a matter of a mere possibility, but rather of a clearly stated intention, especially by the Minister of State for the Economy, Finance, and Privatization in his statement on existing legals texts: 'We will dismantle these regulations as far as they involve exchange controls. They must be abolished in relation to prices, and the law must be set in tune with economic realities.' (*Journal officiel*, 'Débats parlementaires', AN 208.) In doing this, the Government was going in the direction desired by the general reporter of the Finance Committee of the National Assembly, who stated in relation to the *ordonnance* of 1945 on prices: 'The existence of legislation, of regulation, itself brings a temptation to intervene. We must recognize this clearly: from this very human temptation, the new majority is no more preserved than the previous one. We know it so well that we are undertaking to suppress the source of temptation by repealing the *ordonnances* of 1945.' (*Journal officiel*, 'Débats parlementaires', AN 203.)

Thus, there is no doubt as to the intention, repeatedly stated, to repeal purely and simply the *ordonnance* of 1945 relating to prices, without replacing it by other provisions. Now it is right to ask whether it is constitutionally possible to act in this way.

In truth, the repeal of the *ordonnance* of 1945 relating to prices would leave the State completely defenceless in a serious economic crisis. There would only remain the provisions of article 46 §1 of the *loi* of 11 July 1938 on the general organization of the State in time of war, which effectively allows decrees made in the Council of Ministers to 'tax' certain resources. All the same, these powers cannot be used except in a period of war or of 'external tension when circumstances so require'. Taking the 1938 text as a whole, it follows that, even if it permits measures to be taken in peacetime, this can only occur in relation to an actual or potential armed conflict. The notion of 'external tension' can only be interpreted in this sense.

Now, the main characteristic of modern economies is their complexity, due to their interconnection on a planetary scale. From this fact, as the history of the twentieth century has abundantly and dramatically demonstrated, a crisis could break out at any moment, in any part of the world, producing marked economic dislocations, going from a mere price-rise in raw materials to the drying-up of their supply, going from a mere reduction in the purchasing power of consumers to the emergence of explosive social tensions. It is certain that these can occur outside any situation of 'external tension' within the meaning of the *loi* of 1938, and thus do not permit the use of the measures that it would make available without the risk that these would be annulled by a court.

In the same way, the situation envisaged could not come within the scope of the situations envisaged by article 16 of the Constitution [on a state of emergency].

In this situation, France would find itself the only developed country that did not have any power of immediate intervention on prices in case of a serious and sudden crisis. It should be noted that, in reality, all modern countries, however liberal their economic policies, have given the State powers to act at any moment (see R. Savy, 'Les Pouvoirs exceptionnels', *Pouvoirs* 10: 80 ff.), and that this in no way involves a duty to so act, but merely a necessary precautionary measure.

Of course, one could reply that the purpose of the *ordonnance* of 1945 relating to prices was not to deal with exceptional circumstances. Such a remark would be correct, but irrelevant. Given that the Government could intervene on prices in a period of normality, it would be even more justified in doing so in a crisis, and so it was not necessary to refer specifically to this situation among the reasons for the *ordonnance*. But, in permitting its repeal, the enabling law would remove at a single stroke both the ordinary powers, which might be acceptable, and those for use in a crisis, which is not acceptable.

Thus—and, at the end of the day, this lies behind our argument—the intervention of the State in the situations just mentioned is not a mere option, but represents, by contrast, a constitutional obligation. On the one hand, the Preamble to the 1946 Constitution, affirmed by that of the Constitution of 1958, declares that 'The nation proclaims the solidarity and equality of all French men and women before the burdens that result from national calamities', and neither solidarity nor equality can be assured if the State is not in a position to act immediately. On the other hand, the constitutional principle of the continuity of the State only has a meaning if the continuity of national life itself can be secured, and the State especially needs to be able to supervise this in case of necessity.

In such a situation, to give the State the means to intervene on prices at least in a period of crisis is a constitutional requirement, even if the Government abstains from making use of them, which, after all, depends on its decision alone, even given that these provisions are formally restricted to exceptional circumstances.

It is not irrelevant in this context to refer to the decision no. 83–165 DC of 20 January 1984 [*University Professors*], and to consider, *mutatis mutandis*, that the total repeal of the *ordonnance* of 1945 relating to prices, of which certain provisions provide citizens with safeguards in accordance with constitutional requirements that are not replaced by equivalent safeguards, is inconsistent with the Constitution.

Of course, one could consider that a situation of crisis would authorize the Government to take measures that are normally forbidden to it, as the Conseil d'État has accepted through its case-law on exceptional circumstance. But the argument would be irrelevant, since the Conseil d'État has always refused to treat the fixing of the sale price [of goods] as having the purpose of ensuring public order (CE, 6 July 1928, *Syndicat des laitiers*, *Leb*. 548; 11 April 1924, *Groyer*, *Leb*. 382), whereas, by contrast, it has only allowed the intervention of the public authority on prices when this has been expressly made possible by a statutory provision [*une disposition de valeur législative*] (CE, 21 May 1920, *Tison*, *Leb*, 321).

Thus it is certain that, if the *ordonnance* of 1945 relating to prices is eventually repealed, no measure could be taken by a Government faced with a serious crisis without the prior intervention of the legislature, and all the delay that entails. By disregarding the kind of constitutional requirements that it is the State's duty to meet, this provision of the *loi* in question can, in its turn, only be declared as inconsistent with the Constitution . . .

DECISION

On the substance of the loi

13. Considering that, since it is specified in article 38 §1, cited above, that it is for the implementation of its programme that the Government is given the possibility of asking Parliament for authority to take measures that are normally within the province of *loi* by *ordonnance* for a limited period, this text should be understood as requiring the Government to indicate with precision to Parliament the objective of the measures that it proposes to take, and the areas in which they will intervene;

14. Considering that the provisions of an enabling law cannot have as either a purpose or a consequence to dispense the Government from respecting rules and principles of constitutional value when it is exercising the powers conferred on it in application of article 38 of the Constitution;

15. Considering that it is up to the Conseil constitutionnel on the one hand to verify that the enabling law does not contain any provision that permits a breach of those rules and principles, and, on the other hand, not to accept an enabling law as being consistent with the Constitution except on the express condition that it is interpreted and applied with strict respect for the Constitution;

16. Considering that the authors of both references raise complaints of unconstitutionality against each of the articles 1 to 7 of the *loi*;

Concerning article 1 of the loi

17. Considering that article 1 of the *loi* is drafted as follows:

> In order to assure businesses a greater freedom of operation and to define a new competition law, the Government is authorized, within six months of the publication of the present *loi* and under the terms provided for in article 38 of the Constitution, to amend or repeal certain provisions of economic legislation concerning prices and competition, especially those of *ordonnances* no. 45–1483 of 30 June 1945 on prices, and no. 45–1484 of 30 June 1945 on the determination, prosecution, and punishment of economic offences.
>
> In the definition of the new competition law, it shall provide safeguards for the economic actors in the exercise of the powers given to public authorities, and ensure rights to a hearing in their procedures;

18. Considering that the deputies making the first reference argue, first, in support of the claim of unconstitutionality made against these provisions, that the objective of the measures that the Government proposes to take is defined in a manifestly imprecise way; that it is not even established that the Government itself knows what this objective is, as is witnessed by the fact that, during parliamentary debates, it

announced that a senior official would be given the task of producing proposals aimed at defining a new competition law;

19. Considering that, secondly, the deputies making the first reference argue that the repeal of the *ordonnances* of 30 June 1945, announced by the Government during the parliamentary debates, would have the consequence, in cases of an economic or financial crisis, of depriving the Government of any means for immediate action on prices; that, thus, the *loi* submitted for scrutiny by the Conseil constitutionnel permits the abolition of safeguards relating to the respect for constitutional requirements, such as the equality and solidarity of the French before the burdens that follow from national disasters, or the continuity of the life of the nation;

20. Considering that the senators making the second reference formulate criticisms similar in substance, by emphasizing the imprecise character of the terms purporting to define the objectives of the authorization and the scope thus opened for *ordonnances;*

Concerning the argument that the terms of the authorization are insufficiently precise

21. Considering that, even if the Government has to define precisely the objectives of the authorization that it requests in order to achieve its programme, it is not bound to make known the content of the *ordonnances* that it will make by virtue of that authorization, and it is not impermissible for it to make their content depend upon the results of work and studies of which it will only know the details later;

22. Considering that, if article 1 of the *loi* designates as the objective of the *ordonnances* that it authorizes the Government to make the definition of a new competition law and the search for greater freedom of operation for businesses, it does not authorize the Government thereby to amend or repeal the totality of the rules of civil, commercial, criminal, administrative, and social law affecting economic life; that it follows from its terms, clarified by the *travaux préparatoires* and, especially, by the statements of the Government to Parliament, that the authority thus sought involves the amendment or repeal of specific provisions of economic legislation relating to the control of combinations, to competition, and to prices, as well as to the punishment of economic offences contained in the *ordonnances* of 30 June 1945, in *loi* no. 77–806 of 19 July 1977, and in special legislative provisions on prices; that, within these limits, the authority granted by article 1 is not contrary to the terms of article 38 of the Constitution;

23. Considering that the clarification provided by paragraph 2 of article 1 on the safeguards for economic actors and on the observance of rights to a hearing in procedures should not be understood as excluding other safeguards resulting from the principles and rules of

constitutional value, and, in particular, the right to judicial review and the rights of due process; that, furthermore, they should not be understood as excluding the safeguards for human or legal persons not having the status of economic actors;

24. Considering, likewise, that the *ordonnances* should not be contrary to the international obligations of France, in breach of article 55 of the Constitution;

Concerning the possible repeal of the ordonnances *of 30 June 1945 relating to prices*

25. Considering that no principle or rule of constitutional value requires the legislature to enact texts of permanent force conferring special powers on the Government in case of possible eventualities; that, moreover, article 1 of the *loi*, even if it authorizes the amendment or repeal of the *ordonnances* of 30 June 1945 relating to prices, does not permit the amendment or repeal of the rules or principles currently in force giving power to the Government or to agents of a public authority in the case of crisis, exceptional circumstances, or national disaster;

Concerning the whole of article 1

26. Considering that, within the limits of, and conditional upon, the interpretation that has just been given, article 1 is not contrary to the Constitution . . .

III. Parliamentary Procedure

DECISION 13: CC decision no. 59-2 DC of 17, 18, 24 June 1959, *Standing Orders of the National Assembly*, *Rec.* 58; *GD*, no. 3.

Background: The new chambers of the Fifth Republic had to adopt revised standing orders, which were submitted to the Conseil constitutionnel for approval. This decision concerns the first set of standing orders produced by the National Assembly. The National Assembly was required to produce standing orders for its operation under the new Constitution. In order to rationalize Parliament, it was thought best to limit speaking time in debates. To be even-handed, the standing orders proposed that the same time-limits apply to both Government and deputies. In a number of respects, they tried to carry over practices adopted under previous Constitutions. In particular, they wished to continue the practice of passing resolutions that enjoined the Government to take certain actions. The Government objected to this as an infringement of its freedom.

DECISION

The following articles of the standing orders of the National Assembly are declared inconsistent with the Constitution:

Article 31-5, on the ground that the provisions of this text, in so far as they allot five minutes to the Government for speaking [in debate], are contrary to those of article 31 of the Constitution, which states that the members of the Government shall be heard on request, without the length of their intervention being limited.

Articles 81-1 and -4, 82, 86-3 and -4, 92-6, 98-6 and 134-5, in so far as these contain provisions relating to proposals for resolutions, on the ground that, to the extent that such proposals would aim to direct or control governmental action, their use would be contrary to provisions of the Constitution, which, in article 20, entrusting to the Government the determination and conduct of national policy, only provides for the Government's responsibility to be questioned in the situations, and following the procedures, set out in articles 49 and 50;

That, in so far as the proposals for resolutions belong to the right of initiative of members of Parliament in legislative matters, and since that right is defined by the provisions of articles 34, 40, and 41 of the Constitution, and in addition to the fact that they would duplicate private members' bills (*propositions de loi*), the practice of resolutions would be contrary to the letter of the Constitution, especially articles 40 and 41, the formulation of which only covers private members' bills, the passing of which alone can have the consequence of reducing public resources, creating or increasing public charges, and affecting the regulatory power of the Government defined by article 37 or any delegation granted to it by application of article 38;

That it follows from what has been said that, without breaching the Constitution, the above-mentioned articles of the standing orders of the National Assembly relating to legislative procedure and parliamentary control cannot assign to proposals for resolutions a purpose that is different from their own, namely, the formulation of measures and decisions relating to the exclusive competence of the Assembly, in other words, internal measures and decisions concerning the functioning and discipline of the said Assembly, to which it may be appropriate to add only the cases expressly provided for by constitutional and organic texts, such as articles 18 ff. of *ordonnance* no. 59–1 of 2 January 1959, forming the organic law on the High Court of Justice . . .

DECISION 14: CC decision no. 62-20 DC of 6 November 1962, *Referendum Law*, *Rec.* 27; D. 1963, 348, note Hamon; *GD*, no. 14; *SB* 55; Nicholas, '*Loi*, *règlement* and Judicial Review in the Fifth Republic, 259–60, 264–5.

Background: This is explained above, pp. 133–4. The reference was made by the President of the Senate.

DECISION

1. Considering that the competence of the Conseil constitutionnel is strictly limited by the Constitution, as well as by the provisions of the organic law of 7 November 1958 on the Conseil constitutionnel . . . that the Conseil constitutionnel cannot be called upon to rule on matters other than the limited number for which those texts provide;

2. Considering that, even if article 61 of the Constitution gives the Conseil constitutionnel the task of assessing the compatibility with the Constitution of organic laws and ordinary laws, which, respectively, must be submitted to it for scrutiny, without stating whether this competence extends to all texts of legislative character, be they adopted by the people after a referendum or passed by Parliament, or whether, on the contrary, it is limited only to the latter category, it follows from the spirit of the Constitution, which made the Conseil constitutionnel a body regulating the activity of public authorities, that the laws to which the Constitution intended to refer in article 61 are only those *lois* passed by Parliament, and not those which, adopted by the people after a referendum, constitute a direct expression of national sovereignty;

3. Considering that this interpretation follows equally from the express provisions of the Constitution, especially article 60, which regulates the role of the Conseil constitutionnel in referendums, and from article 11, which does not provide for any formality between the adoption of a bill by the people and its promulgation by the President of the Republic;

4. Considering that, finally, this same interpretation is again expressly confirmed by the provisions of article 17 of the above-mentioned organic law of 7 November 1958, which only mentions '*lois* passed by Parliament', as well as article 23 of the said law, which provides that 'when the Conseil constitutionnel declares that the *loi* of which it is seised contains a provision contrary to the Constitution, without finding at the same time that it is not severable from the rest of the *loi*, the President may promulgate it without this provision, or request a new reading from the chambers';

5. Considering that it follows from what has been said that none of the provisions of the Constitution, nor of the above-mentioned organic law applying it, gives the Conseil constitutionnel the competence to rule on the request submitted by the President of the Senate, that it consider whether the bill adopted by the French people by way of referendum on 28 October 1962 is compatible with the Constitution . . .

DECISION 15: CC decision no. 79-110 DC of 24 December 1979, *Finance Law for 1980*, *Rec.* 36; *AJDA* 1980, 357; *GD*, no. 30.

Background: See above, pp. 117–18 and 129–32. The reference was made by the President of the National Assembly and by sixty opposition deputies, on the procedure adopted for passing the finance law for 1980.

DECISION

1. Considering that, under article 40 of the *ordonnance* of 2 January 1959 forming the organic law on finance laws, 'the second part of the finance law for the year cannot be put to debate before a chamber before the vote on the first part';

2. Considering that the meaning of this provision can only be understood with reference to article 1 §1 of the same *ordonnance*, according to which 'finance laws determine the character, amount, and destination of the resources and charges of the State, taking account of the economic and financial equilibrium that they establish';

3. Considering that, in subordinating the discussion of the second part of the finance law—which fixes the global amount of the credits applicable to the appropriations (*services*) voted, and determines the expenditure applicable to new authorizations—to the voting of the first part—which authorizes and quantifies revenue, fixes the ceilings of the major categories of expenditure, and determines the general features of the economic and financial equilibrium—article 40 merely draws the consequences as regards legislative procedure from the fundamental principle declared in article 1; that it aims to guarantee that the basic outline of the previously established equilibrium, as settled by Parliament, will not be infringed during the scrutiny of expenditure;

4. Considering that, even if, for it to be satisfied, this prescription cannot prevent later amendment by the chambers of provisions within the first part of the finance bill, it is necessary that the first part, even if not passed completely, should be adopted as to those provisions which constitute its reason for existence and which are indispensable to the fulfilment of its purpose; that this is the case particularly for the provision that establishes the general features of the equilibrium between receipts and expenditure; that otherwise, and especially in the case of the rejection of this provision, the decisions on the second part relating to expenditure would not have been preceded by the determination of the equilibrium, contrary to what is required, both in its letter and its spirit, by article 40 of the *ordonnance* of 2 January 1959;

5. Considering that it is undisputed that, during the course of the first reading, the National Assembly did not adopt article 25 of the bill (which became article 32 of the finance law submitted to the scrutiny of the Conseil), an article that, within the first part of this law, quantifies

revenue and fixes the ceilings on the charges, establishing thus the general features of the economic and financial equilibrium for 1980; that, in consequence, and although the President of the National Assembly could not but call on the chamber to undertake discussion of the second part, given that the bill had not been withdrawn and its discussion had been kept on the priority agenda, the procedure followed in this first reading was not valid with respect to the provisions of the *ordonnance* of 2 January 1959 forming the organic law on finance laws;

6. Considering that this irregularity follows on the one hand from the fact that the standing orders of the National Assembly do not contain a specific provision to ensure respect for the prescription found in article 49 of the *ordonnance* of 2 January 1959; that, in particular, these standing orders do not permit a reconsideration of the articles in the first part of the finance law before consideration of all the other articles of this law has been completed;

7. Considering that, in the course of the third sitting of the National Assembly, held on 17 November 1979, the Prime Minister requested a second reading, making it clear that this request related to all the articles of the first part, then those of the second part of the bill; that, in accordance with article 49 §3 of the Constitution, he then made a confidence issue 'on the one hand of articles 1 to 25, which form the first part of the finance law, and on the other hand of articles 26 *et seq.*, which constitute the second part, and, finally, of the whole of the text in its original form, as amended by the votes occurring in the first reading and by the amendments laid by the Government at the second reading';

8. Considering that, since the censure motions laid as a result of the Government making the provisions an issue of confidence had been rejected, those provisions were considered to have been adopted, but those of the first part and those of the second part were not passed in a distinct and successive way, as was required by article 40 of the *ordonnance* of 2 January 1959;

9. Considering that, as a result, and although the rest of the procedure was valid, both in the Senate and in the National Assembly, the finance law for 1980 was not adopted in accordance with the provisions of the *ordonnance* of 2 January 1959 forming the organic law on finance laws, provided for under article 47 of the Constitution . . .

DECISION 16: CC decision no. 86-225 DC of 23 January 1987, *Séguin Amendment, Rec.* 13.

Background: This reference was made by Socialist deputies and senators. For full details, see above, pp. 123–4.

DECISION

On the procedure for adopting the whole loi

2. Considering that the deputies making one of the references argue that the *loi* on miscellaneous social measures was adopted improperly; that, in fact, at the public vote on 20 December 1986 on the whole of the *loi* completed by amendment no. 1, the deputies present, with the exception of the Socialist group, voted for their absent colleagues following procedures contrary to constitutional rules, in that, on the one hand, they did not have a proxy power according to the provisions of *ordonnance* no. 58–1066 of 7 November 1958 forming the organic law exceptionally authorizing members of Parliament to delegate their right to vote, and, on the other hand, each of them voted for more than one absent colleague, contrary to the provisions of the final part of the third paragraph of article 27 of the Constitution;

3. Considering that, according to article 27 §2 of the Constitution, 'the right to vote of members of Parliament is personal'; that, according to the third paragraph of the same article, 'In exceptional circumstances an organic law may authorize the delegation of voting. In such a case, no one can be delegated more than one proxy vote';

4. Considering that in applying these provisions, the fact that, in the setting of a public ballot, the number of votes cast in favour of adopting a text is higher than the number of deputies actually present is such as to lead one to think that the proxies used exceeded the limits set out in article 27 cited above, does not render the procedure for adopting this text invalid unless it is established, on the one hand, that one or more deputies have been made to vote contrary to their opinions, or, on the other hand, that without taking such vote or votes into account, the requisite majority would not have been reached;

5. Considering that, whatever the circumstances in which the National Assembly proceeded, during the sitting of 20 December 1986, to a vote by public ballot on the whole of the *loi* currently the subject of a reference, it is not established, nor even alleged, that any of the deputies who appear in the minutes of the sitting as having voted in favour would not have voted that way; that, in these circumstances, the argument based on the breach of the second and third paragraphs of article 27 of the Constitution should be rejected;

On the procedure for adopting article 39

6. Considering that the authors of the reference claim that article 39 of the *loi* was adopted in circumstances inconsistent with the Constitution; that they argue that this article was introduced by way of an amendment during discussion by the chambers of the text adopted by the joint committee, but that it is without any direct relationship to

that text; that, furthermore, this amendment in fact replaced the whole of a bill that the Government had announced would be placed on the agenda for the Council of Ministers on 22 December 1986;

7. Considering that article 39 of the Constitution provides in its first paragraph that 'the right to initiate *lois* belongs concurrently to the Prime Minister and to members of Parliament'; that the first paragraph of article 44 declares that 'the members of Parliament and the Government have the right of amendment'; that, according to the first paragraph of article 45, 'any bill must be debated successively by the two chambers with a view to the adoption of a single text'; that, following the second and third paragraphs of the same article: 'If, following a disagreement between the two chambers, a bill has not been adopted after two readings by each chamber, or, if the Government has declared it urgent, after a single reading by each of them, the Prime Minister has the power to call a meeting of a joint committee to propose a text on the provisions remaining to be discussed. The text drafted by the joint committee can be submitted by the Government for approval by the two chambers. No amendment can be accepted without the Government's agreement';

8. Considering that it follows from these provisions that the adoption by the joint committee of a common text on the provisions remaining to be debated does not prevent the Government, in submitting the text produced by the joint committee for approval by the two chambers, from modifying or complementing it by amendments of its choice, if necessary in the form of additional articles; but, nevertheless, these additions or modifications made to the text in the course of debate cannot be without connection to the latter, nor, in their purpose or importance, can they exceed the limits inherent in the exercise of the right of amendment without breaching articles 39 §1 and 44 §1 of the Constitution;

9. Considering that the amendment from which article 39 of the *loi* originated reproduced in full the provisions of a text drawn up by the Government on the basis of article 2(4) of *loi* no. 86–793 of 2 July 1986, which authorized it to make *ordonnances* for measures necessary to improve employment, and, to this end, 'to amend the provisions of the Labour Code on the duration of work and the flexibility of working hours to permit the adaptation of the operating conditions of businesses to changes in the level of their activity and in general economic conditions, taking account of negotiations between the social partners';

10. Considering that, in fact, the provisions included in article 39 of the *loi*, in the form of twenty paragraphs modifying or completing numerous articles of the Labour Code, provide that the alteration of working hours can be implemented not only by agreement at branch level, but also by agreement at the level of the firm or corporation;

that these remove the previously existing compulsory link between flexibility of working hours and reductions in working time, and leave the social partners with the task of determining by agreement the character and size of the return for the benefit of employees; that they state, however, that, on pain of criminal penalties, agreements at the level of the firm concerning alterations cannot come into force unless, on the one hand, they have not been opposed by unions receiving more than half the votes of the electors registered at the last employee elections, and, on the other hand, they are consistent with the framework laid down by *loi*; that failure to respect the alteration agreement no longer gives rise to a compensatory 50 per cent time off; that, by contrast, changes are made to the rules governing the definition of overtime and the methods for its remuneration; that, in relation to the Sunday rest-day, the provisions included in article 39 give those branches to which a collective agreement so provides the possibility of organizing continuous working for economic reasons and no longer simply for technical reasons; that, finally, special adjustments are made to the provisions of the Labour Code relating to night-working by women;

11. Considering that it follows from what has been said that, by reason both of their scope and importance, the provisions from which article 39 originated exceed the limits inherent in the exercise of the right of amendment; that, therefore, they could not be introduced into the bill on miscellaneous social measures by way of amendment without breaching the distinction established between the bills envisaged in article 39 of the Constitution and the amendments to which these can be subject under article 44 §1; that, in consequence, the Conseil must decide that article 39 of the *loi* referred to it was adopted by an invalid procedure . . .

On article 4

[Article 4 imposed a minimum residence period in France for qualification for certain special social welfare benefits. It was argued, *inter alia*, that this infringed the principle of equality.]

15. Considering that the fixing of a residence qualification for the granting of social benefits does not involve *per se* a discrimination of the kind prohibited by article 2 of the Constitution; that, moreover, neither is it contrary to the principle of the equality of citizens before the law proclaimed by article 6 of the Declaration of the Rights of Man and of the Citizen of 1789 . . .

IV. Decisions onFundamental Freedoms

1. FREEDOM OF THE PERSON

(*a*) *Police Powers and the Criminal Law*

DECISION 17: CC decision no. 76-78 DC of 12 January 1977, *Vehicle Searches, Rec.* 33; D. 1978, 173, note Hamon; *GD*, no. 26; *SB* 65; J. Rivero, *Le Conseil constitutionnel et les libertés*, part I, ch. 5; Nicholas, 169–70.

Background: See above, p. 141. The reference was made by Socialist deputies.

DECISION

1. Considering that individual liberty constitutes one of the fundamental principles recognized by the laws of the Republic and proclaimed by the Preamble to the 1946 Constitution, confirmed by the Preamble to the 1958 Constitution;

2. Considering that article 66 of the Constitution, in reaffirming this principle, confers its safeguard on judicial authorities;

3. Considering that the text submitted for scrutiny by the Conseil constitutionnel aims to give officers of the judicial police, or, on their orders, agents of the judicial police, the power to search any vehicle or its contents on condition only that this vehicle is on a road open to public traffic, and that the search takes place in the presence of its owner or driver;

4. Considering that, provided these two above-mentioned conditions are met, the powers conferred by the provision on officers of the judicial police and or agents acting on their orders could be exercised without restrictions in all cases, without invoking a legal regime of special powers, even when no offence has been committed, and without the *loi* making these controls conditional on the existence of a threat of harm to public order;

5. Considering that, by reason of the scope of these powers, which are not defined elsewhere, conferred on officers of the judicial police and their agents, and as a result of the very general character of the cases in which these powers could be exercised, and the lack of precision in the scope of the controls to which they are capable of giving rise, this text infringes the essential principles on which the protection of individual liberty rests; that, in consequence, it is not consistent with the Constitution . . .

DECISION 18: CC decision no. 80-127 DC of 19, 20 January 1981, *Security and Liberty, Rec.* 15; *GD*, no. 32; L. Philip, 'La Décision *Sécurité et Liberté* des 19 et 20 janvier 1981', *RDP* 1981, 651; J. Rivero, *Le Conseil constitutionnel et les libertés*, part I, ch. 7.

Background: In preparation for the presidential elections in May 1981, the Government proposed a series of measures on law and order. In a number of respects, these provisions were made more severe by amendments tabled by Government supporters (especially the RPR) during the passage of the bill. On the other hand, the *loi* was an advance in the protection of civil liberties, by requiring judicial authorization of detention of a suspect (no longer done by the prosecutor), freeing detainees where there was a procedural irregularity in detention, giving a right for a person to go to court to challenge an expulsion order, and abolishing the system of *tutelle pénale*, which allowed a convict to be kept in prison after his sentence had been served.

The decision was the first major piece of legislation that the Conseil constitutionnel had considered, and marks a watershed in both the character of review over political legislation and in the length of judgments. (It was among the early decisions for which Vedel was a member of the Conseil, and he is thought to have been an important influence in the structure and reasoning of the decision.)

DECISION

The legal basis for crimes and penalties

3. Considering that, in the words of article 8 of the Declaration of the Rights of Man and of the Citizen of 1789, 'no one may be punished except according to a *loi* passed and promulgated prior to the offence, and lawfully applied'; it follows from this, that the legislature must define offences in terms sufficiently clear and precise to exclude arbitrariness . . .

5. Considering that article 24 of the *loi* aims to replace articles 434 to 437 of the Penal Code, and to define various offences consisting in the destruction or wilful damage by various means of moveable objects or immovable property; that the terms 'destroy', 'damage', 'movable objects', 'immovable property', are neither obscure nor imprecise; that the distinctions made both as to the circumstances or the means of destruction or damage, and as to the persons to whose detriment this destruction or damage is committed do not display any ambiguity; that, even if, in the new article 434 of the Penal Code, the legislature excludes from punishment 'minor damage', this provision, created for the benefit of the authors of non-serious acts, and which it is for the competent courts to interpret, does not breach the rule according to which none may be punished except according to a *loi* . . .

The strict and evident necessity of penalties

7. Considering that, according to article 8 of the Declaration of the Rights of Man and of the Citizen of 1789, '*loi* may only create penalties

that are strictly and evidently necessary'; that, according to the authors of both references, it is for the Conseil constitutionnel to censure the provisions of title I of the *loi* submitted for its scrutiny that authorize or impose punishment that is excessive in its eyes either because of the effect of the penalties attached to offences, or of the worsened position of recidivists, or of the limits on the effects of attenuating circumstances, or of restrictions on the conditions for awarding a suspended sentence, or of altering the conditions for implementing sentences;

8. Considering that article 61 of the Constitution does not confer on the Conseil constitutionnel a general power of judgment and decision identical to that of Parliament, but only gives it competence to decide whether the *loi* referred for its scrutiny is consistent with the Constitution;

9. Considering that, within the framework of this mission, it is not for the Conseil constitutionnel to substitute its own judgment for that of the legislature concerning the necessity of the penalties attached to the offences defined by the latter, provided that no provision of title I of the *loi* is manifestly contrary to the principle laid down by article 8 of the Declaration of 1789;

The individuation of sentences

[The *loi* contained provisions increasing the severity of penalties by imposing minimum sentences, creating automatic bars to suspended sentences, and by setting time-limits before which a reduction of a sentence could occur. The opposition deputies and senators argued that such automatic provisions infringed the principle of individuation of sentences implied in article 8 of the 1789 Declaration and fundamental principles recognized by the laws of the Republic].

11. Considering that, on the one hand, if, in the words of article 8 of the Declaration of 1789, cited above, '*loi* must only create penalties that are strictly and evidently necessary', this provision does not imply that the necessity of penalties has to be assessed solely from the point of view of the personality of the convicted person, and even less that, to this end, the judge should be invested with an arbitrary power that article 8 of the Declaration of 1789 specifically wished to proscribe, and which would permit him, at his whim, to dispense persons convicted of *crimes* or *délits* from the criminal law, other than where it so provides;

12. Considering that, on the other hand, if French law has given an important place to the individuation of penalties, it has never conferred on it the character of a unique and absolute principle, prevailing of necessity in all cases over other grounds for criminal punishment; that, thus, even supposing that the principle of the individuation of sentences could, within these limits, be regarded as one of the funda-

mental principles recognized by the laws of the Republic, it could not prevent the legislature from determining rules to ensure the effective suppression of offences, whilst leaving the judge or the authorities entrusted with determining the methods for implementing sentences with a wide power of discretion . . .

Traffic and the right to strike

13. Considering that, according to the authors of the reference, the provisions of articles 16, 17, and 30 of the *loi* submitted for the scrutiny of the Conseil constitutionnel would infringe the exercise of the right to strike and the union rights recognized by the Constitution;

14. Considering . . . that none of the offences created by articles 16 and 17 of the *loi* is committed unless there is a threat to commit a *crime* or a *délit*; that, in these circumstances, there is no possibility that the application of these provisions might, in whatever way, prevent or interfere with the lawful exercise of the right to strike or union action . . .

[Article 30 punished actions that blocked or disrupted rail traffic by placing obstacles in its way.]

16. Considering that the punishment for blocking or disrupting the movement of rail traffic resulting from placing an object on the line is not such as to prevent or interfere in any way with the lawful exercise of the right to strike or of union action . . .

Extension of custody [garde à vue]

[Article 39 of the *loi* permitted an extension of the normal period of twenty-four hours for which a suspect might be held in custody where further investigation was necessary in relation to offences of unlawful detention, hostage-taking, kidnapping of children, or armed robbery. Such an extension had to be authorized by a judge.]

20. Considering that, even if, to be compatible with article 66 of the Constitution, the authorization of an extension of detention in custody in such cases requires the intervention of a judge, no rule or principle of constitutional value requires that this judge must be an investigating magistrate;

21. Considering that the judge, who must necessarily examine the file in order to authorize an extension beyond twenty-four hours of detention in custody, in no way makes a decision on the investigation of the offence or prejudges the guilt of the suspect;

22. Considering that, furthermore, the provisions . . . relating to medical examination of the person held in custody provide additional safeguards for him;

23. Considering that, therefore, article 39 of the *loi* referred for scrutiny by the Conseil constitutionnel is not contrary to the Constitution . . .

Criminal Procedure in the correctional matters

[Articles 47 to 51 of the *loi* gave the *procureur de la République* the power to refer a case either directly to a court for trial, or to an investigating magistrate (*juge d'instruction*) for him to decide whether there was a case to answer. In the former case, the *procureur* could choose between summoning the accused to appear in court, referring the case for immediate trial, or referring the case to the President of the court or to his delegate for a pre-trial review. It was alleged that this allowed unequal treatment of like cases according to the discretion of the *procureur*. Furthermore, it was alleged that the right to due process was breached by not requiring the accused to receive legal assistance when he appeared before the *procureur* for the decision on how the prosecution process was to continue.]

27. Considering that, though by virtue of article 7 of the Declaration of the Rights of Man and of the Citizen of 1789 and of article 34 of the Constitution, the rules of criminal procedure are fixed by *loi*, it is permissible for the legislature to provide different rules of criminal procedure according to the facts, circumstances, and persons to whom they apply, provided that these differences do not result from unjustified discrimination, and that equal safeguards for the accused are ensured;

28. Considering that the institutions of a summons to appear in court, or of a reference for immediate trial, or of a preliminary reference to the President of the court or his delegate have the purpose of enabling cases to be brought to trial without undue delay where an investigation is not necessary; that this objective is consistent with the good functioning of justice and with the freedom of persons liable to be kept in detention;

29. [The Conseil noted that the decision on the choice of the procedure to be followed was given to the *procureur* because of his control of the prosecution process, but that an inappropriate use of the procedure of a reference for immediate trial would lead to the release of the accused or to the court making further inquiries, because of the presumption of innocence in favour of the accused.]

30. Considering that, even if the new article 393 . . . of the Code of Criminal Procedure [created by the *loi*] does not state that a person sent [by the police] to the *procureur de la République* may be accompanied by a lawyer, this is because this member of the judiciary, who only has the right to decide by which route he will conduct the prosecution, is deprived by the new *loi* of the power to issue a warrant for detention, even in the case of a flagrant offence, such a warrant now being issued by a judge . . . [The Conseil noted that the safeguards provided by the *loi* were the same as in other criminal proceedings, and that no

decisions affecting individual liberty could be taken except by a judge. The choices of procedure permitted to the *procureur* were those already available to the investigating magistrate.]

33. Considering that, whatever the option chosen by the *procureur de la République* from the different procedures for prosecution, and without reference to the issue of whether or not there is a preliminary investigation by an investigating magistrate, the trial of the case takes place in the same court; that, informed where necessary by a supplementary investigation, which it can require in any case, it has to decide on the guilt of the accused, who is always presumed to be innocent, according to identical rules of form and substance; that therefore the provisions in question are neither contrary to the right of due process nor of equality before justice . . .

Identity checks

50. Considering that, according to the authors of the references, the provisions of articles 76, 77, and 78 of the *loi* submitted for scrutiny by the Conseil constitutionnel seriously infringe the freedom of movement, both in principle and in the way in which they are to be implemented . . . that, finally, the nature of the operations authorized by the criticized provisions, as well as the insufficiency of the safeguards for the persons subject to them, would permit inevitable abuses against the rights and freedoms of individuals . . .

[Article 76 authorized the judicial police to conduct identity checks in the course of criminal investigations or of the prevention of threats to public order. A person could prove his or her identity 'by any means', and could only be taken to a police station 'if necessary', being held in custody only for as long as was strictly necessary to establish that identity, up to a maximum of six hours. The person detained had to be given the opportunity to notify his or her family or some other person able to establish his or her identity. He or she could also request that the *procureur de la République* be notified.]

54. Considering that the provisions . . . are limited by the rule according to which persons invited to justify their identity can satisfy this invitation on the spot by an appropriate means of their choice, and that they can only be taken to a police station in case of necessity; that strict respect for these prescriptions—the immediate appearance of the person taken to the police station before an officer of the judicial police, the opportunity for him to have his family, or any person able to confirm his identity, informed, or for him to be allowed to do so, the right for him to refer his case to the *procureur de la République*, the duty to detain him only for the period necessary to check his identity, the limitation of his detention to six hours from the initial request to prove his identity—limit the constraints placed on the person who cannot or

does not wish to prove his identity on the spot to those which are necessary to safeguard goals of the public interest of constitutional value, the pursuit of which is the reason for identity checks;

55. Considering that . . . the totality of [the] provisions [for making a record of identity checks and detentions for this purpose] is such as to ensure that competent authorities and courts have the opportunity to check the lawfulness of actions conducted pursuant to article 76 . . .

56. Considering that the pursuit of criminals, and the prevention of threats to public order, especially of threats to the security of persons and property, are necessary for the implementation of principles and rights of constitutional value; that the inconvenience that the application of the provisions of paragraph 1 [of article 76] might represent to the freedom of movement is not excessive, as long as the persons stopped can prove their identity 'by any means', and that, as the text requires, the pre-conditions regarding the legality, the genuineness, and the relevance of the justifications for the operation are met in fact . . .

64. Considering that, with a view to preventing abuses, the legislature has surrounded the procedure for controlling and checking identity that it has created with numerous precautions; that it is up to the judicial and administrative authorities to ensure that they are fully respected, as well as to the competent courts to punish, where necessary, illegalities that are committed, and to provide compensation for their harmful consequences . . .

Retrospectivity of penalties

[Article 100 of the *loi* provided for a reduction in the sentences applicable to a number of criminal offences, but provided that these 'do not apply to offences that were the subject of a judgment on the merits of final resort before the entry into force of this *loi*'.]

71. Considering that these provisions aim to limit the effects of the rule whereby, when it imposes penalties that are less severe than the earlier law, a new penal law should apply to offences committed before its coming into force, and which have not given rise to convictions that are *res judicata*; that, thus, they should be considered as contrary to the principle formulated by article 8 of the Declaration of the Rights of Man and of the Citizen, according to which '*loi* may only create penalties that are strictly and evidently necessary'; that, indeed, the fact that the new, less harsh penal law is not applied to offences committed under the regime of the earlier law effectively permits the judge to pass the sentence provided by the earlier law, which, according to the very assessment of the legislature, is no longer necessary; that, therefore, the second paragraph of article 100 of the *loi*

submitted for the scrutiny of the Conseil constitutionnel is contrary to the Constitution . . .

DECISION 19: CC decision no. 86-213 DC of 3 September 1986, *Terrorism Law*, *Rec*. 122.

Background: As part of the law and order programme of the Chirac Government (the so-called *lois Pasqua*), attempts were made to deal with terrorism. A number of bomb attacks in Paris had heightened the fear of terrorism in France, especially from groups from Iran and the Middle East. The *loi* sought to create a non-jury court for terrorist offences of seven judges, rather than the normal three judges sitting with nine jurors. This was done to prevent threats against lay jurors.

A reference was made by sixty senators on the grounds that (*a*) the *loi* failed to define 'terrorist crimes' adequately; (*b*) it breached equal treatment before the courts; and (*c*) it infringed individual liberty in the extended periods of detention that it permitted.

DECISION

Did loi *define the crimes adequately?*

[The *loi* added a new article 706-16 to the Code of Criminal Procedure that defined 'terrorist offences' by reference to certain specific crimes in the Penal Code, provided that these constituted 'an individual or collective enterprise having as its purpose seriously to disturb public order by intimidation or terror'. This definition was then used to identify the appropriate procedures and penalties for offences (which included an obligatory prohibition on residence in France for at least two years).]

5. Considering that the application of specific rules laid down by the *loi*, both as they relate to the prosecution, judicial investigation (*l'instruction*), and trial, and as they cover the applicable sentences, are limited by two conditions: on the one hand, that the facts considered constitute specific offences defined by the Penal Code or by special *lois*; on the other hand, that these offences relate to an individual or collective enterprise having as its purpose seriously to disturb public order by intimidation or terror;

6. Considering that the first condition fixed by the *loi*, which refers to offences that are themselves defined by the Penal Code or by special *lois* in sufficiently clear and precise terms, satisfies the requirements of the constitutional principle of a basis in *loi* for crimes and penalties; that, equally, the second condition is declared in sufficiently precise terms for the principle not to be infringed; that, thus, the first ground formulated by the authors of the reference cannot be upheld;

Equality before justice

[In derogation of the normal rules, the *loi* provided that, in terrorist cases, the Cour d'assises was to be composed of seven judges and no jury members. Such a court would decide a case on a simple majority, rather than the minimum of eight out of twelve votes necessary in the case of decisions by judges sitting with a jury.]

11. Considering that the authors of the reference claim . . . that there are no offences specific to terrorist activities, even in the intention of the legislature; that a prosecution may relate only to offences already defined and punished by the Penal Code or by special *lois*; that, therefore, in respect of the principle of equality before justice, nothing could justify the trial of these offences by different courts according to whether it is alleged that they were done 'in connection with an individual or collective enterprise having as its purpose seriously to disturb public order by intimidation or terror'; that, whatever the difference in their motives may be, offences defined by the same constituent elements must be tried by the same judges and according to the same rules;

12. Considering that it is permissible for the legislature, in its competence to determine the rules of criminal procedure by virtue of article 34 of the Constitution, to provide different rules of procedure according to the facts, circumstances, and persons to whom they apply, provided that these differences do not result from unjustified discrimination, and that equal safeguards for the accused are ensured, especially respect for the principle of the right to due process;

13. Considering that the difference in treatment established [by the *loi*] between the authors of crimes referred to by the new article 706-16 [of the Criminal Procedure Code] according to whether or not the offences have a relationship with an individual or collective enterprise having as its purpose seriously to disturb public order by intimidation or terror is intended by the legislature to counter the effect of pressures or threats capable of affecting the tranquillity of the trial court; that this difference in treatment does not result from an unjustified discrimination; that, furthermore, by its composition, the Cour d'assises [as composed of seven judges] offers the requisite safeguards of independence and impartiality; that the right to due process is safeguarded before this court; that, in these circumstances, the argument based on breach of the principle of equality before justice has to be rejected . . .

Extension of custody [garde à vue]

[The *loi* inserted article 706-23 into the Code of Criminal Procedure that provided that the *procureur de la République* might request an extension of custody before charge from the President of the local

Tribunal de grande instance, and that the investigating magistrate, when already in charge of the case, might make a similar order. The extension permitted was to be up to forty-eight hours longer than in ordinary criminal cases. The *loi* also provided that a medical examination should take place as a matter of right.]

16. Considering that the authors of the reference argue that, in case of a forty-eight-hour extension of custody, respect for individual liberty requires that the person concerned should appear before a judge, and a medical examination should be conducted daily;

17. Considering that it follows from the new article 706-23 of the Code of Criminal Procedure that the scope of the criticized provisions includes inquiries concerning specific offences that require special investigation because of their relationship with an individual or collective enterprise having as its purpose seriously to disturb public order by intimidation or terror; that this article requires that the extension of custody be subject to a decision by a judge, before whom the person concerned must appear; that, moreover, medical supervision of the person in custody is prescribed; that these provisions are added to safeguards resulting from the rules of general application in the Code of Criminal Procedure, which have the effect of placing custody under the control of the *procureur de la République*, and which require, in conformity with the last paragraph of article 64, a medical examination after twenty-four hours if the person concerned requests it; that, thus, the provisions of the new article 706-23 of the Code of Criminal Procedure do not disregard article 66 of the Constitution;

Offences against the security of the State

[Under the existing law of 21 July 1982, offences against the security of the State were tried by the ordinary courts in peacetime, and not by a special court as hitherto, except where the offences concerned were treason, spying, or other infringements of national security, when a Cour d'assises composed entirely of judges would try the offence. Article 4 of the *loi* on terrorism extended this exception, and the procedures it created for terrorist offences, to bombing, plots and other offences against the authority of the State and the integrity of national territory, *crimes* to disturb the State by way of devastation, and *crimes* involving participation in insurrection.]

22. Considering that the authors of the reference claim . . . that the extension of the rules derogating from the ordinary law, set out in articles 706-17 to 706-25 [on the procedure in terrorist cases] of the Code of Criminal Procedure, to all offences, even those amounting merely to *délits* of negligence, involves a breach of the principle of individual liberty, especially by applying its custody system;

23. Considering that, as was said above, though it is permissible for the legislature to provide different rules of criminal procedure according to the facts, situations, and persons to whom they apply, this is on condition that these differences do not result from unjustified discriminations, and that equal safeguards for the accused are ensured;

24. Considering that the rules of composition and procedure derogating from the ordinary law, which, according to the legislature, are justified by the specific characteristics of terrorism, could not be extended to offences that do not display the same characteristics, and which do not necessarily have any relationship with those mentioned in the new article 708-16 of the Code of Penal Procedure, without infringing the principle of equality before justice; that, therefore . . . article 4 of the *loi* . . . is contrary to the Constitution;

25. Considering that, in this case, there is no need for the Conseil constitutionnel to raise of its own motion any question as to whether the other provisions of the *loi* submitted for its examination are compatible with the Constitution . . .

(*b*) *Right to Health*

DECISION 20: CC decision no. 74-54 DC of 15 January 1975, *Abortion Law*, *Rec.* 19; D. 1975, 529; *GD* no. 23; *SB* 57; J. Rivero, *Le Conseil constitutionnel et les libertés*, part I, ch. 3; Nicholas, 156–62.

Background: One of the reforms supported by Giscard d'Estaing when he became President in May 1974 was the reform of abortion laws. The bill was opposed by a significant number of members of the ruling parties. The reference was made by sixty deputies.

DECISION

1. Considering that article 61 of the Constitution does not confer on the Conseil constitutionnel a general power of judgment and decision identical to that of Parliament, but only gives it competence to decide whether the *loi* referred for its scrutiny is consistent with the Constitution . . .

5. Considering that a *loi* contrary to a treaty is not, as such, contrary to the Constitution . . .

7. Considering that, in these circumstances, when it receives a reference under article 61 of the Constitution, it is not for the Conseil constitutionnel to examine the compatibility of the *loi* with provisions of a treaty or international agreement;

8. Considering that, secondly, the *loi* relating to the voluntary termination of pregnancy respects the freedom of the persons who resort to, or participate in, a termination of pregnancy, whether in a

situation of distress or for a therapeutic reason; that, therefore, it does not infringe the principle of liberty laid down in article 2 of the Declaration of the Rights of Man and of the Citizen;

9. Considering that the *loi* referred to the Conseil constitutionnel does not permit any infringement of the principle of respect for all human beings from the beginning of life, recalled in its article 1, except in cases of necessity and within the terms and limits that it sets out;

10. Considering that none of the derogations set out in this *loi*, as it stands, is contrary to any fundamental principle recognized by the laws of the Republic, nor does it disregard the principle declared in [paragraph 11 of] the Preamble to the Constitution of 27 October 1946, according to which the nation guarantees to the child the protection of its health;

11. Considering that, in consequence, the *loi* . . . does not contradict the texts to which the Constitution of 4 October 1958 refers in its Preamble, nor any of the articles of the Constitution . . .

2. FREEDOM OF ASSOCIATION

See DECISION 1, above.

3. FREEDOM OF EDUCATION

DECISION 21: CC decision no. 77-87 DC of 23 November 1977, *Freedom of Education*, *Rec.* 42; *GD*, no. 27; J. Rivero, *Le Conseil constitutionnel et les libertés*, part I, ch. 6.

Background: See above, p. 152. The reference was made by sixty Socialist senators.

DECISION

1. Considering that, according to article 1 of the *loi* complementing the *loi* of 31 December 1959, as amended by the *loi* of 1 June 1971 on the freedom of education, teachers entrusted with the task of teaching in a private establishment linked to the State by a contract of association are obliged to respect the specific character of that establishment;

2. Considering that, on the one hand, the safeguarding of the specific character of an establishment linked to the State by contract . . . merely implements the principle of freedom of education;

3. Considering that this principle, which is recalled notably in article 91 of the finance law of 31 March 1931, constitutes one of the fundamental principles recognized by the laws of the Republic, reaffirmed by

the Preamble to the 1946 Constitution, and on which the Constitution of 1958 conferred constitutional value;

4. Considering that the affirmation by the same Preamble to the 1946 Constitution, that 'the organization of free and secular public education is a duty of the State', does not exclude the existence of private education, nor the granting of State aid to such education in circumstances defined by the law; that this provision of the Preamble to the 1946 Constitution has no effect on whether the *loi* submitted to the Conseil constitutionnel is consistent with the Constitution;

5. Considering that, on the one hand, according to the terms of article 10 of the Declaration of the Rights of Man and of the Citizen of 1789, 'No one may be troubled on account of his opinions or religion, provided that their expression does not infringe public policy as established by *loi*'; that the Preamble to the 1946 Constitution affirms that 'No one may be harmed in his work or employment on account of his origins, opinions, or beliefs'; that freedom of conscience must be regarded as one of the fundamental principles recognized by the laws of the Republic;

6. Considering that it follows from a comparison of the provisions of article 4 §2 of the *loi* of 31 December 1959, as redrafted by the *loi* submitted for scrutiny by the Conseil constitutionnel, with those of article 1 of the *loi* of 31 December 1959 that the obligation imposed on teachers to respect the specific character of the establishment cannot be interpreted as permitting an infringement of their freedom of conscience, even if it imposes on them a duty of reserve . . .

DECISION 22: CC decision no. 84-185 DC of 18 January 1985, *Education Law*, *Rec.* 36; P. Delvolvé, 'Le Conseil constitutionnel et la liberté de l'enseignement', *RFDA* 1985, 264.

Background: The bill proposed by the Minister of Education, Savary, sought to bring the public and private sectors of education closer, by giving the status of civil servant to all teachers in private schools. This was seen as a take-over bid by the Socialists, directed against religious schools. The proposals were greeted with hostility by the Catholic Church and by the opposition, and they provoked some of the largest street demonstrations seen in France since World War II. The bill was hastily withdrawn by a decision of the President of the Republic, about which Savary only discovered on the television news. He resigned, and new legislation, more moderate in character, was put forward. All the same, it did wish to alter some of the provisions of the *loi Guermeur* of 1977, especially in relation to the duty of reserve of teachers. In its decision on this new *loi Joxe–Chevènement* the Conseil shows how constitutional provisions limit the possibility of repeal of previous legislation. The reference was made by sixty opposition senators and sixty deputies.

DECISION

On article 27-1

4. Considering that the criticisms made against article 27-1 concern the repeal of three amendments to paragraph 2 of article 4 of the *loi* of 31 December 1959, [especially that repeal of] the sentence: 'Teachers providing this teaching are bound to respect the specific character of the establishment provided for in article 1 of this *loi*';

5. Considering that . . . the deputies and senators making the references claim that article 27-1 infringes the specific character of private educational establishments, and, as a consequence, the freedom of education of which that specific character is an expression, in that it repeals article 1 of the *loi* of 25 November 1977, which imposed a duty on teachers taking classes under a contract of association [with the State] to respect the specific character of the establishment . . .

9. Considering that article 1 of the *loi* of 31 December 1959 declares: 'The State proclaims and respects the freedom of education, and guarantees its exercise to all private establishments operating lawfully', and sets out the principles of the organization of this freedom in relation to private establishments in the following terms: 'In the private establishments that have made a contract as provided above, the teaching taking place under the regime of the contract is subject to control by the State. The establishment, while retaining its own specific character, has to provide this teaching in total respect for freedom of conscience. All children shall have access to it, without discrimination as to origins, opinions, or beliefs';

10. Considering that, even if, as those making the references argue, the recognition of the specific character of the private educational establishments merely implements the principle of freedom of education, which has constitutional value, respect for this specific character is affirmed by the last paragraph of article 1 of the *loi* of 31 December 1959 cited above; that, in these circumstances, the scope of the amendments made by article 27-1 [of the *loi* under consideration] to *lois* in force, criticized by the authors of the reference, has to be considered by taking account of the duty imposed by the *loi* to respect the specific character of the establishment;

11. Considering that, thus, the repeal of the provisions of the *loi* of 25 November 1977 imposing on teachers taking classes under a contract of association the obligation to respect the specific character of the establishment does not have the effect of exempting teachers from the duty deriving from the last paragraph of article 1 of the *loi* of 31 December 1959; that such an obligation, even if it cannot be interpreted as permitting an infringement of the freedom of conscience of teachers,

which has constitutional value, requires them to observe a duty of reserve in their teaching . . .

On article 27-2

16. Considering that article 27-2 provides that the making of contracts of association is subordinated, in relation to secondary classes, to the opinion of the relevant department or region, and, in relation to primary classes, to the agreement of the relevant commune, after advice from the communes in which at least 10 per cent of the pupils attending the classes live, with the commune in which the school is sited signing the contract of association with the State and the relevant establishment . . .

18. Considering that, without it being necessary to decide on the issue of whether the provisions of article 27-2 infringe the freedom of education and equality, the said provisions have to be regarded as inconsistent with the Constitution; that, in effect, even if the principle of the free administration of local authorities has constitutional value, it should not lead to the result that essential conditions for the implementation of a *loi* organizing the exercise of a civil liberty should depend on decisions of local authorities, and thus might not be the same throughout the country . . .

4. TRADE-UNION RIGHTS

DECISION 23: CC decision no. 79-105 DC of 25 July 1979, *Strikes on Radio and Television, Rec.* 33; *AJDA* septembre 1979, 46; *GD*, no. 29.

Background: A private member's bill was introduced by a RPR deputy, Vivien, to amend the law on strikes in public broadcasting after a series of strikes in the 'winter of discontent', 1978–9. The main problem was that unions could give a strike notice on each day and then let it slip, making it impossible for the management to know whether a service would operate on a particular day or not. The unions called out only those workers necessary for a minimum service (typically, less than 6 per cent of the work-force), and these would be immediately requisitioned (and paid) by the management to guarantee such a service, so that few would be affected, in terms of salary, by any disruption caused by an actual strike. The *loi* sought to remedy this by setting out more stringent conditions for strike notices, and enabling a decree to establish the categories of personnel required for a minimum service and to empower the management to requisition such employees. Much of the legislation merely put into statutory form existing practice from collective agreements.

Socialist deputies and senators challenged the legislation as being against the right to strike.

DECISION

1. Considering that, according to the Preamble to the Constitution of 27 October 1946, confirmed by that of the Constitution of 4 October 1958, 'the right to strike shall be exercised within the framework of the *lois* that regulate it'; that, in enacting this provision, the drafters of the Constitution intended to recognize that the right to strike is a principle of constitutional value, but within limits, and empowered the legislature to define these by making the necessary reconciliation between the defence of professional interests, of which the strike is a means, and the safeguarding of the public interest, which a strike might infringe; that, especially as far as public-sector strikes are concerned, the recognition of the right to strike cannot prevent the legislature from imposing such limits on this right as are necessary in order to ensure the continuity of the public service, which, just like the right to strike, has the character of a principle of constitutional value; that such limits may go so far as to prohibit the right to strike of those employees whose presence is indispensable to the functioning of those parts of the service whose interruption would harm essential needs of the country;

2. Considering that the provisions contained in paragraph I of article 26 of the *loi* of 7 August 1974, as amended by the *loi* submitted for the scrutiny of the Conseil constitutionnel, are limited to regulating the conditions for lodging a strike notice; that this text is not contrary to any provision of the Constitution, nor any principle of constitutional value . . .

5. But considering that, by providing in the first sentence of paragraph II of the *loi* that: 'when the employees of the national television programming companies are insufficient in number to ensure a normal service, the President of each company may, if the situation requires, requisition categories of employees or agents who must remain in post to ensure the continuity of those parts of the service necessary to carry out the tasks defined in articles 1 to 10' [the general tasks of broadcasting companies], the legislature permits the Presidents of the companies, where a concerted cessation of employment prevents a normal service, and in order to guarantee the generality of tasks entrusted by it to these companies, to restrict the right to strike in those cases where its prohibition does not appear justified in the light of the principles of constitutional value mentioned above; that, therefore, the provisions contained in this sentence have to be considered as incompatible with these principles, in so far as they refer, on the one hand, to carrying out a normal service, and, on the other hand, to the implementation of the tasks defined in articles 1 to 10 of the *loi* of 7 August 1974 . . .

DECISION 24: CC decision no. 82-144 DC of 22 October 1982, *Trade-Union Immunity*, *Rec.* 61; D 1983; 189, note Luchaire; *Droit social* 1983, 185, note Hamon; *SB* 72; M. Forde, 'Bill of Rights and Trade Union Immunities: Some French Lessons' (1984) 13 *Industrial Law Journal* 40, and id., 'Liability in Damages for Strikes: A French Counter-Revolution' (1985) 33 *American Journal of Comparative Law* 447.

Background: See above, p. 162.

DECISION

1. Considering that the deputies making the reference claim that article 8 of the *loi* on the development of institutions to represent of employees is contrary to the Constitution, in that it completes article L521-1 of the Labour Code by a new paragraph, as follows:

> No action can be brought against employees, elected representatives of workers, or union organizations of employees to make reparation for losses caused by, or on the occasion of, a collective labour dispute, except for actions to make good losses caused by a criminal offence and by acts manifestly incapable of being related to the exercise of the right to strike or of union rights. These provisions shall apply to litigation in progress, including that before the Cour de cassation;

2. Considering that it follows necessarily from this text that losses caused by fault, even serious fault, on the occasion of a labour dispute would remain uncompensated either by their authors, or co-authors, or, in the absence of some special provision to that effect, by other physical or legal persons when such losses are related, in however indirect a manner, to the exercise of the right to strike or of union action, and do not follow from a criminal offence;

3. Considering that, since no one has the right to harm another, in principle, every human act that causes loss to another obliges him by whose fault it occurs to make reparation;

4. Considering that, no doubt in certain matters the legislature has created regimes of compensation that derogate partially from this principle, especially in adding or substituting the liability or the guarantee of another physical or legal person for the liability of the author of the loss;

5. Considering that, however, in no area does French law have a regime that removes all liability for losses resulting from civil fault imputable to private physical or legal persons, no matter how serious the fault may be;

6. Considering that, thus, article 8 of the *loi* submitted for scrutiny by the Conseil constitutionnel creates an obvious discrimination to the detriment of the persons to whom it prohibits any action for reparation,

except in the case of a criminal offence; that, in reality, whereas no other person, whether human or legal, public or private, French or foreign, who is the victim of a material or moral loss caused by the civil fault of a private law person is faced with a general prohibition from suing for reparation of that loss, the persons against whom the provisions of article 8 of the *loi* apply . . . would not be able to claim reparation from anyone;

7. Considering that it is true that, according to the preparatory materials, the provisions of article 8 of the *loi* are justified by the desire of the legislature to ensure the effective exercise of the right to strike and of union action, both of which are constitutionally recognized and would be hampered by the abusive threat or commencing of lawsuits against employees, their representatives, or union organizations;

8. Considering that, however, the concern of the legislature to ensure the effective exercise of the right to strike or of union action should not justify the serious infringement by the above provisions of the principle of equality;

9. Considering that, in fact, even though it is for the legislature to define the conditions for the exercise of the right to strike or of union action, as well as of the other rights and liberties that also have constitutional value—and, thus, to define precisely the line dividing lawful acts and behaviour from unlawful acts and behaviour, so that the exercise of these rights cannot be impeded by abusive lawsuits—and, equally, even if it is for it to devise, should the case arise, a special regime of compensation to reconcile the interests at stake, all the same, it cannot deny in principle the right of the victims of fault, who may well be employees, representatives of employees, or union organizations, to equality before the law and before public burdens, even in order to achieve legitimate objectives;

10. Considering that, therefore, article 8 of the *loi* referred to the Conseil constitutionnel, whose provisions are not inseparable from the other provisions of the *loi*, should be declared contrary to the Constitution . . .

DECISION 25: CC decision no. 87-230 DC of 28 July 1987, *Public Serive Strikes*, *Rec.* 48; *RFDA* 1987, 807, note Genevois.

Background: The financial consequences of public-sector strikes have been subject to much controversy. At first, the view taken was that a strike merely suspended the contract of employment, and a pro rata reduction of salary to cover the work not done was appropriate. Legislation of 1961, 1962, and 1977 took a firmer line, imposing a salary reduction equal to that fraction of pay which was treated as indivisible in public accounting (typically, a day). Such a provision was upheld by the Conseil constitutionnel in July 1977 (CC decision no. 77-83 DC of 20 July 1977, *Services Rendered*, *RDP* 1978, 827, note Favoreu).

In 1982 the Socialists amended this by introducing a sliding scale: a strike of less than one hour led to a reduction of 1/160th of the monthly salary; a strike of less than half a day to a reduction of 1/50th; and a strike of one day or less to a reduction of 1/30th. When the right returned to power, various amendments to social legislation were proposed by RPR deputies to reinstate the more severe regime of 1977. The Pelchat amendment, as modified by the Lamassoure amendment, created article 89 of the *loi*, which reimposed the 1977 system of salary retention on the whole of the public sector. Given that the provision merely refers back to clauses of a previous *loi* approved by the Conseil constitutionnel at the time, the way in which the Conseil constitutionnel struck it down in 1987 is important in allowing the Conseil to revoke already promulgated *lois* and to reverse its own decisions.

DECISION

On article 89

3. Considering that the first effect of these provisions is to apply to the employees of each administration or service with a special status, as well as to all those receiving a salary paid monthly . . . the rule according to which the absence of services rendered, during any fraction of the day, gives rise to a retention of salary, the amount of which is equal to the fraction of the salary treated as indivisible by virtue of the said regulation; that, for this rule to be implemented, it is stated that there is no service rendered on the one hand 'when the agent abstains from performing all or part of his hours of service', and, on the other hand, 'when the agent, although carrying out his hours of service, does not perform all or any part of the duties of service that are connected with his functions, such as they are defined as to their nature and modalities by the competent authority . . .';

5. Considering that, it is claimed principally that the provisions of article 89 are contrary to the Constitution by reason of their subject-matter; that they introduce a financial penalty designed to act as a financial disincentive to the use of a right recognized by the Constitution; that they no longer meet a requirement of public accounting as shown by *loi* no. 82-889 of 19 October 1982; that they are not justified by the concern to avoid breaks in the continuity of public service, since this can be ensured by other means, whether by prohibiting certain forms of strike or by requiring a minimum continuous public service; that the authors of the reference also argue that article 89 is unconstitutional by reason of its scope of application; that this article imposes in practice special burdens on private law employees, who do not themselves participate directly in the operation of the public service, and who are not subject to the rules of public accounting; that they thus violate equality between employees;

6. Considering that, according to the seventh paragraph of the

Preamble to the Constitution of 27 October 1946, confirmed by that of the Constitution of 4 October 1958, 'the right to strike shall be exercised within the framework of the *lois* that regulate it'; that, in enacting this provision, the drafters of the Constitution intended the right to strike to be a principle of constitutional value, but within limits, and empowered the legislature to determine these by making the necessary reconciliation between the defence of professional interests, of which a strike is a means, and the safeguarding of the public interest, which a strike might infringe;

7. Considering that, therefore, it is permissible for the legislature to define the conditions for the exercise of the right to strike, and to define the boundaries separating acts and behaviour that constitute a lawful exercise of this right from acts and behaviour that are an abuse; that, in the context of public services, the recognition of the right to strike should not have the effect of preventing the legislature from imposing necessary limitations on this right in order to ensure the continuity of the public service, which, just like the right to strike, has constitutional value; that these limitations may go so far as to prohibit the right to strike of those employees whose presence is essential to the operation of parts of the service whose interruption would harm essential needs of the country . . . ;

12. Considering that, all the same, the mechanism of the automatic retention of the salary of the persons in question that the legislature has adopted to this end, by the generality of the scope of its application, which does not take into account either the character of the various services in question or the harmful effect that concerted terminations of work could have for the community as a whole, could, in a number of cases, cause an unjustified infringement of the exercise of the right to strike, which is constitutionally guaranteed . . . [Article 89 was thus declared unconstitutional.]

5. FREEDOM OF COMMUNICATION

DECISION 26: CC decision no. 86-210 DC of 29 July 1986, *Press Law*, *Rec.* 110; *AJDA* 1986, 538.

Background: The Chirac Government sought to reduce the regulation of the press instituted by the Socialists in 1984, and thereby, *inter alia*, to end prosecutions of the type then pending against its chief financial backer, Hersant, the head of a large press empire.

The case demonstrates the limits of the power of Parliament to repeal provisions of *lois* where the Conseil constitutionnel considers that they offer safeguards for fundamental values. The reference was made by Socialist deputies and senators.

DECISION

1. Considering that the deputies and senators, authors respectively of the first and second reference, claim principally that the *loi* referred for scrutiny by the Conseil constitutionnel, which repeals the *ordonnance* of 26 August 1944 on the organization of the French press, and *loi*, no. 84-937 of 23 October 1984 on the financial transparency and pluralism of press enterprises, substitutes provisions for the texts thus repealed whose scope is unduly limited, and which do not ensure the realization of the objectives of financial transparency and pluralism of the press, which are of constitutional value;

Repeal of previous laws

2. Considering that it is permissible at any moment for the legislature, deciding in the province reserved to it by article 34 of the Constitution, to amend previous texts, or to repeal them and substitute for them other provisions, as the situation requires; that, in order to achieve or reconcile objectives of constitutional value, it is no less permissible for it to adopt new methods, of whose appropriateness it is the judge, and these may include the amendment or repeal of provisions that it considers excessive or unnecessary; that, however, the exercise of this power cannot lead to the removal of legal safeguards for requirements having constitutional value . . .

Financial transparency

16. Considering that, far from being contrary to the freedom of the press, or limiting it, the implementation of the objective of financial transparency tends to reinforce an effective exercise of this freedom by enabling readers to exercise their choice in a truly free way, and to form their opinion by an informed judgment on the kinds of information that are offered by the written press;

17. Considering that, however, it was permissible for the legislature, as was stated above, to incorporate into the *loi* submitted for the scrutiny of the Conseil constitutionnel different methods for achieving the objective of financial transparency from those appearing in previous texts repealed by the said *loi*; that, thus, the fact that the new provisions are less rigorous than the provisions currently in force does not in itself constitute a ground of unconstitutionality;

18. Considering that, even if the provisions of articles 3, 4, 5, and 6 of the *loi*, taken together, do not always permit the public or categories of interested persons to identify immediately those able to exercise control over a specific press publication, their effect is, however, such as to provide essential information, without hiding the fact that the legal persons holding shares in a publishing business and exercising

an influence over it can themselves be dependent on individuals or groups external to that publishing business; that, thus, the judgment made by the legislature of the methods for achieving the objective of transparency is not vitiated by any manifest error;

Pluralism

19. [Article 11 raised the permitted share of the market of daily newspapers of general and political information to 30 per cent of the total publication over the national territory.]

20. Considering that the pluralism of daily newspapers dealing with political and general information is in itself an objective of constitutional value; that, in fact, the free communication of ideas and opinions, guaranteed by article 11 of the Declaration of the Rights of Man and of the Citizen of 1789, would not be effective if the public to which these daily newspapers are addressed were not able to have access to a sufficient number of publications of different leanings and characters; that in reality the objective to be realized is that the readers, who figure among the essential addresses of the freedom proclaimed by article 11 of the Declaration of 1789, should be able to exercise free choice without either private interests or public authorities substituting their own decisions, and without it being possible to make them the subject-matter of a market;

21. Considering that the provisions of article 11 of the *loi* prohibit the exceeding of the threshold of 30 per cent, and, *a fortiori*, the acquisition of an existing daily newspaper by an enterprise that publishes publications of this kind and already exceeds the threshold, only to the extent that this excess profits the acquirer himself; that the text does not provide that this prohibition applies to a legal or human person legally distinct from the acquirer, even though the latter is under its authority or is dependent on it; that, moreover, such an interpretation follows from reading the words of article 11 together with those of article 7 of the *loi*, which, in order to limit the influence of foreign capital, takes into account the effect that can be attached 'directly or indirectly' to certain acquisitions, which article 11 does not do; that, finally, this interpretation is corroborated by the *travaux préparatoires*;

22. Considering that, therefore, unless the ban on front companies set out in article 3 of the *loi*, or any other provision of *loi* or *règlement* is infringed, the provisions of article 11 do not prevent a human person or group from using procedures, perfectly lawful in company law, to make themselves effectively and fully masters of several existing daily newspapers without the threshold of distribution fixed by article 11 being held against them;

23. Considering that, thus, as drafted, far from modifying the

methods for protecting pluralism in the press and, more generally, among the means of communication of which the press is a part, as the legislature may do, the provisions of article 11 do not permit it to ensure that they are effective; that, by their combination with the repeal of previous legislation, they have the effect of depriving a principle of constitutional value of legal protection;

24. Considering that, it follows from that, article 11 has to be declared as not consistent with the Consitution . . .

DECISION 27: CC decision no. 86-217 DC of 18 September 1986, *Freedom of Communication*, *Rec.* 141, *AJDA* 1987, 102.

Background: The Chirac Government sought to reform broadcasting in a number of ways. A major part of the reform was the privatization of the main television station, TF1. The reform also involved a measure of relaxation in the regulation exercised over broadcasting, and the creation of a more independent regulatory authority, the National Commission for Communication and Liberties (CNCL), which replaced the existing regulatory authority and its members.

Private television channels had already been allowed by *loi* no. 85-1317 of 13 December 1985, and the first private channel, Canal 5, had been awarded to Socialist supporters. Before they left office in March 1986, the Socialists had already announced plans for additional private channels.

DECISION

The reform of the regulatory authority

3. Considering that the authors of the reference claim that 'the modern implementation of the freedom of communication proclaimed by article 11 of the Declaration of 1789 presupposes the existence of an independent authority' entrusted with the supervision of respect for constitutional principles in the area of broadcasting; that the independence of such a body implies that the legislature itself cannot put a premature end to the mandate of its members; that, having failed to provide that the members making up the High Authority of Audio-Visual Communication continue in office until the expiry of their mandate, articles 96 and 99 of the *loi* disregard requirements of constitutional value;

4. Considering that it is permissible at any moment for the legislature, deciding in the province that is reserved to it by article 34 of the Constitution, to amend previous texts, or to repeal them and substitute for them other provisions, as the siutation requires; that, in order to achieve or reconcile objectives of constitutional value, it is no less permissible for it to adopt new methods, of whose appropriateness it is the judge, and these may include the amendment or repeal of provisions that it considers excessive or unnecessary; that, however,

the exercise of this power cannot lead to the removal of legal safeguards for requirements having constitutional value;

5. Considering that the replacement of the High Authority of Audio-Visual Communications, created by *loi* no. 82-652 of 29 July 1982, by the National Commission for Communication and Liberties does not in itself remove legal safeguards for constitutional requirements; that, therefore, without disregarding any rule or any principle of constitutional value, the legislature could decide to put an end to the mandate of the members of the High Authority of Audio-Visual Communication at the time of its replacement; that the argument raised can only be rejected;

On the use of broadcasting frequencies

6. Considering that, according to the authors of the reference, the need to secure broadcasting frequencies, combined with the fact that the development of terrestrial television concerns the exercise of public liberties to the highest degree, makes broadcasting frequencies part of the public domain, and this method of communication is, by nature, a public service that corresponds to constitutional requirements . . .

7. Considering that, in the words of article 11 of the Declaration of the Rights of Man and of the Citizen, 'the free communication of thoughts and opinions is one of the most precious rights of man; hence, every citizen may speak, write, and publish freely, save that he must answer for any abuse of such freedom in cases specified by *loi*';

8. Considering that it is up to the legislature to reconcile, in the current state of technology and its control, the exercise of the freedom of communication resulting from article 11 of the Declaration of the Rights of Man with, on the one hand, the technical constraints inherent in the means of audio-visual communication, and, on the other hand, with objectives of constitutional value—the safeguard of public order, respect for the freedom of others, and the preservation of the pluralistic character of socio-cultural currents of expression—which these forms of communication, by reason of their considerable influence, are likely to infringe;

9. Considering that, in order to achieve or reconcile these objectives, the legislature is not bound to place the whole of terrestrial television under the legal regime applicable to public services, nor to adopt a regime of concession; that in fact this means of communication does not constitute a public service activity with a basis in provisions of a constitutional kind; that, in consequence, and whatever the legal status of broadcasting frequencies, it is permissible for the legislature to place the private broadcasting sector under a regime of administrative licensing, provided that the safeguarding of objectives of constitutional value is ensured; that the argument raised cannot be accepted;

Pluralism

[The deputies claimed that the provisions on pluralism were not precise enough and failed to restrict multi-media concentrations of ownership: article 39 only prevented the same person from owning more than 25 per cent of the shareholding in several private companies licensed to operate as public service broadcasters; article 41 prevented ownership of more than one broadcasting company in the same geographical area.]

11. Considering that the pluralism of currents of socio-cultural expression is in itself an objective of constitutional value; that the respect for this pluralism is one of the conditions for democracy; that the free communication of ideas and opinions, guaranteed by article 11 of the Declaration of the Rights of Man and of the Citizen of 1789, would not be effective if the public to which the means of audio-visual communication address themselves were not able to have access, both within the framework of the public sector as well as in the private sector, to programmes that guarantee the expression of tendencies of different characters, respecting the necessity of honesty in information; that in reality the objective to be realized is that the listeners and television-viewers, who are among the essential addressees of the freedom proclaimed by article 11 of the Declaration of 1789, should be able to exercise free choice without either private interests or public authorities being able to substitute their own decisions for it, and without it being possible to make them the subject-matter of a market;

12. Considering that article 1 of the *loi*, which provides that the freedom to exploit and use telecommunication services could be restricted to the extent necessary to safeguard the pluralist expression of currents of opinion, as well as article 3, which creates a National Commission of Communication and Liberties, with the duty of promoting the pluralist expression of currents of opinion, are consistent with the Constitution; that it is necessary to examine whether the means for implementing the principles declared by articles 1 and 3 of the *loi* are equally so; that this implementation depends in part on the rules laid down by the *loi* that are directly applicable, and in part on the rules specified by decree, whose effective application will depend on the intervention of the National Commission of Communication and Liberties . . .

Within the private sector

[It was alleged that the powers of the CNCL were insufficient to protect pluralism in the private sector. The powers of the CNCL were set out particularly in articles 28 to 31 of the *loi*.]

22. Considering that the provisions of articles 28 to 31 have to be interpreted in the light of the principles laid down by the *loi* in articles

1 to 3, which impose a duty on the National Commission of Communication and Liberties to preserve, as a priority, 'the pluralistic expression of currents of opinion'; that, in particular, where there is only a single frequency in a given area, it is for the Commission to impose such obligations on the recipient of a licence as to ensure a free and pluralist expression of ideas and currents of opinion; that the same obligations have to be imposed where the existence of several frequencies, although belonging to different operators, would not be sufficient to guarantee pluralism; that any other interpretation would be contrary to the Constitution, as it would lead to granting the Commission a discretionary power to apply articles 28 to 31 of the *loi*, without requiring it to respect the obligatory framework set out by articles 1 and 3;

23. Considering that, furthermore, the National Commission of Communication and Liberties, like any administrative body, will be subject to judicial review over the exercise of its competence, which can be carried out either by the Government, which is responsible to Parliament for the activities of State administrative bodies, as well as by any person having an interest;

24. Considering that, therefore, articles 28 to 31 of the *loi* are not, in themselves, contrary to the Constitution . . .

[Article 39 prohibited a person from acquiring more than 25 per cent of the share capital of a company licensed to run a national television station. Article 41 prohibited a person from acquiring one or more licences for radio broadcasting unless the population served by him or her was less than 15 million. In addition, it provided that a person licensed to operate a local television station could not acquire another licence to run a similar service in that area.]

31. Considering that article 39 of the *loi* in no way prohibits the same person from becoming the owner of up to 25 per cent of the capital of several private companies, each licensed to provide a television service throughout the whole of the metropolitan territory; that the article imposes no limit on the share that the same person can take in companies licensed to provide a television service in parts of the territory;

32. Considering that neither article 39 nor any other provision of the *loi* provides for any restriction on the awarding of licences for cable television and radio to the same person;

33. Considering that, in the restrictions it imposes, article 41 does not take into account the situation of people with licences for radio broadcasting on longer wavelengths; that it does not preclude the possibility that the same person might hold licences simultaneously for radio broadcasting services and television services; that, in relation to terrestrial television, the second paragraph of article 41 limits itself to prohibiting the same person from cumulating two licences within the

same geographical area, without preventing the same person from being awarded, ultimately, one or more licences at the same time permitting him to serve the whole of the territory, either by way of a national service or by means of a network of local services;

34. Considering that, even if the provisions of article 17 of the *loi*, as well as those of article 41, enable the abuse of a dominant position in the area of communications to be combatted, this circumstance does not suffice, in itself, to ensure respect for the constitutional objective of pluralism;

35. Considering that, according to article 34 of the Constitution, '*loi* shall determine the rules concerning . . . the fundamental safeguards granted to citizens for the exercise of civil liberties'; that, by reason of the insufficiency of the rules contained in articles 39 and 41 with regard to the restriction of concentrations capable of harming pluralism, the legislature has disregarded its competence under article 34 of the Constitution; that, furthermore, from the existence of gaps in the *loi*, there is the risk that situations of concentration will develop, particularly within the same geographical area, not only in the field of broadcasting, but equally in relation to the totality of means of communication, of which broadcasting is one of the essential components;

36. Considering that, as drafted, the provisions of articles 39 and 41 of the *loi* do not in themselves satisfy the constitutional requirement of preserving pluralism either in the area of broadcasting or in that of the media in general; that, in consequence, articles 39 and 41 of the *loi* must be declared inconsistent with the Constitution . . . ;

On the transfer of Télévision française 1 (TF1) to the private sector

[It was argued that the privatization of TF1 could only take the form of a public service concession, not a full transfer of the activity, to the private sector, subject to administrative regulation. This was rejected.]

39. Considering that, as has already been said, it is permissible for the legislature to place the private broadcasting sector under a regime of administrative authorization; that, equally, the legislature might place the national broadcasting company Télévision française 1, once it is transferred from the public sector to the private sector, under a regime of administrative authorization without being bound to resort to a regime of public service concession . . .

On the regulatory powers of the National Commission for Communication and Liberties

[It was argued that the provisions giving 'regulatory status' to the decisions of the CNCL were contrary to article 21 of the Constitution. This was rejected.]

58. Considering that [the provisions of article 21 of the Constitution] confer on the Prime Minister, excepting the powers granted to the President of the Republic, the exercise of regulatory power at a national level; that they do not prevent the legislature from conferring on a State body other than the Prime Minister the task of fixing norms to implement a *loi*, within a precise area and within a framework defined by *lois* and *règlements* . . .

DECISION 28: CC decision no. 88-248 of 17 January 1989, *Conseil supérieur de l'audiovisuel (CSA), RFDA* 1989, 216; *GD*, no. 44.

Background: (See above, p. 191, and J. Chevallier, 'De la C. N. C. L. au Conseil supérieur de l'audiovisuel', *AJDA* 1989, 59.) The CNCL came in for much criticism, particularly from Socialists (including the President), for lacking impartiality in its composition and decisions. A broad consensus of opinion wanted a more obviously independent regulatory authority. The *loi* to adopt a pattern of appointment and tenure very close to that of the Conseil constitutionnel.

Although the CNCL was a regulatory authority, its only enforcement powers were either the Draconian step of suspending a licence, or taking an action in the courts to enforce the terms of the licence or the legislation. The *loi* gave it the power to impose sanctions. Both the grounds of liability and the sanction power itself were challenged by sixty opposition deputies.

DECISION

Grounds of liability to a sanction

8. Considering that in referring to the concept of a 'serious breach' by bodies in the public sector of broadcasting of the obligations imposed on them by virtue of the *loi* of 30 September 1986 as modified, or of the decrees made in the Conseil d'État provided for by article 27, or of the franchise terms (*cahier des charges*), the legislature intended to exclude from non-serious breaches the implementation of an enforcement procedure for national programming companies or the National Broadcasting Institute; that it is for the CSA, subject to judicial review, to comply with the distinction made by the *loi* as to the degree of seriousness of the breach; that, thus, it could not be objected that the legislature underexercised its competence under the Constitution, and especially that provided for in article 34;

Immunity of the President of the Conseil

9. Considering that none may be exonerated by a general provision of a *loi* from all personal liability, whatever the nature or seriousness of the act attributed to him; that, thus, the provisions of the last sentence of article 13 of the *loi* of 30 September 1986, as reformulated by article 8 of the referred *loi*, must be declared contrary to the constitutional

principle of equality, in that it provides that: 'Actions taken to implement these decisions cannot in any case give rise to the personal liability of the President of the Authority'. . .

The regulatory powers of the CSA

14. Considering that the first two paragraphs of article 21 of the Constitution are written thus: 'The Prime Minister directs the operation of the Government. He is responsible for national defence. He ensures the implementation of *lois*. Subject to the provisions of article 13, he exercises the power to make regulations, and makes appointments to civil and military services. He may delegate certain powers to ministers';

15. Considering that these provisions confer on the Prime Minister, excepting the powers granted to the President of the Republic, the exercise of regulatory power at a national level; that they do not prevent the legislature from conferring on a State body other than the Prime Minister the task of fixing norms to implement a *loi*, provided that this authorization concerns only measures limited in scope both by their area of application and by their content;

16. Considering that the *loi* empowers the CSA to determine by way of *règlement* not only the rules of conduct relating to advertising, but equally all the rules relating to institutional communication, sponsorship, and practices analogous thereto; that, by reason of its excessively broad scope, this authorization disregards the provisions of article 21 of the Constitution . . .

Administrative sanctions

[It was claimed that the powers conferred on the CSA by the new articles 41-1 and 42-11 added to the *loi* of 30 September 1986 to impose sanctions breached both the separation of powers and the freedom to communicate (article 11 of the 1789 Declaration).]

26. Considering that it is up to the legislature to reconcile, in the current state of technology and its control, the exercise of the freedom of communication resulting from article 11 of the Declaration of the Rights of Man with, on the one hand, the technical constraints inherent in the means of audio-visual communication, and, on the other hand, with objectives of constitutional value—the safeguard of public order, respect for the freedom of others, and the preservation of the pluralistic character of socio-cultural currents of expression—which these forms of communication, by reason of their considerable influence, are likely to infringe;

27. Considering that, in order to achieve these objectives of constitutional value, it is permissible for the legislature to impose a regime of administrative authorization on the different categories of broadcasting

service; that it is equally permissible for it to give an independent administrative body the task of supervising respect for constitutional principles in the area of broadcasting; that, without infringing the principle of the separation of powers, *loi* may give the independent authority charged with safeguarding the exercise of the freedom of broadcasting the powers of sanction that, within limits, are necessary for accomplishing its mission;

28. Considering that it is for the legislature to attach to the exercise of these powers measures designed to safeguard constitutionally guaranteed rights and freedoms;

29. Considering that, in accordance with the principle of respect for the right to due process, which constitutes a fundamental principle recognized by the laws of the Republic, no sanction may be imposed without the holder of a licence having been enabled both to present observations on the facts that are alleged against him and to have access to the file concerning him . . . [The Conseil then noted various provisions of the *loi* that, *inter alia*, provided for a *procédure contradictoire* (a procedure whereby both sides can put their point of view), conducted by a judge of the administrative courts, to be followed, that ensured that penalties were not automatic and were proportionate both to the seriousness of the breach committed and to the financial benefits gained thereby, and that administrative and criminal penalties were not cumulative. In addition, the body imposing penalties was to be independent.]

31. Considering that it is right to note equally that any decision imposing a sanction may be subject to an appeal to the Conseil d'État, as article 42-8 provides; that this process suspends the implementation of the withdrawal of the licence as mentioned in article 42-3; that, in other cases, suspension of the implementation of the challenged decision may be requested [under the normal rules of procedure of the administrative courts]; that exercise of the right of appeal, being restricted to the person penalized, may not, in accordance with general principles of law, lead to an aggravation of his or her situation;

32. Considering that, since they concern breaches of the obligations attached to an administrative licence, and having regard to the safeguards provided, which, moreover, apply to both contractual penalties and to sanctions liable to be imposed by [the various provisions of the *loi*], [the articles granting power to the CSA to impose sanctions] are not contrary, in principle, to articles 11 and 16 of the Declaration of the Rights of Man and of the Citizen . . .

On the penalties imposed

[It was claimed that the provisions of the *loi* authorizing the imposition of sanctions by the CSA breached article 8 of the 1789 Declaration and article 34 of the Constitution, in that *loi* had not fully defined the terms of the offences].

35. Considering that it follows from these provisions, as well as from the fundamental principles recognized by laws of the Republic, that a penalty may not be imposed unless the following are respected: the principle that offences and penalties must be based on a *loi*, the principle of the necessity of penalties, the principle of the non-retroactivity of more severe incriminating penal laws, and the principle of the right to due process;

36. Considering that these requirements not only relate to penalties imposed by criminal courts, but extend also to any sanction having the character of a punishment, even if the legislature has left the task of imposing it to a body of a non-judicial kind;

37. Considering that, however, applied outside criminal law, the requirement that penalized offences be defined is satisfied, in administrative matters, by reference to an obligation that the holder of a licence was placed under by virtue of *lois* and *règlements;*

38. Considering that [the sanction powers of the CSA may only be invoked after a default notice is served on the licensees, requiring them 'to respect the obligations that are imposed on them by *lois* and *règlements*, and by the principles set out in article 1 [of the *loi*]']; that the obligations capable of giving rise to a sanction are only those resulting from the terms of the *loi*, or those whose respect is expressly imposed by a licensing decision taken in application of the *loi* and of the *règlements* which fix the general principles defining the obligations of the different categories of broadcasting services within the framework set out by the legislature;

39. Considering that, subject to the reservations of interpretation mentioned above, articles 42-1 and 42-2 are not contrary to the provisions either of article 8 of the Declaration of the Rights of Man, nor article 34 of the Constitution which defines the competence of the legislature . . .

6. FREEDOM OF PROPERTY

DECISION 29: CC decision no. 81-132 DC of 16 January 1982, *Nationalizations*, *Rec.* 18; *GD*, no. 33; *SB* 67; L. Favoreu, 'Les Décisions du Conseil constitutionnel dans l'affaire des nationalisations', *RDP* 1982, 377; L. Favoreu, *Nationalisations et Constitution*; J. Rivero, *Le Conseil constitutionnel et les libertés*, part I, ch. 8.

Background: See DECISION 2.

DECISION

On the principle of nationalization: see DECISION 2

Unequal treatment

21. Considering that, as far as the nationalization of banks is concerned, article 13 of the *loi* declares first, in paragraph I the general rule according to which the companies covered by the nationalization are identified, as well as the exceptions made to this rule, then, in paragraph II, it draws up a list of the companies nationalized . . . [The grounds of complaint were the exclusion of (i) banks with credit of under 1 milliard francs; (ii) mutualist and co-operative banks; and (iii) banks the majority of whose capital was held by foreigners, since they infringed equality.]

28. Considering that the principle of equality is no less applicable between legal persons as between physical persons, because, legal persons being groupings of physical persons, breach of the principle of equality between the former would be equivalent to breach of equality between the latter;

29. Considering that the principle of equality does not prevent a *loi* from establishing non-identical rules with respect to categories of persons who are in different situations, but this can only be the case where the non-identity is justified by a difference in situation and is not incompatible with the purpose of the *loi*.

30. Considering that the exclusion of banks with the status of commercial or industrial property companies or the status of discount houses is not contrary to the principle of equality, as certain of the features of the status of these establishments are specific to them;

31. Considering that even if banks whose majority capital belongs directly or indirectly to physical persons not resident in France, or to legal persons not having their registered office in France, have the same legal status as other banks, the legislature may exclude them from nationalization without breaching the principle of equality, because of the difficulties that nationalizing these banks might cause at an international level, which eventuality would, in its view, harm the public interest that the objectives pursued by the *loi* on nationalization would serve;

32. Considering that, by contrast, the exception made for banks whose majority share capital belongs directly or indirectly to companies of a mutualist or co-operative character does breach the principle of equality; that, in fact, it is justified neither by the specific character of their status, nor by the nature of their activity, nor by potential difficulties in applying the *loi* that would work against the public interest goals that the legislature intends to pursue;

33. Considering that, thus, the following provisions of article 13-I of the *loi* submitted for scrutiny by the Conseil constitutionnel must be declared incompatible with the Constitution: 'Banks the majority of whose share capital belongs directly or indirectly to companies of a mutualist or co-operative character' . . .

Compensation

[The method of compensation was to give former shareholders in the nationalized companies some interest-bearing bonds, which could be redeemed after a period of seven and a half years.]

43. Considering that, by virtue of the provisions of article 17 of the Declaration of the Rights of Man and of the Citizen, the deprival of a property right for public necessity requires a just and prior indemnity . . .

53. Considering that it is true that, according to the [provisions of the *loi*], the reference to the average stock-market price for the years 1978, 1979, and 1980 only forms 50 per cent of the calculation of the exchange value of the shares, and is complemented as to 25 per cent by reference to the net accounting position, and 25 per cent by reference to ten times the net average profit;

54. Considering that, according to the intentions of the legislature, the appeal to criteria other than the average stock-market price was precisely to correct the imperfections in the reference to the average stock-market value . . .

55. But considering that this goal is unequally achieved by the provisions currently examined; that, in particular, the reference to the net accounting position, excluding the assets of subsidiaries, as well as the reference to the net average profit, excluding the profits of subsidiaries, lead to very different results for the companies in question, determined not by a difference in objective economic and financial facts, but by the diversity of the management techniques and methods of presenting accounts followed by the companies, which, in itself, should have no bearing on the assessment of compensation;

56. Considering that, furthermore, as a necessary consequence, the provisions of the articles currently examined deprive the former shareholders of the dividends that they would have received for the 1981 financial year, and which are not made up for by the interest that the bonds given in exchange will produce in 1982;

57. Considering that, overall, as far as the shares of the companies quoted on the stock exchange are concerned, the method of calculating the value of exchange leads to inequalities in treatment, the extent of which is not justified alone by practical considerations of speed and simplicity; that the inequalities of treatment are compounded, in a

number of cases, by a substantial underestimate of the said exchange value; that, finally, the refusal to give former shareholders the benefit of dividends relating to the financial year 1981, or to provide them with an equivalent advantage in an appropriate form, removes without justification the compensation to which the former shareholders are entitled;

60. Considering that it follows from what has been said that articles 6, 18, and 32 of the *loi* submitted for scrutiny by the Conseil constitutionnel are incompatible with the requirements of article 17 of the Declaration of the Rights of Man and of the Citizen as far as the just character of the compensation is concerned . . .

DECISION 30: CC decision no. 86-207 DC of 25, 26 June 1986, *Privatizations*, *Rec.* 61; *AJDA* 1986, 575; *GD*, no. 41; *SB* 82.

Background: See DECISION 12.

DECISION

As far as articles 4 and 5 of the loi and the list annexed thereto are concerned

[Article 4 provided that, by 1 March 1991 at the latest, the majority shareholding, held directly or indirectly by the State, in the enterprises listed in the annex should be transferred to the private sector. Article 5 enabled the Government to make *ordonnances* within six months, fixing, *inter alia*, the rules for the valuation of the enterprises to be privatized, the legal and financial forms of transfer, the terms of payment, the restrictions on the acquisition or disposal of rights in the newly privatized companies, the terms for protecting the national interest, the conditions for developing popular and worker shareownership, and the tax regime of the transfers. The list annexed to the *loi* contained the names of sixty-five enterprises belonging to the public sector.]

On the principle of privatization

50. Considering that article 34 of the Constitution places within the domain of *loi* 'the rules concerning . . . the nationalization of undertakings, and the transfer of property in undertakings from the public sector to the private sector';

51. Considering that, though this provision leaves the legislature to judge the appropriateness of transfers from the public sector to the private sector, and to determine the property or enterprises to which such transfers should relate, it does not dispense it, in the exercise of this competence, from respect for the principles and rules of constitutional value, which are imposed on all organs of the State;

52. Considering that the deputies making the first reference claim

that the provisions of article 4, and the statements in the list of undertakings annexed to the *loi* disregard the provisions of the ninth paragraph of the Preamble to the 1946 Constitution, according to which 'any property or business whose exploitation has or acquires the character of a national public service or a *de facto* monopoly should become the property of the community'; that it follows from this that the transfer from the public sector to the private sector of certain enterprises figuring on the list annexed to the *loi* whose exploitation has the character of a national public service or a *de facto* monopoly would be contrary to the Constitution;

53. Considering that, though the necessity of certain national public services is derived from principles or rules of constitutional value, the determination of those other activities which should be turned into a national public service is left to the judgment of the legislature or of the regulatory power, as the case may be; that it follows that the fact that the activity has been turned into a public service by the legislature, without the Constitution requiring it, does not prevent this activity, and the enterprise running it, from being the object of a transfer to the private sector;

54. Considering that none of the enterprises that figure on the list mentioned in article 4 can be regarded as exploiting a public service whose existence and operation would be required by the Constitution; that, in particular, even if, as the deputies making the first reference allege, the legislature had intended, by nationalizing a collection of banks, to create a public service of credit, this creation, which did not follow from any constitutional requirement, could not prevent the return to the private sector of certain credit activities, and the banks providing them, as a result of a new *loi*;

55. Considering that the notion of *de facto* monopoly cited in the ninth paragraph of the Preamble to the 1946 Constitution should be understood as taking into account the whole of the market within which the enterprises carry out their activities, as well as the competition that they face in that market from other firms; that one should not take account of the advantageous position that any enterprise occupies momentarily, or, with respect to production, that only represents a part of its activities; that, taking account of these considerations, it is not established on the facts that it is by a manifest error in evaluation that the enterprises figuring on the list annexed to the *loi*, as well as their subsidiaries, have been considered as not amounting to *de facto* monopolies;

On the conditions for the transfer

57. Considering that the deputies making the reference claim that it would be impermissible for enterprises from the public sector being

transferred to the private sector to be disposed of at a price below their real value; that a disposal at such a price would fundamentally disregard the principle of equality by providing the acquirers with an unjustified advantage, to the detriment of the totality of citizens; that, however, no provision of the enabling law provides guarantees against the disposal at inadequate prices of the enterprises mentioned in article 4 of the *loi*; that, furthermore, the unconditional obligation imposed on the Government to proceed to the transfer of the totality of the State's majority shareholdings in these enterprises before 1 March 1991 could have the effect of disposing of important parts of the national patrimony at a low price, if, as might be expected, this massive influx exceeds the capacity of the market, without excluding the possibility of transfers into foreign hands, which would be harmful to national independence;

58. Considering that the Constitution opposes the transfer of property or enterprises forming part of public property to persons pursuing private-interest objectives for prices lower than their value; this rule follows from the principle of equality invoked by the deputies making the reference to this Conseil; that it is also based on the provisions of the Declaration of the Rights of Man of 1789 concerning the right of property and the protection due to it; this protection concerns not only the private property of individuals, but also, on an equal footing, the property of the State and of other public persons;

59. But considering that article 4 of the *loi* provides, in its second paragraph, that the transfers shall be made by the Government in conformity with the rules laid down by the *ordonnances* mentioned in article 5; that article 5 of the *loi* provides that *ordonnances* shall fix the rules for the valuation of the enterprises and for the determination of the offer price, which would forbid the transfer of the enterprises mentioned in article 4 if the price for which they could be transferred was lower than their real value; that the preparatory materials for the *loi* show that the Government has committed itself to proceed to a valuation by independent experts, and not to transfer the businesses mentioned in article 4 of the *loi* at a price below their real value; that the safeguards that should preserve national independence will follow equally from the *ordonnances* provided for by article 5;

60. Considering that it follows from what has been said above that article 4 of the *loi* has to be understood as providing the final date of 1 March 1991 for the implementation of the transfers only at a price that conforms to the patrimonial interests of the State and while respecting national independence, it being understood that the transfers that, at that date, could not have been undertaken or completed could only take place or be completed under the authority of a new provision of a *loi*; that any other interpretation would be contrary to the Constitution;

61. Considering equally that the provisions of article 5, to which the second paragraph of article 4 refers, have to be understood as requiring the Government to make provision by way of *ordonnance* for the valuation of the enterprises to be transferred to be made by competent experts, totally independent of the potential buyers; that it should be conducted according to objective methods currently practised for the total or partial transfer of assets of companies, taking account, with appropriate weight in each case, of the stock-exchange value of their shares, the value of assets, the profits made, the existence of subsidiaries, and future prospects; that, further, the *ordonnance* should forbid a transfer if the price proposed by the purchasers does not exceed, or at least is not equal to, this valuation; that the choice of purchasers should not be preceded by any special treatment; that national independence must be safeguarded; that any other interpretation would be contrary to the Constitution . . .

DECISION 31: CC decision no. 89-256 DC of 25 July 1989, *TGV Nord*, *RFDA* 1989, 1009, comment P. Bon.

Background: In order to expedite the construction of the high-speed train (TGV) link to the Channel Tunnel, an amendment (article 9) was moved to the *loi* on planning and new towns that empowered the Government to invoke the procedure of expropriation in extreme urgency set out in articles L15-9 ff. of the Code of Expropriation. This procedure had been enacted by *ordonnance* no. 58-997 of 23 October 1958, modified by the *loi* of 4 August 1962, and codified by the only article of the *loi* of 23 December 1970. The procedure departed from that usually applied in expropriation, in that (i) the public authority could be authorized to take possession of the property not by a judge, but by a decree issued after a favourable opinion from the Conseil d'État, and (ii) compensation could be paid as an interim amount assessed by the Services of Public Lands (or according to the offer made by the public authority, if higher) before expropriation was ordered, though the final quantum was to be fixed by the judge in proceedings begun within a month thereafter.

DECISION

Legislative procedure

[It was argued that the amendment either had no connection with the text under discussion, or exceeded the scope of the right of amendment. The Conseil rejected this on the ground that infrastructural works such as railways had relevance to the planning and land policies of public bodies.]

Challenging enacted lois

9. Considering that, according to the authors of the reference, article 9 of the *loi* is contrary to the Constitution, in that it extends the

scope of application of the provisions of article L15-9 of the Code of Expropriation for Public Utility, which are themselves unconstitutional;

10. Considering that the validity with respect to the Constitution of the terms of a promulgated *loi* may be properly contested on the occasion of the submission to the Conseil constitutionnel of provisions of *lois* that modify it, complement it, or affect its scope;

11. Considering that it is thus the task of the Conseil constitutionnel to ensure that the terms both of the single article of *loi* no. 70-1263 of 23 December 1970, codified under article L15-9 of the Code mentioned above, and of the texts of legislative force to which article L15-9 refers are not contrary to the Constitution . . .

Substantive validity of article 9

17. Considering that article 2 of the Declaration of 1789 lists property among the rights of man; that article 17 of the same Declaration equally proclaims: 'Property being an inviolable and sacred right, none can be deprived of it except when public necessity, legally ascertained, evidently requires it, and on condition of a just and prior indemnity';

18. Considering that the objectives and conditions for the exercise of the right of property have undergone an evolution characterized by the extension of its range of application to new areas, and by the limitations required in the name of the public interest, that it is in relation to this evolution that the reaffirmation by the Preamble to the Constitution of 1958 must be understood;

19. Considering that to conform to these constitutional requirements, a *loi* may authorize the expropriation of immovable property or real rights only in order to carry out an operation whose public utility has been lawfully ascertained; that the entry into possession by the expropriator must be conditional upon the prior payment of compensation; that, in order to be just, compensation must cover the totality of the direct, material, and certain harm caused by the expropriation; that, in case of disagreement on the determination of the quantum of compensation, the person expropriated must have an appropriate method of redress;

20. Considering nevertheless that the payment by the expropriating body of an interim award representing the compensation due is not incompatible with respect for these requirements, if such a mechanism responds to imperative reasons of the public interest, and provides safeguards for the rights of the owners concerned;

21. [The Conseil noted that article L15-9 was limited to major works of national importance, and could only be invoked when very localized difficulties were likely to delay the works being carried out. Furthermore, expropriation was conditional upon payment of compensation to the owner, or, if this was impeded, upon payment of an interim

award, with the final amount being assessed by the judge, who may award a special compensation to take account of the speed of the procedure.]

22. Considering that since its scope of application is narrowly circumscribed, and the totality of safeguards provided for the benefit of the owners involved, article L15-9 of the Code of Expropriation for Public Utility is not contrary to article 17 of the Declaration of the Rights of Man and of the Citizen;

As far as the argument based on Breach of a fundamental principle recognized by the laws of the Republic is concerned

23. Considering that, though it permits an anticipated taking of possession of unbuilt land in the circumstances analysed above, the procedure governed by article L15-9 in no way prevents the intervention of the civil judge to fix definitively the amount of the compensation; that, thus, in any case, the importance of the functions conferred on the judicial authority in relation to immovable property by the fundamental principles recognized by the laws of the Republic is not disregarded . . .

As far as the argument based on breach of the principle of equality is concerned

[The Conseil] found that, in determining the fundamental principles of property law, the legislature might provide for different procedures for expropriation in different circumstances, provided that no unjustified discrimination occurred, and the owners were given equivalent safeguards. Article L15-9 satisfied both requirements, and its extension to railway routes was not contrary to the Constitution.]

V. EQUALITY

1. THE GENERAL PRINCIPLE

DECISION 32: CC decision no. 73-51 DC of 27 December 1973, *Ex Officio Taxation*, *Rec.* 25; Hamon, D. 1974 Chr. 83; *GD*, no. 21; Nicholas, 100–1; J. Rivero, *Le Conseil constitutionnel et les libertés*, part I, ch. 2.

Background: The finance law for 1974 sought to take steps to reduce tax evasion. Article 62 of the *loi* permitted the Fisc to fight against incomplete tax returns in relation to income tax. (At that time France did not operate a pay-as-you-earn system, but required all individuals to make a tax return annually, on which assessment would be based.) Article 180 of the General Code of Taxation permitted the Fisc to decide ex officio the amount of tax due when no return was made, or where the taxpayer refused to reply to demands to explain or justify items in his or her return. The provision also applied where the return appeared inadequate. The parliamentary committee sought to amend this by allowing the taxpayer to demonstrate that circumstances did not give rise to a presumption that he or she had illegal or undeclared resources. The Minister of Finance, M. Giscard d'Estaing, succeeded in having this amendment qualified by the addition of a final proviso to the paragraph, stating that it did not apply in those cases where the relevant tax band for the income in question was at least 50 per cent lower than the limit for the highest band of income tax, thereby excluding large defrauders of the Fisc from the possibility of having the ex officio tax demand set aside. The President of the Senate referred to provision to the Conseil.

DECISION

1. Considering that the provisions of article 62 of the finance law for 1974 aim to add to article 180 of the General Code of Taxation so as to enable the taxpayer, taxed ex officio on income tax in the circumstances laid down by the latter article, to obtain discharge from the amount assigned to him in this respect if he establishes, subject to review by the tax judge, that circumstances do not give rise to a presumption of 'the existence of illegal or hidden resources, or of behaviour intending to evade the normal payment of the tax';

2. Considering that, however, the final proviso of the paragraph added to article 180 of the General Code of Taxation by article 62 of the finance law for 1974 aims to create a discrimination between citizens with regard to the possibility of presenting evidence against a decision of ex officio taxation by the administration concerning them; that, thus, the said provision infringes the principle of equality before the law contained in the Declaration of the Rights of Man of 1789 and solemnly reaffirmed by the Preamble to the Constitution . . .

DECISION 33: CC decision no. 89-269 DC of 22 January 1990, *Miscellaneous Social Measures*, *AJDA* 1990, 471; *RFDA* 1990, 406.

Background: A number of parliamentary amendments sought to make changes in the Code of Social Security and Health. Two issues were raised by the Conseil of its own motion: the scope of article 17 on the contract governing payments to doctors, and the rights of foreigners.

DECISION

18. Considering that, according to article 34 of the Constitution, '*loi* shall determine the fundamental principles . . . of social security'; that, among the fundamental principles belonging within the competence of the legislature is that whereby the tariff for medical fees for care provided to those on social insurance is fixed by means of an agreement between practitioners or their representative organizations, or, in default, by imposition; that, by contrast, the fixing of the means of implementing the fundamental principles laid down by the legislature

belongs to the competence of the regulatory power; that it follows, therefore, that article 17 of the *loi* referred does not infringe the provisions of article 34 of the Constitution;

19. Considering that, by virtue of the article 21 of the Constitution, the Prime Minister ensures the implementation of *lois* and, subject to the provisions of article 13, exercises regulatory power; that he may delegate certain of his powers to ministers;

20. Considering that these provisions do not prevent the legislature from conferring on a public body other than the Prime Minister the task for fixing the norms for the implementation of the principles laid down by *loi*, as long as such delegation only concerns measures of limited scope both as regards their field of application and their content;

21. Considering that the entry into force of one or other of the agreements provided for by article L162-5 of the Code of Social Security is subject to ministerial approval; that this approval has the effect of conferring a regulatory character on the provisions of the agreement that comes within the area of the provisions of article L162-6 of the Code just mentioned; that this mechanism for implementing the principles laid down by *loi*, whose sphere of application and scope are narrowly circumscribed, is not contrary to article 21 of the Constitution.

Relationship to 1946 Preamble

[Objection was taken to different agreements being made according to whether charges were reimbursed to specialist practitioners or general practices.]

23. Considering that, by virtue of the eleventh paragraph of the Preamble to the Constitution of 27 October 1946, confirmed by the Constitution of 4 October 1958, the nation 'guarantees to all, especially to the child, the mother, and aged workers, the protection of health, material security, rest, and leisure';

24. Considering that it is incumbent upon the legislature as well as on the regulatory authority, within their respective competences and respecting the principles laid down by paragraph 11 of the Preamble, to determine their concrete modes of application; that, in particular, their role is to determine the appropriate rules designed to realize the objective defined by the Preamble; that, in this respect, recourse to an agreement to govern the relationships between the primary funds for sickness insurance and doctors aims to reduce the portion of medical fees that remains to be paid by persons covered by social insurance, and, in consequence, to permit the effective application of the principle laid down by the cited provisions of the Preamble; that the possibility of arranging, by special agreements, the relationships between the

primary funds for sickness insurance and, respectively, general practitioners and specialists has, as its purpose, making the conclusion of such agreements easier; that, in this situation, it could not be objected against article 17 of the *loi* that it disregards the provisions of the eleventh paragraph of the Preamble of the 1946 Constitution;

Article 24 on the National Solidarity Fund

30. Considering that article 24 of the *loi* redrafts article L815-5 of the Code of Social Security, under which 'the additional benefit is due to foreigners in application of Community regulations or of reciprocal international agreements';

31. Considering that, with regard to foreigners, the legislature may make specific provisions on condition that it respects international agreements to which France is a party and fundamental freedoms and rights of constitutional value recognized to all who reside in the territory of the Republic;

32. Considering that the additional benefit from the National Solidarity Fund is granted to aged persons, especially those who have become unable to work, who do not have a sum available, from resources of whatever origin, to assure them a minimum for living; that the allocation of this benefit is subject to a period of residence in France;

33. Considering that to exclude foreigners lawfully resident in France from the enjoyment of the additional benefit if they cannot rely on international agreements or regulations made on that basis, disregards the constitutional principle of equality;

34. Considering that it follows, thus, that article 24 of the referred *loi* has to be declared contrary to the Constitution . . .

2. EQUALITY IN VOTING AND REVERSE DISCRIMINATION

DECISION 34: CC decision no. 82-146 DC of 18 November 1982, *Feminine Quotas*, *Rec.* 66; *AJDA* 1983, 128.

Background: A parliamentary amendment by Socialist deputies was inserted into a *loi* amending the Electoral Code for local elections. The provision in question sought to require that the lists of candidates drawn up by the parties, from which the electors would choose, should contain a maximum of 75 per cent of persons of the same sex. In other words, women would be guaranteed a minimum of 25 per cent of the places on the list of candidates. The Government is understood to have indicated to the Conseil that it did not attach much importance to the provision, and the Conseil, seised of the *loi* on other grounds, declared the provision unconstitutional.

DECISION

1. [The Conseil refused to admit a reference by a private citizen, M. Alain Touret, on the ground that he was not among the persons listed in article 61 of the Constitution.] . . .

4. Considering that, by virtue of article 4 of the *loi* submitted to the Conseil, municipal councillors of towns of 3,500 inhabitants or more are elected by voting from a list; that voters can alter neither the content nor the order of presentation on the lists, and, by virtue of article L260 *bis*: 'the lists of candidates cannot include more than 75 per cent of persons of the same sex';

5. Considering that, according to article 3 of the Constitution:

> National sovereignty belongs to the people, which shall exercise it by its representatives and by means of referendum.
>
> No section of the people, nor any individual, may arrogate its exercise to itself.
>
> Suffrage may be direct or indirect, under the conditions provided by the Constitution. It shall always be universal, equal, and secret.
>
> Within the terms settled by *loi*, all adult French nationals of both sexes, enjoying their civil and political rights, are voters.

and according to article 6 of the Declaration of Rights and Man and of the Citizen: 'All citizens, being equal [in the eyes of the law], are equally eligible for all public dignities, positions, and employment according to their abilities, and without distinction other than that of their virtues and talents';

6. Considering that it follows from a comparison of these texts that the status of citizen itself gives rise to the right to vote and to be eligible on identical terms to all who are not excluded by reason of age, incapacity, or nationality, or for any reason designed to protect the freedom of the voter or the independence of the person elected; that these principles of constitutional value oppose any division of voters or eligible candidates into categories; that this applies for all political elections, especially for the election of municipal councillors;

7. Considering that it follows from what has been said that the rule that, in establishing the lists to be submitted to voters, includes a distinction between candidates by reason of their sex is contrary to the constitutional principles mentioned above; that, thus, article L260 *bis* of the Electoral Code, as provided under article 4 of the *loi* submitted for scrutiny by the Conseil constitutional, must be declared contrary to the Constitution; . . .

DECISION 35: CC decision no. 86-196 DC of 8 August 1985, *Elections in New Caledonia*, *Rec.* 63; *GD*, no. 40; D. 1986, 45, note Luchaire.

Background: As part of the attempt to resolve the political situation in the colony of New Caledonia, it was proposed to institute a new electoral system and a new Congress with greater powers. The electoral system was to meet the aspirations of the native population (now a minority) by weighting representation in favour of their region. A challenge was made to these provisions by the right-wing parties, who drew their support from the settler community.

DECISION

On the principle of equality

12. Considering that the deputies making one of the references claim that articles 3 to 5 of the *loi* are contrary to the principle of equality, as, according to them, they tend 'to confer the majority within the Congress to an ethnic group that is not majoritarian in number in the population of the territory'; that they consider, in fact, that 'by the over-representation of some regions and the reduced representation of another', the criticized provisions disregard both the principle of equality in voting and of equality before the law without distinction as to origin, race, or religion, declared respectively by the third paragraph of article 3 and by the first paragraph of article 2 of the Constitution . . .

14. Considering that, by the terms of article 2 §1, cited above, of the Constitution, the Republic 'ensures equality before the law to all citizens, without distinction as to origin, race, or religion'; that, according to article 3 §3, suffrage 'shall always be universal, equal, and secret'; that article 6 of the Declaration of the Rights of Man and of the Citizen of 1789 provides that the law 'most be the same for all, whether it protects or punishes. All citizens, being equal in its eyes, are equally eligible for all public dignities, positions, and employment according to their abilities, and without distinction other than that of their virtues and talents';

15. Considering that these provisions do not prevent the legislature, in accordance with article 74 of the Constitution, from being able to create and delimit regions within the framework of: the specific organization of an overseas territory, taking account of all relevant matters, especially the geographical division of populations; that, in doing that, article 3 of the *loi* does not breach article 2 of the Constitution;

16. But considering that, in order to be representative of the territory and its inhabitants in accordance with article 3 of the Constitution, the Congress, whose role as the deliberating organ of an overseas territory is not limited to the mere administration of the territory, must be elected on an essentially demographic basis; that, even if it does not follow that representation must be necessarily proportional to the population of each region, nor that other requirements of the public interest cannot be taken into account, these considerations can only be brought in, however, to a limited extent, which, in this case, is manifestly exceeded . . .

17. Considering that, thus, the statement of the number [of seats for each region] appearing in paragraph 2 of article 4 of the *loi* must be declared to be inconsistent with the Constitution; that, in consequence, paragraph 2, inseparable from the statement of these numbers, must be declared incompatible as a whole with the Constitution . . .

3. EQUALITY BEFORE JUSTICE

DECISION 36: CC decision no. 75-56 DC of 23 July 1975, *Single Judge, Rec.* 22; Nicholas, 162–5; J. Rivero, *Le Conseil constitutionnel et les libertés,* part I, ch. 4.

Background: In order to improve the speed of justice in criminal cases, especially minor ones, resort to a single judge was to be permitted not only in the Tribunal de police, which deals with *contraventions,* but also in the Tribunal correctionnel, which deals with *délits.* A reference was made by sixty opposition senators on the ground that this deprived the accused of essential safeguards by giving the power to award sentences of up to five years' imprisonment to a single judge sitting without a jury.

DECISION

2. Considering that the new provisions of article 398-1 of the Code of Criminal Procedure give the President of the Tribunal de grande instance the power, in all matters within the jurisdiction of the Tribunal correctionnel except press offences, to decide, at his discretion and without appeal, whether that court will be composed of three judges, in conformity with the rule laid down by article 398 of the Code of Criminal Procedure, or only of one such judge exercising the powers conferred on the President [of the court];

3. Considering that, thus, matters of the same kind may be tried either by a collegial court or by a single judge, as the President of the court decides;

4. Considering that, especially when a criminal *loi* is involved, by conferring such a power, article 6 of the *loi* referred to the Conseil constitutionnel . . . infringes the principle of equality before justice that is included in the principle of equality before the law proclaimed by the Declaration of the Rights of Man of 1789 and solemnly reaffirmed by the Preamble of the Constitution;

5. Considering that proper respect for this principle forbids citizens in similar situations and being prosecuted for the same offences from being tried by courts composed according to different rules;

6. Considering that, finally, article 34 of the Constitution, which reserves to *loi* the task of fixing the rules determining criminal procedure, forbids the legislature from conferring on another body the exercise . . . of the tasks set out in the disputed provisions of article 6 of

the *loi* referred to the Conseil constitutionnel where a matter so fundamental as the rights and liberties of citizens are concerned . . .

8. Considering that these provisions must be considered as contrary to the Constitution . . .

4. EQUALITY BEFORE PENALTIES

DECISION 37: CC decision no. 89-271 DC of 11 January 1990, *Electoral Expenses*.

Background: Legislation had been passed in March 1988, prior to the presidential elections, to ensure that the support given to political parties and candidates was open to scrutiny. All the same, a significant number of irregularities in connection with companies came to light. This caused a political scandal, but it also gave rise to proposals for the public funding of political expenses.

The provision on public funding gained widespread support, but was criticized as being against the interests of smaller parties. The provision on amnesty did give rise to considerable controversy. The RPR and Communists voted against it, the UDF and the UDC abstained, leaving the Socialists to vote in the provision amnestying those who had been involved in irregularities in relation to the funding of political parties, particularly in the 1988 presidential and parliamentary elections. Given this controversy, a reference was made by the Prime Minister under article 61 §2 of the Constitution.

DECISION

1. Considering that the text submitted to the Conseil constitutionnel consists of a total of twenty-seven articles under four different titles; that the Prime Minister does not raise any particular ground with respect to them; that, nevertheless, it is for the Conseil constitutionnel to note of its own motion any provision of the *loi* referred to it that disregards rules or principles of constitutional value . . .

10. Considering that article 11 of the *loi* referred . . . defines the mechanisms for allocating State aid [to political parties]; that the new first paragraph of article 9 of the *loi* of 11 March 1988 specifies that: 'The first portion of the aid provided under article 8 is attributed to parties and political groups that have presented candidates in at least seventy-five constituencies in the most recent National Assembly elections . . . The allocation is made in proportion to the number of votes obtained in the first round by each of the parties or political groups in question. Only results equal or superior to 5 per cent of the votes cast in each constituency are taken into account' . . .

11. Considering that, by the terms of article 2 §1 of the Constitution, the Republic 'ensures equality before the law for all citizens, without distinction as to origin, race, or religion'; that article 3 of the Consti-

tution declares, in its first paragraph, that 'National sovereignty belongs to the people, which shall exercise it by its representatives and by means of referendum', and, in its third paragraph, that suffrage 'shall always be universal, equal, and secret'; that, finally, article 4 of the Constitution provides that 'Parties and political groups contribute to the expression of suffrage. They may form and exercise their activity freely. They must respect the principles of national sovereignty and democracy';

12. Considering that these provisions do not prevent the State granting financial aid to political parties or groups that contribute to the expression of suffrage; that the aid granted must comply with objective criteria if it is to be compatible with principles of equality and freedom; that, furthermore, the mechanism of aid adopted must not result in nor establish the dependence of a political party on the State, nor compromise the democratic expression of diverse currents of ideas and opinions; that, even if the granting of aid to parties or groupings on the sole basis that they present candidates at National Assembly elections may be subordinated to a requirement that they satisfy a minimum of support, the criteria adopted by the legislature must not lead to disregard for the requirement of the pluralism of currents of ideas and opinions, which constitutes the foundation of democracy;

13. Considering that articles 10 and 11 of the *loi* referred satisfy these constitutional requirements to the extent that they provide that State aid is granted not only to parties and groups represented in Parliament, but equally to political parties and groups 'in proportion to their results, in the National Assembly elections'; that it is not contrary to the Constitution to lay down the principle that, in the latter case, aid shall be distributed 'in proportion to the number of votes obtained in the first round by each of the parties or political groups' that, subject to provisions specific to overseas departments and territories, have presented candidates in 'at least seventy-five constituencies at the most recent National Assembly elections';

14. Considering that, on the other hand, the fact that, for the purposes of allocating State aid to parties in proportion to their results in the elections, only those 'results equal to or higher than 5 per cent of the votes cast in each constituency' are taken into account is, by reason of the chosen threshold, capable of hindering the expression of new currents of ideas and opinions; that, thus, article 11 of the *loi* referred, in so far as it imposes this condition, must be declared contrary to the provisions of article 2 and 4 of the Constitution taken together . . .

Amnesty

19. Considering that the first paragraph of article 19 provides that 'Other than where there has been personal gain by their authors, all

offences committed before 15 June 1989 in relation to the direct or indirect financing of electoral campaigns, or political parties, or groups are amnestied, except for offences provided for by articles 132 and 175 to 179 of the Criminal Code, and those committed by a person holding on that date, or on that of the offence, a mandate as member of Parliament' . . .

20. Considering that, according to article 34 of the Constitution: '*Loi* shall determine the rules concerning . . . amnesty';

21. Considering that, by virtue of this competence, the legislature may, for the purpose of political or social peace, remove all criminal character from certain facts punishable by the criminal law, by prohibiting all proceedings with respect to them, or removing the condemnations that have been pronounced; that it is for it, then, to assess to which offences and, as necessary, to which persons the benefit of the amnesty should apply; that the principle of equality does not prevent it thus setting the boundaries for the scope of the amnesty, provided that the categories adopted are defined in an objective manner . . .

23. Considering that, in principle, the criteria taken into account by the legislature when determining the scope of the amnesty that it adopted are not contrary to the Constitution; that, nevertheless, the implementation of these criteria could lead to the exclusion of members of Parliament from the benefit of the amnesty only if they held this status on 15 June 1989 and found themselves thereby called to exercise the powers conferred on Parliament in relation to amnesty by article 34 of the Constitution; that, on the contrary, by adopting, for all cases, the status of the persons concerned at the date of the criminal events, even though they had ceased to be members of Parliament on 15 June 1989, the legislature has introduced a discrimination between the authors of identical acts in relation to the amnesty that is not justified by the objective of political and social peace pursued by the *loi*; that it follows, thus, that the words, 'or that of the events', in the text of article 19 of the *loi* referred must be declared contrary to the Constitution.

BIBLIOGRAPHY

This bibliography contains some of the principal works available in English and French on this topic. It is not comprehensive. In particular, commentaries on individual cases have been omitted. (The editions to works given here are sometimes more up-to-date than those referred to in the text.) Fuller lists can be found, notably in Genevois, and in L. Favoreu and L. Philip, *Les Grandes Décisions du Conseil constitutionnel*, cited below.

Works in English

General Works on French Government

FORDE, M., 'Bill of Rights and Trade Union Immunities: Some French Lessons' (1984) 13 *Industrial Law Journal* 40.

—— 'Liability in Damages for Strikes: A French Counter-Revolution' (1985) 33 *American Journal of Comparative Law* 447.

GRAHAM, C., and PROSSER, T., *Privatizing Public Enterprises* (Oxford, 1991).

HAYWARD, J. E. S., *Governing France: The One and Indivisible Republic* (2nd edn., London, 1983).

ROSS, G., HOFFMAN, S., and MALZACHER, S., *The Mitterrand Experiment* (Cambridge, 1987).

WRIGHT, V., *The Government and Politics of France* (3rd edn., London, 1989).

Works on the Conseil constitutionnel

BEARDSLEY, J. E., 'The Constitutional Council and Constitutional Liberties in France' (1972) 20 *American Journal of Comparative Law* 431.

DAVIS, M. H., 'The Law/Politics Distinction: The French Conseil constitutionnel and the US Supreme Court' (1986) 34 *American Journal of Comparative Law* 45.

HARRISON, M., 'The French Constitutional Council: A Study in Institutional Change' (1990) 38 *Political Studies* 603.

KEELER, J. T. S., and STONE, A., 'Judicial–Political Confrontation in Mitterrand's France: The Emergence of the Constitutional Council as a Major Actor in the Policy-Making Process', in Ross, Hoffman, and Malzacher, *The Mitterrand Experiment*, ch. 9.

MORTON, F., 'Judicial Review in France: A Comparative Analysis' (1988) 36 *American Journal of Comparative Law* 89.

NEUBORNE, B., 'Judicial Review and Separation of Powers in France and the United States' (1982) 57 *New York University Law Review* 360.

NICHOLAS, B., '*Loi, règlement* and Judicial Review in the Fifth Republic' [1970] *Public Law* 251.

—— 'Fundamental Rights and Judicial Review in France' [1978] *Public Law* 82, 155.

PROSSER, T., *The Privatisation of Public Enterprises in France and Great Britain* (Working Paper no. 88/364, European University Institute, Florence).

—— 'Constitution and Economy: Privatisation of Public Enterprises in France

and Great Britain' (1990) 53 *Modern Law Review* 304.
STONE, A., 'In the Shadow of the Constitutional Council: The "Juridicisation" of the Legislative Process in France' (1989) 12 *West European Politics* 12.
TALLON, D., 'The Constitution and the Courts in France' (1979) 27 *American Journal of Comparative Law* 567.
TUNC, A., 'The Fifth Republic, the Legislative Power, and Constitutional Review' (1960) 9 *American Journal of Comparative Law* 335.
VON MEHREN, A. T., and GORDLEY, Y., *The Civil Law System* (2nd edn., Boston, 1977), chs. 4, 5.
VROOM, C., 'Constitutional Protection of Individual Liberties in France: 'The *Conseil constitutionnel* since 1971' (1988) 63 *Tulane Law Review* 266.

Works in French

General Works on French Government

AVRIL, P., *La Ve République, histoire politique et constitutionnelle* (Paris, 1987).
DREYFUS, F., and D'ARCY, F., *Les Institutions politiques et administration de la France* (3rd edn., Paris, 1989).
DUVERGIER, M., *La Cinquième République* (Paris, 1959).
—— *La Cohabitation des français* (Paris, 1987).
FOURNIER, J., *Le Travail gouvernemental* (Paris, 1987).
MAUS, D., *Les Grands Textes de la pratique institutionnelle de la Ve République* (2nd revised edn., Paris, 1985).
—— *Les Institutions politiques françaises* (Paris, 1990).
Two periodicals are of general interest here: *Pouvoirs*; *Revue française de science politique* (esp. vol. 34 (1984)).

Decisions of the Conseil constitutionnel

All decisions are reported in the *Recueil des décisions du Conseil constitutionnel*, which is published at the end of each year. Selected decisions are published with commentaries in the following periodicals: *Actualité juridique, droit administratif*; Recueil Dalloz-Sirey; *Revue de droit public; Revue française de droit administratif*. Shorter summaries are to be found in the *Revue française de droit constitutionnel*. An analytical summary is to be found in the *Table analytique des décisions, 1959–1984*, published by the Conseil constitutionnel.

The principal case-book on the subject is L. Favoreu and L. Philip, *Les Grandes Décisions du Conseil constitutionnel* (5th edn., Paris, 1989). Selected important decisions are found in Kahn-Freund, Lévy, and Rudden, *A Source-Book of French Law* (3rd edn. by B. Rudden, Oxford, 1991).

Major Books on Constitutional Law

AVRIL, P., and GICQUEL, J., *Droit parlementaire* (Paris, 1988).
Avis et débats du Comité consultatif constitutionnel (Paris, 1961).
Conseil constitutionnel, *La Déclaration des droits de l'homme et du citoyen et la jurisprudence* (Paris, 1989).
Conseil constitutionnel et Conseil d'État (Colloquium at University of Paris 2, Paris, 1988).

DEBBASCH, C., PONTIER, J.-M., BOURDON, J., and RICCI, J.-C., *Droit constitutionnel et institutions politiques* (3rd edn., Paris, 1990).
FAVOREU, L. (ed.), *Le Domaine de la loi et du règlement* (2nd edn., Paris, 1981).
—— (ed.), *Nationalisations et Constitution* (Paris, 1982).
—— *Les Cours constitutionnelles* (Paris, 1986).
—— *La Politique saisie par le droit* (Paris, 1988).
—— and JOLOWICZ, J. A., *Le Contrôle juridictionnel des lois* (Paris, 1986).
—— and PHILIP, L., *Le Conseil constitutionnel* (Paris, 1980).
FRANCK, C., *Les Fonctions juridictionnelles du Conseil constitutionnel et du Conseil d'État dans l'ordre constitutionnel* (Paris, 1974).
GENEVOIS, B., *La Jurisprudence du Conseil constitutionnel* (Paris, 1988).
GICQUEL, J., HAURIOU A., and GÉLARD, P., *Droit constitutionnel et institutions politiques* (10th edn., Paris, 1989).
HAMON, L., *Les Juges de la loi* (Paris, 1987).
LUCHAIRE, F., *Le Conseil constitutionnel* (Paris, 1980).
—— *La Protection constitutionnelle des droits et libertés* (Paris, 1987).
—— and CONAC, G., *La Constitution de la République française* (2nd edn., Paris, 1987).
MORANGE, J., *Droits de l'homme et libertés publiques* (2nd edn., Paris, 1989).
PHILIPPE, X., *Le Contrôle de proportionnalité dans les jurisprudences constitutionnelle et administrative françaises* (Paris, 1990).
RENOUX, T., *Le Conseil constitutionnel et l'autorité judiciaire* (Paris, 1984).
RIVERO, J., *Le Conseil constitutionnel et les libertés* (Paris, 1984).
—— *Les Libertés publiques* (vol. i, 5th edn., Paris, 1987; vol. ii, 4th edn., 1989).
ROBERT, J., *Les Libertés publiques et droits de l'homme* (4th edn., Paris, 1988).
TURPIN, D., *Contentieux constitutionnel* (Paris, 1986).

Major Articles on French Constitutional Law

BACOT, G., 'La Déclaration de 1789 et la Constitution de 1958', *RDP* 1989, 685.
BOCKEL, A., 'Le Pouvoir discrétionnaire du législateur', in G. Conac *et al.* (eds.), *Itinéraires: Études en l'honneur de L. Hamon* (Paris, 1982), 43.
BOUDÉANT, J., 'Le Président du Conseil constitutionnel', *RDP* 1987, 443.
CHAZELLE, R., 'Continuité et tradition juridique au sein de la seconde chambre: Le Sénat et le droit parlementaire coutumier', *RDP* 1987, 711.
COLLY, F., 'Le Conseil constitutionnel et le droit de propriété', *RDP* 1988, 135.
DELVOLVÉ, P., 'Le Conseil constitutionnel et la liberté de l'enseignement', *RFDA* 1985, 624.
FAVOREU, L., 'Le Droit constitutionnel jurisprudentiel', *RDP* 1983, 333; *RDP* 1986, 395; *RDP* 1989, 399.
—— 'Le Conseil constitutionnel et l'alternance' (1984) 34 *Revue française de science politique* 1002.
—— 'Les Cents premières Annulations prononcées par le Conseil constitutionnel', *RDP* 1987, 589.
—— 'Ordonnances ou règlements d'administration publique?', *RFDA* 1987, 686.
—— '"Les Règlements autonomes n'existent pas"', *RFDA* 1987, 871.
—— 'Le Droit constitutionnel, droit de la Constitution et constitution du droit', *RFDC* 1990, 71.

FERSTENBERG, J., 'Le Contrôle, par le Conseil constitutionnel, de la régularité constitutionnelle des lois promulgées', *RDP* 1991, 339.

FLAUSS, J.-F., 'Les Droits sociaux dans la jurisprudence du Conseil constitutionnel', *Droit social* 1982, 645.

GARRIGOU-LAGRANGE, J.-M., 'Les Partenaires du Conseil constitutionnel ou de la fonction interpellatrice des juges', *RDP* 1986, 647.

GOGUEL, F., 'Le Conseil constitutionnel', *RDP* 1979, 5.

HAMON, L., 'Contrôle de constitutionnalité et protection des droits individuels', D. 1974 Chr. 83.

—— 'Le Droit du travail dans la jurisprudence du Conseil constitutionnel', *Droit social* 1983, 155.

LEBEN, C., 'Le Conseil constitutionnel et le princie d'égalité devant la loi', *RDP* 1982, 295.

LUCHAIRE, F., 'De la méthode en droit constitutionnel', *RDP* 1981, 275.

—— 'Un Janus constitutionnel: L'égalité', *RDP* 1986, 1229.

OBERDORFF, H., 'A propos de l'actualité de la Déclaration de 1789', *RDP* 1989, 665.

PHILIP, L., 'Le Développement du contrôle de constitutionnalité et l'accroissement des pouvoirs du juge constitutionnel', *RDP* 1983, 401.

—— 'Bilan et effets de la saisine du Conseil constitutionnel' (1984) *Revue française de science politique* 988.

Pouvoirs, 13: *Le Conseil constitutionnel* (revised edn., Paris, 1986).

RIALS, S., 'Les Incertitudes de la notion de Constitution sous la Ve République', *RDP* 1984, 587.

RIVERO, J., 'Les Notions d'égalité et de discrimination en droit public français' (1965) XIV *Travaux de l'Association Henri Capitant* 343.

ROBERT, J., 'L'Aventure référendaire', D. 1984 Chr. 223.

—— 'Conseil d'État et Conseil constitutionnel: Propos et variations', *RDP* 1987, 1151.

ROMI, R., 'Le Président de la République, interprète de la Constitution', *RDP* 1987, 1265.

SAVY, R., 'La Constitution des juges', D. 1983 Chr. 105.

TERNEYRE, P., 'La Procédure législative ordinaire dans la jurisprudence du Conseil constitutionnel', *RDP* 1985, 691.

—— 'Droit social et la Constitution', *RFDC* 1990, 339.

13emes journées juridiques franco-italiennes (Paris, 8–10 October 1987): 'L'Effet des décisions des juridictions constitutionnelles à l'égard des juridictions ordinaires', *Journées de la Société de législation comparée* 1987.

VIER, C.-L., 'Le Contrôle du Conseil constitutionnel sur les règlements des Assemblées', *RDP* 1972, 165.

INDEX